Property of
Andy
Schuster

# *Introduction to Offshore Structures*
### *Design*
### *Fabrication*
### *Installation*

# Introduction to Offshore Structures

## Design
## Fabrication
## Installation

**W.J. Graff**
*Professor of Civil Engineering*
*University of Houston*

Gulf Publishing Company
Book Division
Houston, London, Paris, Tokyo

## Introduction to Offshore Structures
### Design, Fabrication, Installation

Copyright ©1981 by Gulf Publishing Company, Houston, Texas. All rights reserved. Printed in the United States of America. This book, or parts thereof, may not be reproduced in any form without permission of the publisher.

**Library of Congress Cataloging in Publication Data**
Graff, W. J. (William J.)
   Introduction to offshore structures.

   Includes bibliographical references and index.
   1. Drilling platforms—Design and construction.
2. Offshore structures—Design and construction.
I. Title.
TN871.3.G67                627'.98               81-6259
ISBN 0-87201-694-3                             AACR2

# Contents

**Preface** . . . . . . . . . . . . . . . . . . . . . . . . . . ix

## Part I — Steel Template Structures

**Chapter 1**    **Offshore Oil Operations** . . . . . . . . . . . . . . . . 1
Exploration. Exploratory Drilling. Development Drilling.
Production and Product Transport. Worker Transportation.

**Chapter 2**    **Early Platform History** . . . . . . . . . . . . . . . . 4
The Beginning. Postwar Boom. Santa Barbara Channel,
California. The British Petroleum Company. Gulf of Mexico
Accident Reports.

**Chapter 3**    **Platforms, Catwalks, and Heliports** . . . . . . . . . . . 21
Drilling/Well-Protector Platforms. Tender Platform.
Self-Contained Template Platforms. Tower-Type Template
Platforms. Production or Treatment/Reinjection Platforms.
Quarters Platform. Flare Jacket and Flare Tower. Auxiliary
Platforms. Catwalks. Helicopters and Heliport Design.

**Chapter 4**    **An Overview of Engineering Procedures** . . . . . . . . . 43
Operational Criteria. Environmental Criteria. Foundations
Design. Structural Design. Construction and Installation.

**Chapter 5**    **The World's Tallest Platform—*Cognac*** . . . . . . . . . 48
Announcement. Description. Deck Structure. Oceanographic
Criteria. Structural Design. Fabrication. Installation. Cost.

**Chapter 6**    **Design Loads and Forces** . . . . . . . . . . . . . . . . 66
Wind Forces. Wave Forces. Current Force. Dead Load. Live
Load. Impact. Other Forces. Deck Floor Loads.

| Chapter 7 | **Pile Foundations** . . . . . . . . . . . . . . . . . . . 79<br>Axial Loads. Safety Factors. Installing Undrivable Piles. Lateral Pile Load Analysis. Battering of Piles. Group Piles. |
|---|---|
| Chapter 8 | **Design for an Eight-Leg Jacket** . . . . . . . . . . . . . 92<br>General Description. Environmental Conditions. Design. Materials. Laminations and Lamellar Tearing. Weldability. |
| Chapter 9 | **Computer Methods for Static & Dynamic Analyses** . . . . 132<br>Steps Involved in Static Analysis. Typical Static/Elastic Analysis Program. Dynamic Analysis. Typical Offshore Structure Dynamic Analysis Program. |
| Chapter 10 | **Tubular Joint Design and Fatigue Analysis** . . . . . . . 149<br>**Tubular Joints.** Types of Joints. Multiplane Connections. Brief History of Tubular Joints. Parameters of an In-Plane Tubular Joint. Elastic Stress Distribution. Punching Shear Stress. Overlapping Braces. Stress Concentration. Chord Collapse and Ring Stiffener Spacing. Stiffened Tubes. **Fatigue of Tubular Joints.** Fatigue Behavior. |
| Chapter 11 | **Fabrication and Installation** . . . . . . . . . . . . . . 201<br>Jacket Fabrication. Deck Substructure. Piles. Jacket Load-out and Installation. Piles and Well Conductors. Deck Substructure Installation. |
| Chapter 12 | **Corrosion** . . . . . . . . . . . . . . . . . . . . . . . 239<br>Corrosion Mechanism. Galvanic Corrosion. Atmospheric Corrosion of Steel. Principles of Cathodic Protection. Offshore Structure Corrosion Zones. Biological Corrosion. Stress Corrosion. Corrosion Fatigue. |

## Part II — Concrete Gravity Structures

| Chapter 13 | **The Gravity Platform** . . . . . . . . . . . . . . . . . 260<br>General Features. |
|---|---|
| Chapter 14 | **Environmental Loads** . . . . . . . . . . . . . . . . . 270<br>Wave Loads. Importance of Morison's Equation. Wind Loads. Current Forces. |
| Chapter 15 | **Geotechnical Design** . . . . . . . . . . . . . . . . . 275<br>General Aspects. Foundation Stability. Skirts. Foundation Failure Modes. |

| Chapter 16 | **Structural Design** . . . . . . . . . . . . . . . . . . 285 |
|---|---|
| | Limit State Design. Prestressing. |
| Chapter 17 | **Manifold Platforms** . . . . . . . . . . . . . . . . . 308 |
| | General Description. Base Slab Design. Solid Vertical Wall Design. Perforated Vertical Wall Design. *Petrobras* Gravity Platforms. |
| Chapter 18 | **Tower Platforms** . . . . . . . . . . . . . . . . . . . 320 |
| | General Description. Tremie Method of Concrete Placement. *Andoc* Platform. |
| Chapter 19 | **Materials, Corrosion, and Fatigue** . . . . . . . . . . . 336 |
| | Concrete. Corrosion. Fatigue Behavior. |
| Chapter 20 | **Deck Structures** . . . . . . . . . . . . . . . . . . . 344 |
| | General Description. Design Loads. Integrated vs. Modular Design. Transition Pieces. Steel Grades. |
| Chapter 21 | **Construction and Installation** . . . . . . . . . . . . 354 |
| | Construction. Towing. Installation. Instrumentation. |
| **Index** | . . . . . . . . . . . . . . . . . . . . . . . . . . 364 |

*To my wife, Ruby*

# *Preface*

This book discusses the design, fabrication, and installation of fixed offshore platforms used to drill for and produce petroleum from beneath the sea. It is written as an overall view for the seasoned engineer working in a restricted or specialized aspect of the subject and for the young engineer or student facing the opportunities of a career in the offshore industry for the first time. The book also should be useful for in-house company training programs and to management and/or business people regularly concerned with hardware, processes, or terms related to offshore structures.

Drilling for petroleum beneath the sea has increased tremendously in the last 25 years. Shortly after World War II, wells were drilled in shallow, protected water using onshore methods, although these were inadequate. These inadequacies led to the growth of an entirely new industry which, with its many attendant service industries, has today become one of the world's most intensive concentrations of manpower, capital, and technology.

I became involved in the offshore industry when structural engineering graduate students began asking for a course to acquaint them with this broad new area in which they found themselves working. (At the same time, two events began the development of this book, although writing a book was not contemplated at that time: my summer employment in the marine engineering division of Brown & Root, Inc., and attendance at a week-long short course on offshore structures at the University of California, Berkeley.) With as much enthusiasm as the students who kept inquiring about and encouraging some action, I initiated a "current topics" course. It was well received, later formalized, and then offered several times over the following years. Each repetition of the course was a learning experience for the teacher as well as the students.

In 1970 I was invited to become a member of the American Welding Society committee on tubular joints. I readily accepted because, essentially, the only tubular joints then in use were those used on offshore jackets. My activity and interest in offshore structural design mushroomed during the next several years. Engineering assignments with several companies engaged in offshore activity helped to broaden my viewpoint and increase my capability and confidence.

The academic year 1977-1978 was spent as a visiting professor at a university in Denmark. The opportunity to organize weekly seminars and a one-day conference on offshore structures resulted in my becoming acquainted with knowledgeable people involved in the North Sea offshore activities. The Danish government was preparing a comprehensive treatise on offshore oil and gas exploration to acquaint and encourage more Danish companies to enter the field. I was asked to write a chapter on steel template platform design for the treatise. Thus this book evolved; one day I realized that many portions of it were at hand in the form of class notes, published articles, and reports of various kinds. The effort of rewriting all of this material into a unified presentation was underestimated, but that is how many things are accomplished.

Offshore structures are universally designed for functional purposes based on rational and empirical design data; the whole is judged primarily on economic grounds. Nevertheless, economics are not addressed at all in this book. Economics are a major consideration in any engineering design, but their impact is usually best appreciated *after* the technical aspects have been thoroughly studied. While this book is written from the practical viewpoint of what is actually done, there undoubtedly are gaps in the presentation: the reasons why some things are done have simply not been published by the doers; or it may be that I am unaware of the rationale for some aspects of the overall effort.

The material in this book is by no means original. It has been gleaned from many sources, sometimes conflicting. These sources were scattered; the details at times incomplete. This is understandable in so new a subject; each published paper emphasizes its rather specific topic. My contribution has been in assimilating, organizing, closing gaps, and presenting a unified treatment.

Extensive illustrative material has been included, while mathematics has deliberately been held to a minimum. This presentation is consistent with the philosophy that one must first learn to identify the chessmen before coping with their specific moves. Extensive illustrative material broadens the understanding of the descriptive material. The references at the end of each chapter are intended to introduce the reader to literature providing greater depth in the various specific topics. It is hoped that the book will serve as a reference after it has accomplished its purpose of introducing the reader to the new and interesting subject of offshore platforms.

Experienced designers may note that the discussion of some specific item in this book may not be as directly to the point as it could have been. In those instances I would appreciate receiving a note that will sharpen the presentation should the book merit future revision. Obviously, I would like to know of errors so that they may be corrected in any later printings.

The subject of offshore structures is so broad that arbitrary deletions or abridgments of material have had to be made to reach an end to the manuscript. I would appreciate being informed of topics that readers believe should receive more adequate treatment in an introductory text.

The book is divided into two parts: Part I, consisting of Chapters 1-12, discusses steel template platforms. This technology was developed years ahead of that employed in designing and constructing the concrete gravity platforms discussed in Part II. A feeling is occasionally expressed in some quarters that concrete gravity platform designs may not be as viable as steel template designs for water depths much beyond the depths of present concrete structures. In Part I, Chapters 6, 7, 8, and 11 should be of particular interest to the reader with only limited time. In Part II the same reader should select Chapters 17, 18, 20, and 21. In a few instances mention is made of the same subject in more than one place in the book. This repeated information adds emphasis and accommodates the reader who chooses to read selectively.

Preparation of the book could not have been accomplished without the preprints and proceedings of the Offshore Technology Conference. The cooperation of the Offshore Technology Conference staff with regard to the use of illustrative materials is indeed appreciated.

As students took the offshore structures design course over the years, they have significantly contributed to the subject matter presented here. The author gratefully acknowledges their part in the development of this book.

The assistance of several individuals was significant during the initial stages of manuscript preparation. I am grateful to the many people who gave encouragement to the effort. Still, several specific acknowledgments with thanks are in order: to Dr. Jay B. Weidler, Brown & Root, Inc., and Mr. Peter W. Marshall, Shell Oil Company, for comments on early draft material; to Mr. Olav Furnes, Det Norske Veritas; to Dr. Torgier Moan of the Norwegian Institute of Technology and Dr. Palle Thoft-Christensen of Aalborg University Centre for furnishing copies of many published papers and reports originating in European meetings or publications.

Acknowledgment with thanks is due to the University of Houston Civil Engineering Department and the Cullen College of Engineering Educational Media Center for preparation of the many line drawings. Two former Chairmen of the Civil Engineering Department gave considerable encouragement to me: Dr. George Pincus and Dr. Ardis White. The assistance of two other members of the University of Houston Civil Engineering Department is also gratefully acknowledged: Dr. Michael O'Neill and Dr. Osman Ghazzaly.

Lastly, I wish to acknowledge and give special thanks to my wife, Ruby, for typing the manuscript. Her readiness to help, constant encouragement, and promptness in typing draft after draft materially expedited the overall project.

It is sincerely hoped the book will be as interesting to read as it was to write. I firmly believe that a tremendous future waits in what is called "the offshore."

*W. J. Graff*
*March 1981*
*Houston, Texas*

# Part I

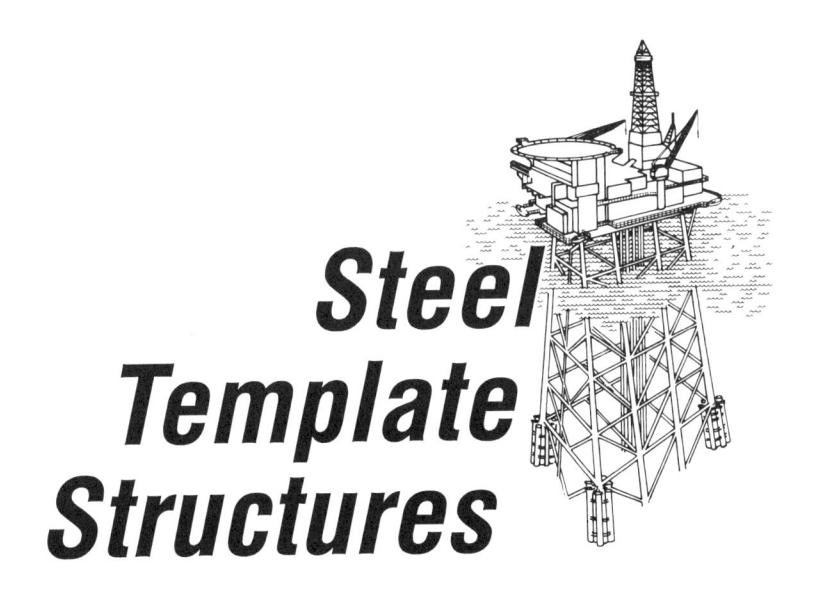

# *Steel Template Structures*

# Chapter 1

# *Offshore Oil Operations*

Over the past 20 years, two major categories of fixed platforms have been developed: the *steel template* type pioneered in the Gulf of Mexico, and the *concrete gravity* type first developed for the North Sea. A third fixed platform type, the *tension leg* platform, is now being developed. It relies on many tightly anchored cables to hold its floating structure fixed in position.

The number of offshore platforms in the bays, gulfs, and oceans of the world today approaches 10,000. As a background to all this activity, a brief description of offshore oil operations follows. The subject is very broad, beginning with the geologist and geophysicist hunting for potential oil-bearing strata within the ground under the ocean floor and ending with transportation of the oil from the offshore site to the shore location.

Oil operations conducted offshore can be broken into five main areas: exploration, exploration drilling, development drilling, production operations, and transportation.

### Exploration

This phase consists of trying to locate oil-bearing strata within the ground underneath the ocean. Geologists and geophysicists are responsible for this activity. Geology is the science of rocks. The geologist looks at and studies surface formations and drilled-out core samples to describe the geometry of the earth's faults and strata. The geophysicist uses methods of remote data gathering such as seismic exploration and instrumentation for measuring gravity fields to make interpretations as to the possible presence of oil-bearing strata. Within a designated area of the ocean, geophysicists perform seismic surveys from specially equipped boats, systematically charting the ground structure underneath the ocean. When an area thought to be favorable is found, cores are drilled from core-drilling ships. These specialized ships can remain dynamically positioned on location and drill in seas with 30 ft (9 m) waves and in depths approaching 4000 ft (1200 m).

## 2    Introduction to Offshore Structures

### Exploratory Drilling

Once it is decided that an area may contain oil-bearing strata, an exploratory well must be drilled to confirm or deny the presence of hydrocarbons. The formation can yield gas rather than oil, or both. Exploratory wells are drilled with a mobile drilling rig—one mounted on a ship or some form of movable platform. Large, self-contained stationary platforms as discussed in this book are *not* used for exploratory drilling. The *jack-up mobile rig* is used in water depths of 50-250 ft (15-76 m). In very shallow water, less than 50 ft (15 m), a *submersible* unit is used. The submersible rig is towed to location, flooded, and rests on the ocean floor during drilling. The jack-up rig floats to the location with its legs raised high in the air. On location the legs are jacked-down into the water and, as they penetrate the mud of the ocean floor, the bargelike drilling deck and derrick are gradually raised out of the water.

For drilling exploratory wells in water deeper than 250 ft (76 m), *floating drilling rigs* are used. Floating drilling rigs are classified as either semi-submersible or ship-shape hulls. The *semi-submersible* has open framework through which the waves can pass easily. After arriving on location, the semi-submersible is flooded to a deep draft; thus, although it is floating, it forms an extremely stable platform for drilling. The *ship-shape hull* can be moved from one drilling site to another more easily than the semi-submersible, but the vessel rolls and pitches in moderately high seas, causing more shut down time than a semi-submersible.

### Development Drilling

Development drilling is the process of drilling holes into known accumulations of oil so that the oil reserves may be exploited in the most profitable manner. Generally, development drilling is done from a self-contained platform. The platform is of adequate size to contain all the necessary equipment and supplies. Efficient development drilling requires that many wells be drilled from one surface location using *directional drilling*. This form of drilling has the tremendous advantage of having the flow from many wells converge to one surface location for treatment and storage prior to shipment to shore. Early self-contained platforms permitted drilling 8-10 wells; present designs allow 32-40 wells from one platform. As will be discussed in Chapter 5, the *Cognac* platform has been designed for drilling 62 wells.

While the number of decks and the location of equipment on the decks of a self-contained platform vary from one platform to another, the lowest level usually is the one on which cement units and reserve supplies of drilling mud are located. The intermediate decks house the electrical power, pumping, and primary processing equipment, and the top deck supports the quarters, the drilling rig, and communications.

Not all development drilling is done from self-contained platforms. In water depths to about 50 ft (15 m), a mobile drilling unit may be used to drill the well, and a well-protector jacket will be placed around the well riser to protect it from environmental forces. The well jacket also serves as a place from which to run flowlines for production.

Another development drilling method uses a tender ship and a *tender type platform*. The tender type platform is large enough to support only the drilling derrick and its attendant power equipment. The drilling crew quarters, drilling mud, and other drilling supplies, etc., are located on the moored tender ship adjacent to the platform.

## Production and Product Transport

Once development drilling has been completed, production of the well is begun. In deep water the production and processing equipment is put on the same self-contained platform used for development drilling. In shallow water drilling platforms are usually quite small and are thus converted to well-protector platforms when production is started. A separate platform adjacent to the well-protector platform is constructed for the processing or treatment equipment.

Storage of the oil is a main concern in offshore operations. Frequently, after all of the drilling has been completed, the drilling platform (if it is large enough) becomes a well-protector platform and a storage platform. Large oil storage tanks ranging in capacity from 10,000 to 30,000 barrels are placed on it. If the platform is in shallow water, the oil is shipped to shore by barges or pipeline. Often, in deeper water, a tanker ship is anchored next to the treatment platform and serves for storage as well as transport, or the oil may be transferred to a second tanker for movement to shore to save raising anchors.

## Worker Transportation

Worker transportation is one of the most basic problems associated with offshore operations. Transportation is accomplished either by boats or helicopters. High-speed crew boats transport work crews when time is available and the distance is less than about 50 miles (80 km). Helicopters transport crews and other personnel over long distances and/or when time is important. The transportation of equipment to offshore rigs is accomplished with work boats. These boats generally are about 30 ft (9 m) wide and 140 ft (43 m) long; they are versatile, high-powered, and essential to offshore operations. Thus, all platforms must be provided with mooring bitts, bumpers, cranes, stairs, etc., for use with work boats and crew boats.

# Chapter 2

# *Early Platform History*

### The Beginning

The offshore oil industry began off the coast of California in the late 1890s. H.L. Williams bought property on the California coast with significant oil deposits indicated beneath it. In 1887 he completed his first wells onshore near the beach. Evidence of gas pockets on the beach led Williams to believe that oil deposits could be found offshore. The land operation was extended out over the water by means of a pier, and in 1887 the first hole over water was drilled. Eleven piers were constructed and, by 1900, drilling was being conducted in the water 500 ft (150 m) from the shoreline.[1] Figure 2-1 shows how this first marine oilfield looked in 1903 at Summerland in Santa Barbara County.

As early as 1909 or 1910, wells were being drilled in Ferry Lake in Caddo Parish, Louisiana. Wooden derricks were erected on hastily built wooden platforms constructed on top of cypress tree pilings.[2]

In 1922 oil was discovered beneath the waters of Lake Maracaibo, Venezuela. Drilling was begun from timber platforms erected in the shallow water of this inland lake. By 1930, the lake was a dense array of drilling rigs and production platforms as shown in Figure 2-2. Underwater transmission lines (pipelines) were used to transport crude oil to the lake shore. Oil development in and around Lake Maracaibo has been intense; it is estimated there are over 6000 platforms in the lake today.

California's offshore development was pioneered by the Signal Oil and Gas Company. Around 1930, the company was drilling wells in the tidelands at Elwood, California, and in 1938 began drilling in the area of the turning basin of Long Beach Harbor.

Marine drilling in the Gulf of Mexico began in the 1930s with wells placed in the swamps and marsh areas of Louisiana. Timber platforms were used and canals were dredged so that supply barges could reach the drill sites. The first attempt to drill in the Gulf proper was in 1933, off Creole in Cameron Parish,

**Figure 2-1.** *Summerland, California oil field in 1903: the first marine oilfield developed in the U.S. (Source: D.B.J. Thomas, International Symposium on the Integrity of Offshore Structures, Institution of Engineers and Shipbuilders in Scotland, Glasgow, April 1978.)*

Louisiana. A wooden platform was constructed in 12 ft (3.5 m) of water 3000 ft (915 m) off the coast. Development in the 1930s was greatly suppressed due to the inaccessibility of the producing areas, but soon swamp buggies and new road construction opened the way for geophysicists and their equipment to conduct geophysical surveys.

In 1937 evidence of oil-bearing formations was found west of the town of Creole, with indications that the find extended out into the Gulf. Superior Oil Company and Pure Oil Company leased the property (7000 acres onshore and 33,000 acres offshore) and commissioned Brown & Root, Inc., to design a platform to be placed in 14 ft (4.3 m) of water about a mile (1.6 km) from the coastline. The platform was designed with emphasis on the ability to withstand high seas and hurricanes. This platform was the first to be constructed in the Gulf in an area remote from the shore. It was constructed from timber piles and had a 100 x 300-ft (30 x 90-m) base from which conventional land drilling was performed. The well began producing in the spring of 1938, resulting in the opening up of the Creole Field.

**6   Introduction to Offshore Structures**

*Figure 2-2. View of Lake Maracaibo oilfield in 1930. (Source: D.B.J. Thomas, International Symposium on the Integrity of Offshore Structures, Institution of Engineers and Shipbuilders in Scotland, Glasgow, April 1978.)*

Early in 1938, approximately one mile (1.6 km) off the coast of McFaddin Beach, Texas, a 50 x 90-ft (15 x 27-m) timber platform was constructed in water about 10-15 ft (3-4.5 m) deep. This platform is shown in Figures 2-3 and 2-4. Notice the untreated timber piles and the pile groups used as docking fenders.

**Figure 2-3.** Close view of timber platform one mile off the coast of McFaddin Beach, Texas, in 1938. (Copyright © 1968 Offshore Exploration Conference.)

Approximately 25 wells were drilled from pile foundations off the Gulf coast from 1937 to 1942.[3] These truly offshore operations demonstrated the many difficulties that had to be overcome if drilling beyond the shoreline was to continue with any efficiency. None of the oil companies or service industries were prepared for this offshore operation. The nearest supply base to the first Creole operaton was 13 miles (21 km) away in Cameron, Louisiana. Because of the unavailability of radios, all equipment orders had to be dispatched with the first boat going in to shore. All heavy equipment had to be transported by flattop barges pulled by chartered shrimp boats, and the drilling crews were transported every shift by round-bottom shrimp boats. When a fog set in, the rig could only be located by turning off the ship's engines and listening for the rig operations. Problems also resulted from marine borers and hurricane damage.

### Postwar Boom

World War II brought a temporary halt to offshore development, but the advances in technology, as a result of the war, allowed the industry to get a firm grasp on a strong future.

**8    Introduction to Offshore Structures**

*Figure 2-4.* Overall view of 1938 McFaddin Beach, Texas, timber platform. (Copyright © 1968 Offshore Exploration Conference.)

In 1946 Magnolia Oil Co. (Mobil Oil Co.) constructed a platform in 14 ft (4.3 m) of water and approximately five miles (8 km) offshore. Although it was still within sight of land, it was the first operation carried out this far from shore. The platform was 174 x 77 ft (53 x 23 m) and stood 19 ft (5.8 m) above the mean high-tide level. Construction was done entirely on the site and took about 60 days. A total of 338 steel piles supported the derrick. Communications were maintained by radio. Working crews were housed on a boat anchored at Eugene Island, Louisiana. Two crew boats were used for continuous transportation between the rig and the quartersboat.

The platform was designed to withstand hurricane winds of 150 mph (67 m/sec) and a maximum wave height of 18 ft (5.5 m). This rig was the first to provide for the drilling of three wells by the mounting of the derrick on skids. It also represented the first use of steel piles, and from this time on, wooden piles became a thing of the past.[4] The relentless attack by the teredo, or shipworm, resulted in a loss of confidence in the use of timber piles. Though the Magnolia well proved to be a dry hole, it nevertheless represented the first major effort to drill in waters far from shore in the Gulf of Mexico.

The year 1947 saw the construction of two platforms that became the design standard for many years. Superior Oil Company made a radical design change in their platform so that it could operate 18 miles (29 km) offshore in 20 ft (6 m) of water. The total platform size was 173 ft (53 m) long by 108 ft (33 m) wide. It was a completely self-contained system, including drilling rig, equipment, pipe racks, and all supporting facilities. Living quarters were on a separate platform connected to the drilling platform by a bridge. (See Figure 2-5.) The new design called for six steel templates or jackets fabricated onshore and carried to the site by barge. They were then lowered into the water by a crane and fixed to the bottom using 268, 8 and 10-inch (20 and 25-cm) steel piles driven through the jacket legs.[2]

*Figure 2-5.* Superior Oil Company 1947 platform in Vermilion Block 71 of the Gulf of Mexico. (Copyright © 1968 Offshore Exploration Conference.)

The term "template" derives from the fact that the jacket legs serve as guides for the tubular piles. This construction method allowed the actual placement of the structure in the water to be completed in about nine days as opposed to two months' installation time by the common method of onsite construction. The new design also allowed for the use of bracing below the water line. The old type platforms were braced only above the water line and provided very limited lateral resistance to wave impact forces. Previously, above-the-water bracing had been sufficient only because of the shallow water depth. This new method of construction allowed platforms to be placed into much deeper water.[5]

The summer of 1947 in the Gulf of Mexico also saw the construction of a much smaller drilling platform of 2700 sq ft (250 sq m) area as opposed to the 20-30,000 sq ft (1850-2780 sq m) of earlier platforms. This structure was placed in 18 ft (5.5 m) of water 10.5 miles (17 km) offshore, out of sight of land. The platform held only the derrick and some basic machinery and was accompanied by a converted war surplus LST ship which housed the crew quarters as well as supplies and other necessary equipment. The small drilling platform with the accompanying tender ship system became very popular due to the great reduction in construction costs. On September 9, 1947, this well became the first offshore well to begin production. Ten days later the strongest hurricane of the season came through, producing winds up to 90 mph (40 m/sec) at the platform. Everything survived, and the first large oil deposit in open, unprotected water began to produce.

Offshore production was significantly hampered by the lack of support equipment, but the obvious growth of a new oilfield led to the design and construction of better equipment made especially for offshore work. In 1949 the first derrick barge designed for offshore work was commissioned. Pile diameters increased, and lightweight derricks for offshore use were designed. Designs improved as a result. Platforms had fewer piles, but they were of larger diameter, and there was more space in the cross-bracing. By the mid-1950s, the average-size pile was 30 inches (76 cm) in outside diameter (OD).[2] In 1969, the average-size pile had increased to 48 inches (122 cm) OD. The early piles had wall thicknesses as thin as 3/8 inches (9.5 mm), yet in 1969 variations in wall thickness were common, the heaviest wall being at the mudline. Wall thickness ranged from 5/8 inches (16 mm) to about 1.25 inches (32 mm).[6]

By May of 1949, there were 10 offshore platforms in the Gulf of Mexico and 25 tender rigs. Later that year a hurricane swept through the Texas Gulf, causing major damage to one platform and minor damage to a tender rig. This event caused a reevaluation of the design parameters at that time.

The development of mobile drilling units dates from about 1949. In that year there was only one barge-mounted rig capable of drilling in water as deep as 20 ft (6 m).

## Early Platform History 11

Significant construction in offshore areas of the Gulf of Mexico was slight in 1950 due to a controversy with the federal government over ownership of offshore properties. This dispute was settled by the Submerged Lands Act of 1953. From this time on, construction in the Gulf of Mexico accelerated tremendously.

In 1955 the first platform in over 100 ft (30 m) of water was in operation, with a deck size of 220 x 106 ft (67 x 32 m). This new platform, built by Shell Oil Company, introduced the use of skirt piles. The trusswork consisted of three jackets of eight piles each. (See Figure 2-6.) All of the deck pieces and the jackets were installed by lifting them in upright position from transportation

**Figure 2-6.** Shell Oil Company 1955 Platform in Grand Isle Block 47 of the Gulf of Mexico. (Copyright © 1968 Offshore Exploration Conference.)

barges with a 250-ton derrick barge. Each of the jackets was fabricated on its side, set into the water, rotated into the vertical, then placed on a barge for transportation to the site.

Careful consideration of deck size began in 1956, and through new compact platform designs, the necessary deck area was reduced to an average size of 110 x 140 ft (33 x 43 m). In 1957 a new type of barge that allowed a controlled jacket launch was introduced, eliminating the need of a derrick barge for launching. This development led to a significant reduction in installation time.[7] Since that year, many jackets have been transported to the site lying horizontally on barges and rotated into the vertical only after being launched from the end of the barge at or near the offshore location.

In 1959 a platform had been installed in the Gulf of Mexico in more than 200 ft (60 m) of water, and more than 200 platforms had been placed in the Gulf. The early 1960s saw the development of the packaged structure. By reducing the amount of onboard supplies and redesigning to reduce space, a self-contained platform was designed that could operate with a deck size of 66 x 118 ft (20 x 36 m). Platforms continued to go into deeper and deeper water. A platform was placed in 285 ft (87 m) of water in 1965, and another platform was placed in 340 ft (104 m) of water in 1967. In the early 1970s Shell Oil Company installed a platform in the Gulf in 373 ft (114 m) of water, and the Tenneco Corporation placed a platform 130 miles (208 km) off the Louisiana coast in 375 ft (114 m) of water. This platform is over 400 ft (122 m) tall and weighs over 8000 tons (7256 tonnes).

Recently, the Shell Oil Company installed a platform in 1020 ft (310 m) of water on the continental slope, as distinct from the continental shelf, about 100 miles (160 km) southeast of New Orleans, Louisiana. This platform is described in Chapter 5.

### Santa Barbara Channel, California

In the years following World War II the growth of drilling off the California coast developed in much the same fashion as that in the Gulf of Mexico. As platforms were placed in deeper water, they became more complex in their functional requirements and structural configurations. To illustrate the development in the Santa Barbara Channel, two platforms constructed 10 years apart are described.

The first, constructed in 1966, was a joint venture of the Mobil Oil Company and the Atlantic Richfield Company. The eight-pile template or jacket type platform was located approximately 2.5 miles (4 km) southwest of Goleta, California, in 211 ft (64 m) of water; it was named *Holly*.[8] *Holly* was designed as a 30-well drilling and production platform. Figure 2-7 shows a sketch of the structure, and Figure 2-8 is a picture of how the finished platform looked. The state of development within the offshore industry at that time is

Early Platform History 13

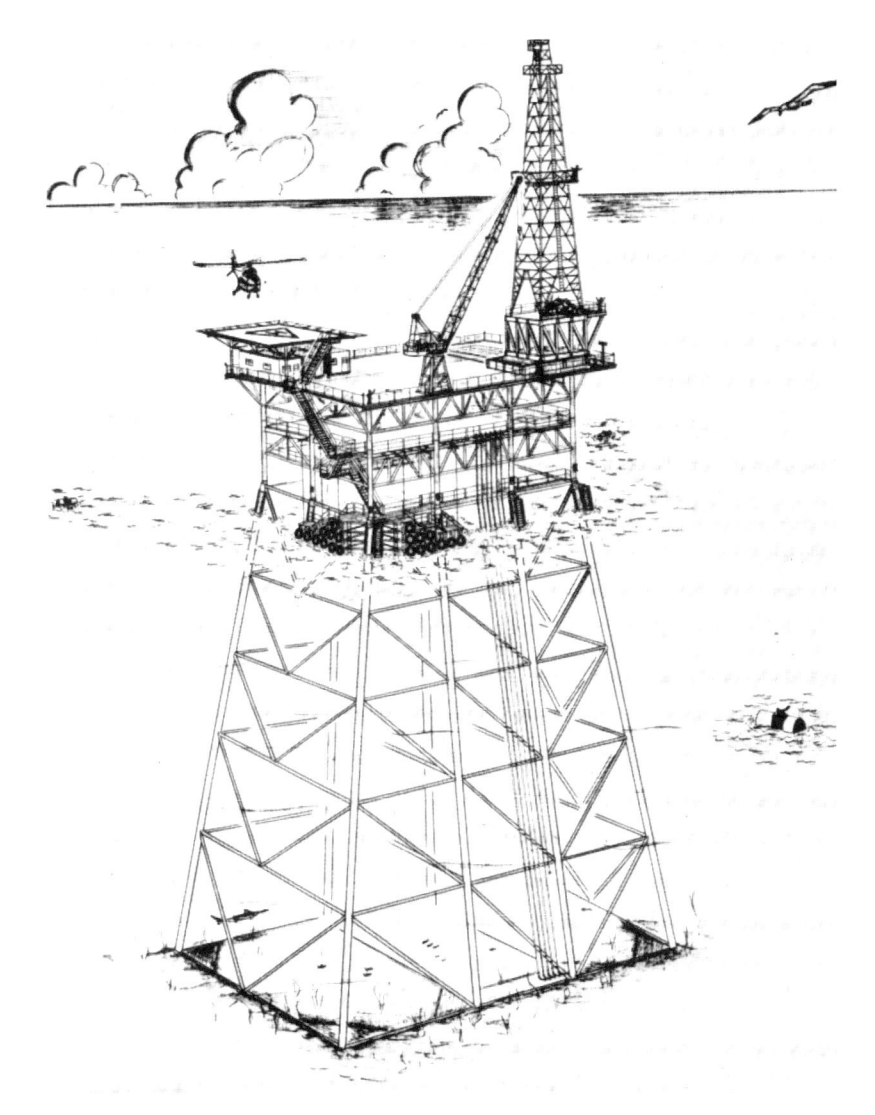

*Figure 2-7.* Template or jacket type offshore structure named Holly, a 30-well drilling and production platform in water depth of 211 feet. (Source: Mashburn, M.K. and Hubbard, J.L., "An Ocean Structure," Proceedings of the First Civil Engineering in the Oceans Conference, 1968, p. 185.)

**14    Introduction to Offshore Structures**

**Figure 2-8.** Aerial view of Holly. (Source: Mashburn, M.K. and Hubbard, J.L., "An Ocean Structure," Proceedings of the First Civil Engineering in the Oceans Conference, 1968, p. 184.)

expressed by the following quotation: "The relatively deep water and multiplicity of wells required a structure of sizable dimensions which subsequently proved to be a challenge to the ingenuity of the designers and builders because of the limitations of the available erection equipment."[8] For economic reasons, the platform was fabricated at a construction yard in the Gulf of Mexico and the completed components were towed over 4000 miles (6400 km) via the Panama Canal to the offshore site.

Platform *Holly* had two decks, a production deck 60 x 100 ft (18 x 35 m) at an elevation of 38 ft (12 m) above mean lower low-water level, and a drilling deck 80 x 125 ft (24 x 38 m) at an elevation of 60 ft (18.3 m) above mean lower low-water level. The decks were fabricated of solid steel plate placed over wide-flange beams supported on tubular trusses. Above the drilling deck, a 45 ft (14 m) square heliport was located at an elevation of 80 ft (24 m). This heliport could accommodate a 10-passenger helicopter.

Two boat landings were provided. Each landing had three levels at 3.5 ft (1.1 m) difference in elevation to allow for an 8-ft (2.4-m) tidal range. The eight 36-inch (0.9-m) OD piles were driven to an average penetration of over 100 ft (31 m). Five 6-inch (15-cm) diameter pipelines were laid on the sea floor to transfer the produced crude oil to a shore facility.

Decks for platforms the size of *Holly* were installed in one of two ways. *Holly's* decks were divided into nine segments so that no single segment would weigh more than 100 tons (90 tonnes), and the segments were lifted into place with a 100-ton (90-tonne) derrick barge. The deck structure could have been cut into four or five segments and lifted into place with a 250-ton (227-tonne) derrick barge. While derrick barges of 250-ton (227-tonne) capacity were available, the decision to use 100-ton (90-tonne) lifts was based on the lower daily rental rate of the smaller derrick barge.

Exxon's 850-ft (259-m) platform in Santa Barbara Channel was installed in 1976.[9] The platform is located approximately 25 miles (40 km) west of the city of Santa Barbara about 5 miles (8 km) offshore in 850 ft (259 m) of water. It is an excellent example of a self-contained, deepwater platform for combined drilling and production activities combined with crew housing. The design made provision for drilling 28 wells. The platform's name is *Hondo*.

This platform, typical of platforms along the California coast, was designed primarily to resist earthquakes. The three basic requirements were: (1) to withstand all loads expected during fabrication, transportation, and installation; (2) to withstand the loads resulting from severe storms and earthquakes; (3) to function safely as a combined drilling, production, and housing facility.

The specific criteria were: (1) to avoid structural damage when subjected to an earthquake spectrum represented by a ground acceleration of 0.25 g; (2) to be safe against collapse when subjected to an earthquake represented by a ground acceleration of 0.5 g; (3) to withstand without loss of structural integrity plastic strain equal to 2.0 times the deformation of the 0.25 g earthquake, or 1.25 times the deformation of the 0.5 g earthquake.

The design storm condition was chosen as that with a recurrence interval of 400 years. Specifically, the wave height was 44 ft (13.4 m) trough to crest, the storm tide was 8 ft (2.4 m), and the storm wind velocity was 100 mph (45 m/sec).

Figure 2-9 shows elevation sketches of the jacket. Figure 2-10 shows the completed deck structure. The deck segments were installed in five lifts with a 500-ton (454-tonne) derrick barge. There are three deck levels 86 × 170 ft (26.2 × 51.8 m) each, and a flare boom 220 ft (67 m) long.

## 16  Introduction to Offshore Structures

EL.(+)85'-0"

JOINT

EL.(-)397'-0"

EL.(-)850'-0"

SOUTH ELEVATION          EAST ELEVATION

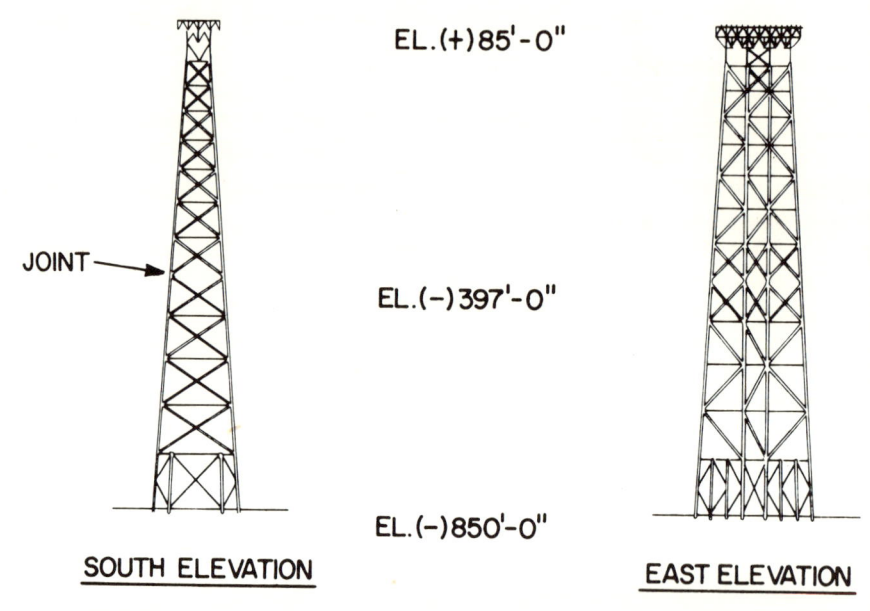

**Figure 2-9.** Exxon Company 850-foot platform in Santa Barbara Channel, California. (Source: Ruez, W.J., "Exxon's 850-foot platform in the Santa Barbara Channel," Proceedings of the Third Civil Engineering in the Oceans Conference Vol. II, 1976; p. 805.)

**Figure 2-10.** Deck Structure for Exxon Company 850-foot platform in Santa Barbara Channel. (Courtesy Exxon Company.)

# Early Platform History

The eight legs of the jacket were framed with X type and diagonal type bracing. In addition to the eight 48-inch (1.2-m) OD piles driven through the jacket legs, there were twelve 54-inch (1.4-m) OD skirt piles, four on each side, and two on each end of the jacket. (See Figure 2-9.)

The jacket was fabricated in two sections so that it could be transported on barges no longer than 450 ft (137 m). After reaching the offshore site and being launched endwise from the barges, the two jacket pieces were mated in the water and joined by welding. Preinstalled access tubes enabled welders to enter dewatered habitats and make the full-penetration groove welds from within the jacket legs.

## The British Petroleum Company

Around 1950, while the developments mentioned previously were taking place in the Gulf of Mexico and in the Santa Barbara Channel, the British Petroleum Company was engaged in similar exploration operations at Umm Shaif off the coast of Abu Dhabi in the Persian Gulf. Here, as in the Gulf of Mexico, the drilling was taking place in water depths less than 100 ft (30 m), and the problems of offshore technology were essentially the same. The Umm Shaif operation has grown steadily over the years. Figure 2-11 shows the complex as it was in 1976. In the foreground is a power and water treatment/injection platform. The quarters platform is shown to the right in Figure 2-11. Well-protector and processing platforms are shown in the distance.

## Gulf of Mexico Accident Reports

In the 1960s hurricanes in the Gulf of Mexico caused serious reevaluation of platform design criteria. Wave heights of 42 ft (13 m) and wind gusts up to 200 mph (89 m/sec) in hurricane Hilda, the 100-year storm that struck in 1964, destroyed 13 platforms in the Gulf. The next year another storm of 100-year recurrence probability, hurricane Betsy, destroyed three platforms and damaged many others. It became evident that the 25-year-recurrence design storm that had been used in the early days was insufficient to protect offshore investments. The reasoning was: If a cluster of platforms designed for a 25-year storm were placed in one localized area of the Gulf, the chances against their destruction would be good, but in spreading the platforms throughout the Gulf, the chances for survival of the same number of platforms would be greatly reduced.

With the occurrence of two 100-year storms, designers abandoned the 25-year and 50-year conditions and began designing for storms with a recurrence interval of 100 years. As a result, platform elevations increase, structural members became larger, and welded tubular joints were investigated more closely for high, localized static stresses and fatigue behavior.

**18** Introduction to Offshore Structures

**Figure 2-11.** British Petroleum Company oil complex at Umm Shaif in the Persian Gulf, 1976. (Source: D.B.J. Thomas, International Symposium on the Integrity of Offshore Structures, Institution of Engineers and Shipbuilders in Scotland, Glasgow, April 1978.)

It is difficult to reconstruct the history of accidents involving platforms in the Gulf of Mexico. The U.S. Geological Survey has compiled a list of accidents dating from 1956.[10] This list consists of five extensive tables identified as follows:

1. Blowouts
2. Explosions and fires
3. Pipeline breaks and leaks
4. Significant pollution incidents
5. Major accidents

In each table the location, date, type of accident, how controlled, volume of oil spilled, and the extent of damage are given.

According to the U.S. Geological Survey list, 19 platforms were completely destroyed or lost between 1956 and December 1976, 10 of these as the direct result of hurricanes in 1964 and 1965. Industry magazine articles have carried stories describing several Gulf of Mexico hurricanes; some articles have had location maps of lost and damaged platforms. While generally kept as unpublished information, several articles give brief damage reports evaluating the reasons for failure.[11-21]

Between 1956 and December 1976, the U.S. Geological Survey report lists 13 other instances in which platform structural damage was extensive, although the platforms were not lost. Blowouts—that is, uncontrollable flow of gas and/or oil from the wellhead and usually accompanied by fire—were by far the most common cause.

Lloyd's of London Press, Ltd. publishes weekly book-form reports of all marine casualties. From these reports, all incidents relating to drilling rigs have been extracted for the period January 1966 to December 1976. This consolidated report is available for about $500-$600. The report lists name and type of rig, name and type of incident, date, location, year of construction, owner, cost of repairs, injuries, deaths, and damage to the environment.

### References

1. Londenberg, R., "Man, Oil, and the Sea," *Offshore,* Oct. 1972, p. 54.
2. Lee, G.C., "Offshore Structures: Past, Present, Future, and Design Considerations," *Proceedings of the Offshore Exploration Conference* (OECON), 1968, pp. 169-196.
3. "First Well in Gulf Just 25 Years Ago," *Offshore,* Oct. 1963, p. 17.
4. "The Drilling Well At Sea," *Oil and Gas Journal,* Nov. 2, 1946, p. 63.
5. Short, E.H., Jr., "Prefabricated Offshore Drilling Platform Set Up in Record Time," *Oil and Gas Journal,* Aug. 9, 1947, p. 82.

6. McClelland, B.; Focht, J.A., Jr.; Emrich, W.J., "Designing and Constructing Heavily Loaded Piles," *Ocean Industry,* Jan. 1969, pp. 56-59.
7. Lee, G.C., "Offshore Platform Construction Extended to 400 Foot Water Depths," *Journal of Petroleum Technology,* Apr. 1963, pp. 383-388.
8. Hubbard, J.L. and Mashburn, M.K., "An Ocean Structure," *Proceedings of the Civil Engineering in the Oceans Conference,* Sept. 6-8, 1967, pp. 183-202.
9. Ruez, W.J., "Exxon's 850 Foot Platform for Santa Barbara Channel," *Proceedings of the Third Civil Engineering in the Oceans Conference,* June 5-12, 1975, Vol. II, pp. 801-817.
   Bandgette, J.J., "A Deep Water Platform in the Santa Barbara Channel," *Journal of Petroleum Technology,* April 1978, pp. 498-506.
10. "Accidents Connected with Federal Oil and Gas Operation on the Outer Continental Shelf," U.S. Geological Survey, Conservation Division, Jan. 1977.
11. Cooper, G.W., "Hurricane Damage to Structures," 22nd ASME Annual Petroleum Mechanical Engineering Conference (paper), Philadelphia, PA, Sept. 17-20, 1967.
12. "New Study Shows Why Fixed Platforms Fail During Storms," *Oil and Gas Journal,* Oct. 16, 1967, pp. 101-103.
13. "Hilda's Visit is Brief But Costly for Industry," *Oil and Gas Journal,* Oct. 12, 1964, pp. 104-106.
14. "Hilda's Damage May Hit $100 Million," *Offshore,* Nov. 1964, pp. 15-22.
15. Rodgers, L.C., "Hilda Kicks Operating Costs Up," *Oil and Gas Journal,* June 21, 1965, pp. 154-158.
16. "Oil Installations Hit Hard by Camille," *Oil and Gas Journal,* Aug. 25, 1969, pp. 40-41.
17. "Camille Knocks Out 300,000 Barrels per Day and Costs Industry," *Offshore,* Sept. 1969, pp. 33-35.
18. "Giant Shell Platform Lost in Hurricane Camille," *Ocean Industry,* Sept. 1969, p. 13.
19. Davis, W.J., "Camille's Impact," *Ocean Industry,* Oct. 1969, pp. 11-17.
20. Gray, E.M., Jr., "Sun Designs and Erects Platform in 10 Months," *Offshore,* Nov. 1969, pp. 71-74.
21. Sterling, G.H., "The Failure of the South Pass 70 Platform B in Hurricane Camille," *Journal of Petroleum Technology,* Mar. 1975, p. 263.

# Chapter 3

# *Platforms, Catwalks and Heliports*

There are several types of steel template offshore platforms. Economics control the specific choice of platform to be placed at a given location. In deep water (approaching 400 ft or 122 m) all functions are combined on one multilevel structure called a self-contained platform. (See Figures 3-1 and 2-7.) In more shallow water it is advantageous to separate the functions and have several separate platforms. This chapter discusses several special purpose platforms. These are:
1. Drilling/well-protector platforms
2. Tender Platforms
3. Self-contained template platforms
4. Self-contained tower platforms
5. Production platforms
6. Quarters platforms
7. Flare jacket and flare tower
8. Auxiliary platforms

Catwalks and heliports are also discussed.

## Drilling/Well-Protector Platforms

Platforms built to protect the risers on producing wells in shallow water are called well-protectors or well jackets. Usually a well jacket serves from one to four wells. Such a platform may be either one large pipe or caisson, or an open lattice truss template structure.

There are two main types of well jackets: the slip-over type and the development type. Both types protect the well (or wells) from ship collisions and environmental forces, and serve as the support for aids to navigation devices, meter-run equipment for calibration of meters, wireline units, helicopter pads, flowline risers, and conductor tubes.

## 22 Introduction to Offshore Structures

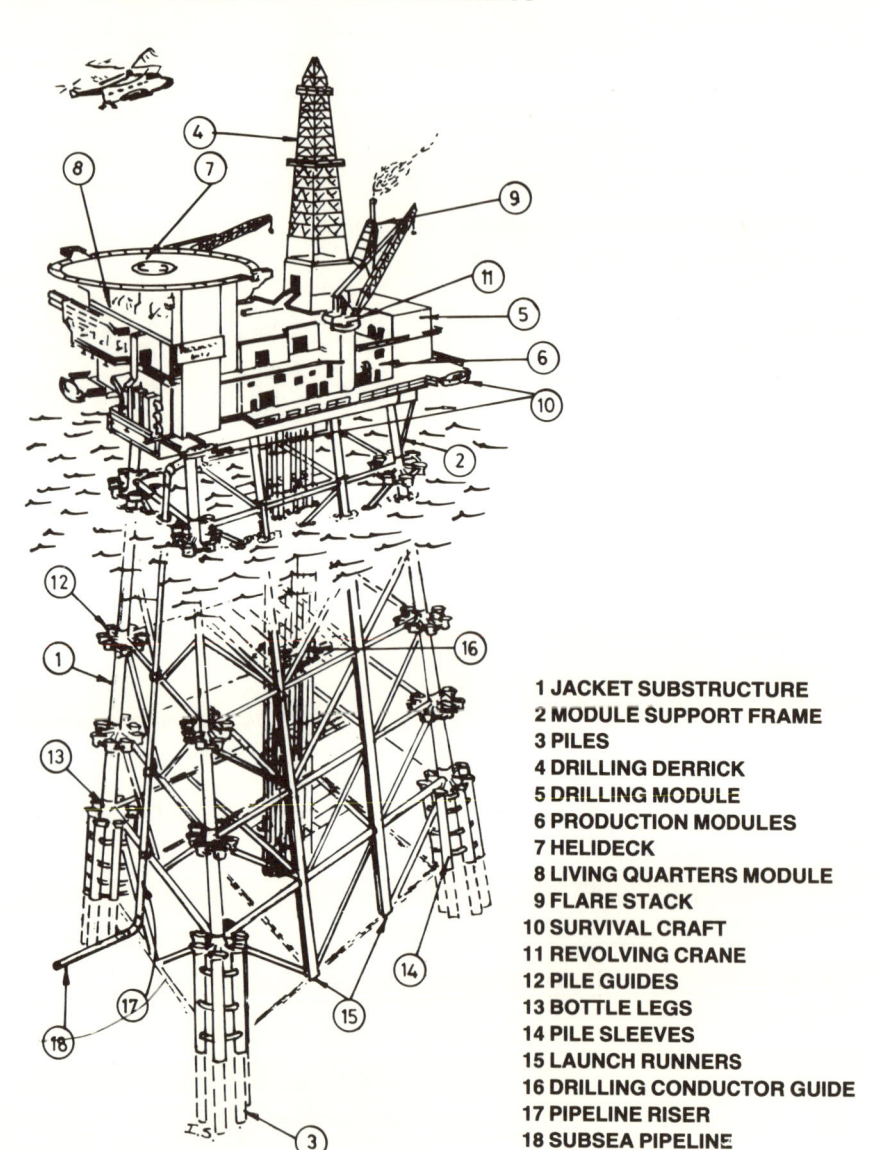

1 JACKET SUBSTRUCTURE
2 MODULE SUPPORT FRAME
3 PILES
4 DRILLING DERRICK
5 DRILLING MODULE
6 PRODUCTION MODULES
7 HELIDECK
8 LIVING QUARTERS MODULE
9 FLARE STACK
10 SURVIVAL CRAFT
11 REVOLVING CRANE
12 PILE GUIDES
13 BOTTLE LEGS
14 PILE SLEEVES
15 LAUNCH RUNNERS
16 DRILLING CONDUCTOR GUIDE
17 PIPELINE RISER
18 SUBSEA PIPELINE

**Figure 3-1.** Typical North Sea steel-piled self-contained drilling/production platform for water depths of more than 400 feet. (Source: J.G. Timar paper, Lectures on Offshore Engineering, Institute of Building Technology and Structural Engineering, Aalborg University Centre, Aalborg, Denmark, spring 1978.)

## Platforms, Catwalks and Heliports

The slip-over well jacket is used for exploratory wells in water 50-100 ft (15-30 m) deep. A large caisson is first driven into the mud through which the well is drilled. The caisson protects the well initially and, within a few months after drilling, a well jacket is slipped over the caisson. The slip-over well jacket is a four-legged structure with one side left open. After it is slipped around the caisson, additional bracing is added on the open side. The slip-over jacket ordinarily serves one well. Piles are driven through the jacket legs into the ocean floor to secure the unit.

The development well jacket may accommodate several wells, depending on the design conditions. This type of jacket is installed prior to drilling. In shallow water with depth to approximately 15 ft (4.5 m), the mobile exploratory drilling vessels are small, and the slot in the vessel which permits the vessel to locate the drilling rig over the well jacket is narrow. Thus, well jackets with a plan size of about 8 x 16 ft (2.4 x 4.8 m) are used; usually such jackets are for one well. In water depths of 15-45 ft (4.5-14 m) a 16 x 20 ft (4.8 x 6 m) four-well jacket is usually used. At this water depth, the slot width in the drilling vessel may still be the limiting factor on jacket size. In water depths of 45-100 ft (14-31 m) the well jackets range from 20 to 30 sq. ft (1.9 to 2.8 sq. m). Mobile drilling rigs for such water depths are large, and slot size is not a limitation.

For water depths of more than 100 ft (31 m), drilling/well-protector platforms become more complicated. For example, if it has been decided to put the living accomodations on one platform and the processing or treatment facilities on another platform, the drilling platform can be supported by a jacket with only four legs (piles). (See Figure 3-2.) During the drilling phase, this platform contains the derrick and substructure, drilling mud and drill water, fuel, mud and water storage tanks, primary power plant, and several pumps. Usually, the pumps are all mounted on one skid as a pump package.

The total weight of the drilling equipment is around 3000-4000 kips (a kip is a thousand pounds), or $13.4 \times 10^6$ N-$17.8 \times 10^6$ N. The various component weights are:

| | |
|---|---|
| Derrick and drilling rig | total = 1346 kips ($5.99 \times 10^6$ N) |
| Set back (lengths of stored pipe) | 225 kips ($1.00 \times 10^6$ N) |
| Substructure beneath drilling rig | 160 kips ($0.71 \times 10^6$ N) |
| Substructure support mat | 60 kips ($0.27 \times 10^6$ N) |
| Drilling derrick | 80 kips ($0.35 \times 10^6$ N) |
| Tools | 15 kips ($0.07 \times 10^6$ N) |
| Drawworks | 78 kips ($0.35 \times 10^6$ N) |
| Drawworks motors | 12 kips ($0.05 \times 10^6$ N) |
| Traveling blocks and equipment | 50 kips ($0.22 \times 10^6$ N) |
| Hook load | 666 kips ($2.96 \times 10^6$ N) |

**24** Introduction to Offshore Structures

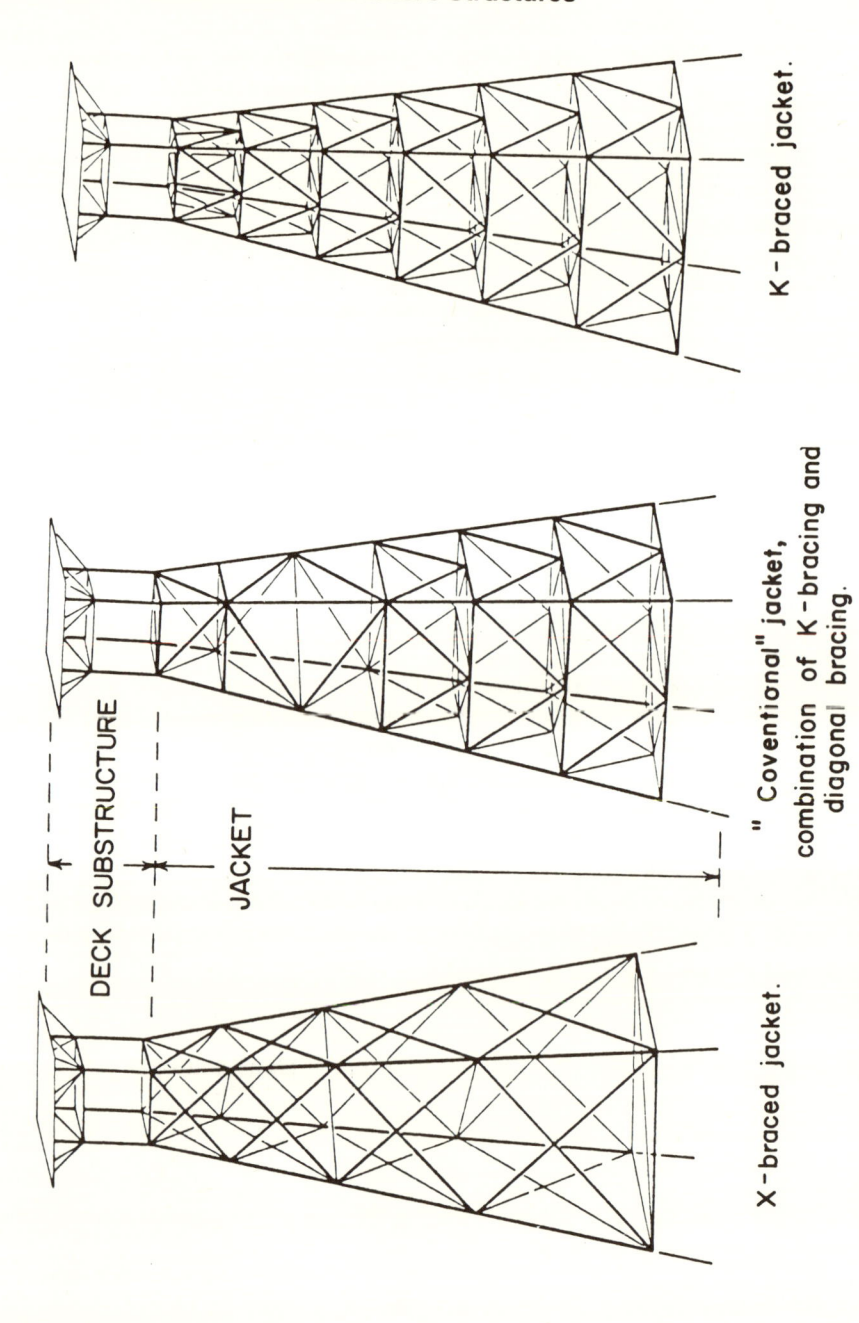

*Figure 3-2. Drilling/well-protector jackets with four legs. (Copyright © 1972 Offshore Technology Conference.)*

| | |
|---|---|
| Storage tanks | total, full = 1090 kips ($4.85 \times 10^6$N) |
| Drilling water (two tanks) | 294 kips ($1.31 \times 10^6$N) |
| Potable water | 147 kips ($0.65 \times 10^6$N) |
| Fuel oil | 128 kips ($0.57 \times 10^6$N) |
| Reserve mud tank with desander, degasser, and shale slaker | 272 kips ($1.21 \times 10^6$N) |
| Active mud | 249 kips ($1.11 \times 10^6$N) |
| Pump package | total = 265 kips ($1.18 \times 10^6$N) |
| Rotating equipment | 32 kips ($0.14 \times 10^6$N) |
| Reciprocating pumps | 132 kips ($0.59 \times 10^6$N) |
| Other | 101 kips ($0.45 \times 10^6$N) |
| Power plant | total = 274 kips ($1.22 \times 10^6$N) |
| Well logging unit | total = 25 kips ($0.11 \times 10^6$N) |

After wells have been drilled (usually from four to nine) and the drilling equipment is removed, other equipment is installed so that the platform may protect the wells while producing crude petroleum. The new equipment consists of the Christmas trees of valves on the wells, a manifold for collecting the production of the wells so that it can go to the processing or treatment platform in one pipeline about 6 inches (152 mm) in diameter, fire safety equipment, navigation warning lights, and a well-kill system.

A well-kill system controls a runaway well. Well control is accomplished by pumping mud and brine down into the well. The weight of the long column of heavy fluid (mud) is greater than the upward force of the crude petroleum from the pressure in the ground. A well-kill system consists of a mud storage tank, a diesel-driven mud pump, and a brine storage tank. There may also be mixing tanks for manufacturing the brine and for mixing the chemicals for the mud.

### Tender Platform

The tender platform is not used as commonly now as it was 20 years ago. In terms of size and operation it falls between the well jacket and the self-contained platform. Generally, the derrick and substructure, drilling mud, primary power supply, and mud pumps are placed on the platform. The drilling crew quarters, remaining equipment, and supplies are located on the tender ship moored adjacent to the platform. The two are usually connected by a long walkway. Figure 3-3 shows a 1959 Gulf of Mexico tender platform operation in 200 ft (60 m) of water.

**26** Introduction to Offshore Structures

*Figure 3-3.* Gulf of Mexico tender platform operation in 1959: The South Timbalier Block 131 location in a water depth of 200 feet. (Copyright © 1968 Offshore Exploration Conference.)

## Self-Contained Template Platforms

The self-contained platform is a large, usually multiple-decked, platform which has adequate strength and space to support the entire drilling rig with its auxiliary equipment and crew quarters, and enough supplies and materials to last through the longest anticipated period of bad weather when supplies cannot be brought in. The bad weather period is usually three to four days in the Gulf of Mexico, so the platform should accommodate supplies and materials for approximately twice this time.

There are two types of self-contained platforms: the template type and the tower type. Actually, the tower platform is also a template structure; however, the piles are driven in a different manner. Template and tower platforms are described in this chapter in addition to other platform types.

## Platforms, Catwalks and Heliports

The self-contained template platform consists of a large multilevel deck structure supported by long piles driven deep into the ocean floor. The template, also called a jacket, is a three-dimensional welded frame of tubular members and is used as a guide for driving piles through the hollow legs of the jacket. The jacket also holds the piles together so that they act as a single unit against lateral forces. The well jacket used in shallow water is a template structure, but because of its simple function compared to a self-contained platform, it is not considered part of this discussion.

Self-contained template platforms have been designed and constructed in many sizes and shapes. Early template structures had many legs and a multiplicity of horizontal and diagonal braces. The first Shell Oil Company self-contained platform, built in 1955 in 72 ft (22 m) of water, had 53 pilings. Today, in the Gulf of Mexico most platforms fit one or the other of two classifications: those with 10 or 12 piles and those with eight piles. In the early days of exploration in the Gulf of Mexico the diameters of tubes available for jackets were limited, and more piles were needed to get sufficient soil support. More recently, with the availability of very large tubes, the trend is toward the eight pile type of platform. This latter type has been used in water depths approaching 400 ft (122 m).

In plan view the jacket legs form a rectangle. The jacket dimensions vary from company to company, but generally speaking, the sizes are about the same. The leg spacings in a horizontal plane are usually given for about 10-15 ft (3-4.5 m) above the mean water line. At this elevation, a 10-pile jacket has four legs along each side, each spaced 40-45 ft (12-14 m) apart. In some cases the center bay along the side is as much as 60 ft (18 m). In the narrow direction the center leg of the three legs is used for piling in addition to the four legs on each side.

In plan view the rectangle formed by the jacket legs in an eight-pile platform is more narrow, but the spacings at the 10-15-ft (3-4.5-m) elevation above the mean water line are similar. The length of the center bay is closer to 60 ft (18 m) while that of the shorter bays along the side is about 45 ft (14 m). Across the narrow direction of the jacket, the leg spacing is 45 ft (14 m). Jackets constructed in more recent years frequently have larger-diameter corner legs; typical pile sizes of Gulf of Mexico eight-pile jackets of today are 60-inch (1.5-m) OD for corner legs and 48-inch (1.2-m) OD for in-between legs. Allowing one-inch annular clearance between the pile and the inside of the leg means that the jacket legs have internal diameters of about 62 inches (1.6 m) and 50 inches (1.3 m), respectively.

Jacket legs are not vertical. From the plan view rectangular size at 10-15-ft (3-4.5-m) elevation (seadeck level), the legs flare out, or are said to have "batter," as follows: batter of one foot in seven or eight for legs on the long sides, and batter of one foot in 10 or 12 for legs in the narrow dimension of the jacket.

Today, the eight-pile platform configuration is adapted to increased water depths and various soil conditions by adding skirt piles. Because of the batter of the legs, there is increased separation between the legs at the ocean bottom end of the jacket. The skirt piles are usually placed in-between the legs almost in the plane of the side of the jacket. Their purpose is to assist in resisting the overturning moment on the structure. Skirt pile guides usually extend vertically only between the two lowest levels of horizontal bracing in the jacket. In plan view these guides must be slightly outside the plane of the side of the jacket so that the pile extensions necessary for above-water driving can be supported by ring guides at various levels up the side of the jacket. In still more recent platform designs the lower portions of the legs are constructed of very large-diameter tubes so that several piles may be driven through pile guide tubes provided in the large-diameter legs. These enlargements of the lowers ends of the jacket legs are called *bottles*.

The wells from a template platform are drilled through conductor tubes driven into the ocean floor. These tubes are positioned vertically within the jacket by several levels of horizontal framing containing guide rings for the passage of the tubes. Conductor guides hold conductors laterally only.

Although the template type jacket is designed to float, it does so with almost all of its members submerged, and only a small portion of the structure at or above the water surface. (See Figure 3-4.) Thus, the jacket is transported to the erection site on a barge, is launched from the barge, and is subsequently upended into position. The piles are driven after the jacket has been positioned. The deck structure, divided into pieces appropriate for the lifting capacity of the derrick barge, is brought to the site on transportation barges and is then lifted into position on the tops of the piles extending up from the jacket legs. The deck structure is welded to the pile ends, and the piles are welded to the upper ends of the jacket legs. Some companies fill the annular space between the piles and the inside of the jacket legs with grout; others do not. Filling the annular spaces with grout enables the piles and jacket to resist lateral forces as a single, rigid structure. Those who leave this space open do so on the premise that salvaging of the jacket for use at another drilling site will be easier without the weight and complication of the grout. (Very few jackets have actually been relocated.)

## Tower Type Template Platforms

The self-contained template platform of the tower type is characterized by relatively few large-diameter, non-battered legs and fewer diagonal braces of larger size than those used in regular template type structures. The tower type jacket was conceived to eliminate the need to launch the structure from a barge; it can be floated to location using the buoyancy of its larger-diameter legs. The tower platform was originally designed to supply deepwater structures for the Pacific coast and for Cook Inlet, Alaska.

*Figure 3-4. Deepwater jacket after launching and before upending. (Courtesy J. McDermott Incorporated.)*

The regular template structure has many cross braces, horizontal and diagonal. The tower structure has relatively few braces, and none extend into the splash zone above the mean low-water level. This reduces the lateral resistance the structure offers to the passage of large storm waves and eliminates having bracing members in the path of ice floes in cold climates.

The pile foundation for a tower platform usually consists of several groups of piles, frequently four. Each group of piles is driven through one of the large-diameter legs with each pile subsequently serving as a conductor through which a well may be drilled.

There are from 8 to 12 cylindrical tubular piles per leg. The piles are arranged in a circle around the inside of the large-diameter leg. The guide tubes for the piles are structurally positioned using bulkheads with an inner cylinder inside that is concentric with the outer shell of the leg. Cement grout is used to fill the annulus between the outer cylindrical shell and the inner cylinder. This large usage of grout keeps the outer cylindrical legs from failing by local buckling and also bonds the piles to the inner and outer cylinders to achieve composite action.

**30    Introduction to Offshore Structures**

The four-legged tower platforms in Cook Inlet have legs ranging in diameter from 14 to 17 ft (4.3 to 5.2 m). The bracing members, all of which are below the level of formation of maximum ice thickness, range in diameter from 48 to 74 inches (1.2 to 1.9 m).

The piping and valves required for flooding the legs for upending of the tower are located inside the legs, as are the conduits for cathodic protection anodes, the flowline risers, pumps, and instrumentation. Upending and positioning of the tower structure in the water can be accomplished without the use of a derrick barge, although one is needed for placing the various segments of the deck structure on the upended tower.

Figure 3-5 shows a tower platform in Cook Inlet, Alaska. A similar Cook Inlet platform, Figure 3-6, is shown with ice crushing against the legs. A tower

**Figure 3-5.** Cook Inlet four-legged tower platform. (Copyright © 1968 Offshore Exploration Conference.)

## Platforms, Catwalks and Heliports 31

**Figure 3-6.** *Ice crushing against the legs of a four-legged Cook Inlet platform. (Copyright © 1968 Offshore Exploration Conference.)*

## 32  Introduction to Offshore Structures

type jacket is shown being towed in Figure 3-7. The small-diameter braces at the jacket end nearest the tugboat were placed there temporarily just for the towing phase.

### Production or Treatment/Reinjection Platforms

Production platforms support buildings, compressors, storage tanks, treating equipment, and other attendant facilities.

A production or treatment/reinjection platform is basically a platform for separating the oil-gas-water mixture of the produced crude petroleum into crude oil, natural gas, and water, and treating each of these by simple in-the-field processes prior to transport, disposal, or reinjection into the earth. The processing facilities on a particular treatment platform may vary depending on what is to be done with the natural gas and whether the crude oil is to be off-loaded into tankers or pumped ashore through a pipeline. Sometimes there are facilities on the treatment platform for injecting pressurized water into the oil-bearing strata within the earth by means of an injection well, or wells, to improve the production of petroleum from other wells.

If there are two pipelines from the treatment platform to shore, one for crude oil and one for natural gas, the platform must also have two sets of metering and shipping equipment. Pumps send the crude oil through strainers, then through meters, and into the liquid pipeline. Compressors send the natural gas through strainers, then meters, and into the gas pipeline. For each system, there are meter calibration or test loops, recording instruments, and sphere launchers to force cleaning devices through the pipelines. In simplest form these cleaning devices, sometimes called pigs, are spheres of hard rubber somewhat larger in diameter than the pipeline through which they are to pass. They are inserted into the pipeline through by-pass loops in the line and are forced along by the pressure of the compressed gas or liquid behind them.

If there is no gas pipeline, the natural gas is usually burned off through a flare tower remote from the treatment facilities on the platform. For some deepwater self-contained platforms, the flare tower is part of the deck substructure, extending far away from and above the decks proper. When water depths permit, it is safer to mount the flare tower on a separate jacket remote from the treatment platform.

Personnel safety is of primary importance on any platform. On a production or treatment platform, there should be safety systems to provide for gas leak detection and fire protection. The water supply system, or systems, must be adequate for utilities and fire protection. Protection must also be provided against those fires which cannot be fought with water.

The production and treatment facilities include equipment for separation of the natural gas from the liquid mixture and for separation of the water and

Platforms, Catwalks and Heliports 33

**Figure 3-7.** Jacket for four-legged tower platform under tow. Copyright © 1968 Offshore Exploration Conference.)

### 34   Introduction to Offshore Structures

entrained sand from the crude oil. Some of the separated natural gas powers gas turbines to drive generators for production of electricity. The pumps and compressors on the platform are electric-driven. The various pieces of process equipment on such a platform cannot be identified in a general way because systems vary; in some cases the separations go through two stages. There is a manifold for incoming crude petroleum from the drilling/well-protector platforms, pressure storage tanks for gas to be flared, fuel gas storage tanks, surge drums, and sometimes crude oil storage tanks.

As an alternate to flaring, the natural gas is often compressed to around 5000 psi ($34.48 \times 10^6$ Pa), or more, and reinjected into the oil-bearing strata of the earth. The reciprocating piston compressors develop primary and secondary shaking forces which are transmitted into their foundations. If reinjection of gas into the ground is anticipated, the platform should be designed to withstand these shaking forces.

### Quarters Platform

The living accommodations platform for offshore workmen is commonly called a quarters platform. For deepwater self-contained platforms approaching 400 ft (122 m), the quarters are made an integral part of the single platform for economic reasons. In more shallow water the living quarters may be separated from the drilling and treating activities as a matter of crew safety. The quarters platform is built near enough to the drilling or production platform to enable the two to be connected by a bridge. In some instances where it is desirable to increase the equipment load capacity of an older self-contained platform, a separate quarters platform is constructed, moving that portion of the weight off of the first platform to permit the installation of heavier operating machinery.

A drilling crew consists of approximately 18 men. The crew leader is called a driller. There are five skilled laborers called roughnecks, a derrickman, a motorman, a diesel engine operator, a pump operator, a mud man, a crane operator, and six ordinary laborers called roustabouts.

Two crews are always on the offshore platform. Each crew works a 12-hour day, from 1100 to 2300 and from 2300 to 1100. This arrangement gives each crew some daylight working hours. There are a number of additional people on the platform, for example: the overall supervisor called a tool pusher, a customer company representative, a welder, a maintenance man, two cooks and two galley employees (one for 0600 to 1800 and one for 1800 to 0600), and a bedroom steward. The bedroom steward makes the beds, changes the linens (towels each day, bed linens each week), and cleans the living quarters. The tool pusher, company representative, welder, and maintenance man work from 0600 to 1800. The galley and bedroom people are usually provided through a subcontract with a hotel service company. In addition, there may be onboard several engineers of special expertise: a petroleum/environmental specialist, a directional drilling engineer, one or two chemical mud specialists, a gas turbine specialist, etc.

Altogether, the number of people who require housing is between 50 and 75. Crew personnel usually work seven days and are ashore seven days, or they work 14 days and are ashore 14 days. Four meals are served each day: 0400-0600, breakfast; 1000-1230, lunch; 1600-1830, dinner; and 2200-0030, night meal. Only one 30-minute eating period is included in the 12-hour working day.

The tool pusher and the company representative each has an office of about 110 sq ft (10.2 sq m); each has a bedroom of about 110 sq ft, and they share a bathroom between their bedrooms. The welder and maintenance man share a bedroom of about 110 sq ft. They share the bathroom with another bedroom for two people usually reserved for visiting specialists.

There is a hospital room (actually an isolation room equipped for only simple medication) with two to four beds and a private bath. When someone becomes seriously ill or has an accident, a helicopter is brought out from shore and the individual is removed to a regular hospital. There is a radio and communications room, including ship-to-shore microwave telephone and side band radar.

There are four large rooms in the living quarters. These are: dayroom (TV and lounge), galley, kitchen, and change room. These each have areas of about 700 sq ft (65.1 sq m) except for the kitchen, which is about two-thirds the size of the galley (eating area). There are usually two hot food buffet counters in the galley. Sometimes a game room is provided with billiard tables. Showers, toilets, and lockers are available for work clothes in the change room; sometimes washing machines and dryers also. Each man is responsible for his own work clothes. On smaller platforms without washers and dryers, each man brings several sets of work clothes to the platform when he reports for work. Each man also brings ordinary street clothes to wear when he is not working.

A large walk-in freezer is adjacent to the kitchen. In the kitchen there are at least two large refrigerators, a large cooking stove with oven, coffee urns, a grill, industrial-size food mixer, dishwasher, disposal, and sinks.

Most of the bedrooms on the platform are made to accommodate four people. Each room contains four single beds arranged as upper and lower bunks, sometimes all on one wall and sometimes each pair on opposite walls. Each room contains four clothes lockers and a mirror. There is a reading light by each bunk in addition to a ceiling light. The area of a bedroom is about 120 sq ft (11.2 sq m). Frequently, the bedrooms are grouped about a community bathroom with several individually enclosed toilets and several showers. There is also a large walk-in linen storage room adjacent to the bedrooms, and a janitor's closet.

## 36  Introduction to Offshore Structures

Living accommodations can be arranged in many ways. On some platforms they comprise three floors: the four-man bedrooms, bathroom, and linen storage on the bottom floor; kitchen, galley, TV lounge, and change room on the middle level; supervisory offices, radio room, and bedrooms for the supervisory people on the top floor. The heliport is mounted on the top level above the quarters.

The number and type of people needed to operate a production or treatment platform is different from that needed for drilling. Even so, the total number of people who require housing is still in the range of 50-75.

If a separate quarters platform is to be built, the space limitations listed previously can be relaxed somewhat. However, the number of berths might be twice as many.

There are other considerations connected with living accommodations. There must be a sewage system and septic tank. There must be storage tanks for potable water and utility water. Usually, water is brought from shore by supply boats and pumped into the storage tanks. Fire safety water is pumped from the sea.

Sometimes, when there are personnel from two cultures on the same offshore platform, there will be two galleys, two kitchens, and a separation within the bedroom-bathroom area insofar as possible.

If the offshore operation is less than 50 miles from the coastline, the crews are usually transported by boat; if farther out at sea, helicopters are usually used.

### Flare Jacket and Flare Tower

A flare jacket is a triangular-shaped tubular steel truss structure that extends from the mudline to approximately 10-13 ft (3-4 m) above the mean water line (MWL). It is secured to the ocean bottom by driving tubular piles through its three legs. The pile tops extending above the tops of the jacket legs are cut off level at an elevation about two feet above the jacket top. The flare tower is mounted on top of these pile ends.

The flare tower is fabricated the same as the flare jacket and extends from the three to six meter elevation above MWL to about one-third the height of the derrick on the drilling platform, or 10-20 ft (3-6 m) higher than the heliport landing surface on top of the quarters platform. The columns of the flare tower may stab directly into the pile tops. To facilitate the stabbing, curved shims are inserted in the jacket legs and welded to the sides of the piles extending out of the jacket legs. The flare tower columns are welded into the tops of the piles, making a single monolithic structure. The catwalk frames into the side of this monolithic structure. Sometimes, a small working deck is welded to the tops of the piles at the proper elevation to provide a support surface for the catwalk and to permit the flare tower to have a larger size base. The flare tower is mounted on this working deck. All of the components are welded together to form a single structure.

## Platforms, Catwalks and Heliports 37

The flare jacket is usually constructed with either K or X bracing throughout most of its length. (See Figure 3-8.) The top two or three levels of bracing may be diagonals. The flare tower may be constructed with diagonal bracing, and the flare pipes are mounted inside the framework. Another design is also shown in Figure 3-8. In this case the main flare pipe is made a part of the flare tower; diagonal bracing is used.

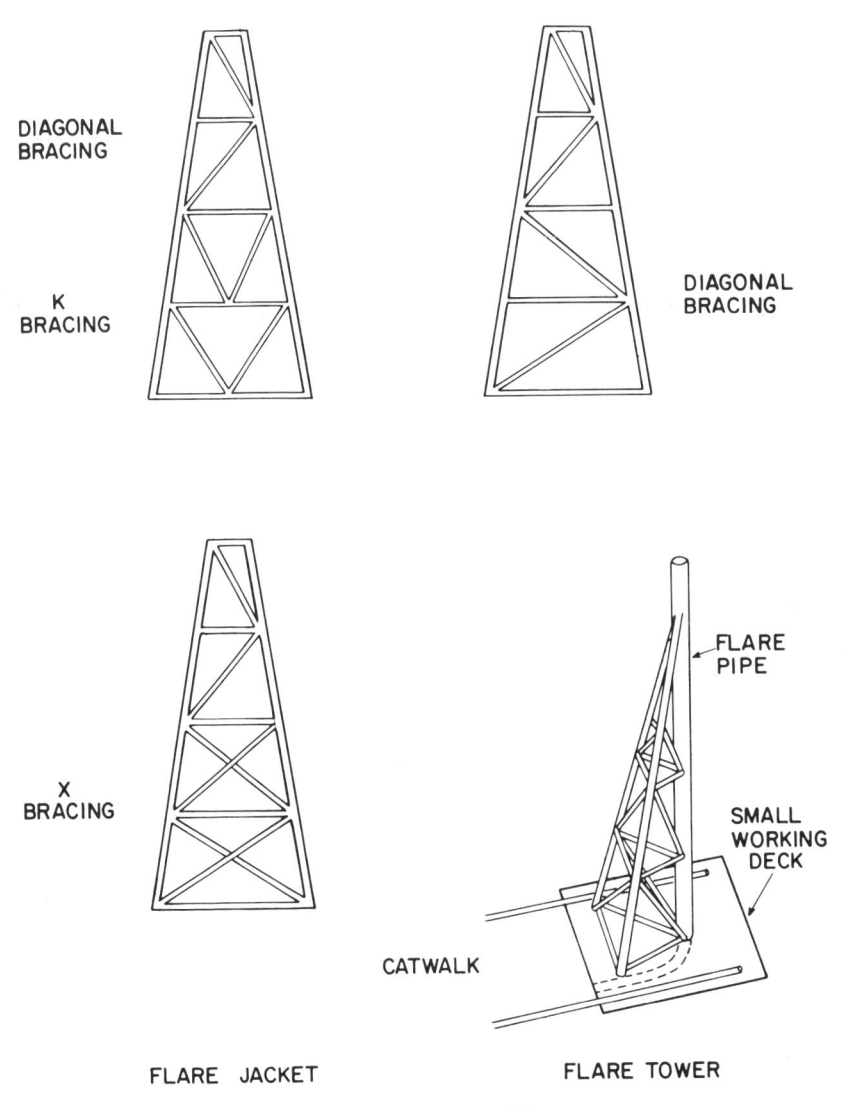

*Figure 3-8.* Flare jackets and flare towers.

**38    Introduction to Offshore Structures**

Usually, three pipes extend vertically within the flare tower: the main gas line, the gas line for the pilot flame, and the flame igniter line (flame-front generator). Another pipeline coming from the producing well may also be part of the flare tower. This pipeline is for emergency use if it becomes necessary to temporarily flare the entire crude petroleum production from the well.

### Auxiliary Platforms

Sometimes small platforms are built adjacent to larger platforms to increase available space or to permit the carrying of heavier equipment loads on the principal platform. Such auxiliary platforms have been used for pumping or compressor stations, oil storage, quarters platforms, or production platforms. Sometimes they are free standing; other times they are connected by bracing to the older structure.

### Catwalks

A catwalk is a bridge 100-160 ft (30.4-48.7 m) in length that connects two neighboring offshore structures. A catwalk may serve one or all of the following functions: supporting structure for pipelines, pedestrian movement, or a bridge for material-handling. Usually, a particular catwalk serves a combination of these functions. Catwalks are straight, single-span, tubular steel truss bridges. Length, width, elevation, and truss type vary with each catwalk.

The pedestrian walkway may be located on either the top or bottom horizontal truss for a rectangular cross section catwalk. Rectangular cross section catwalks are usually made of four Warren or four Pratt bridge trusses. The braces are usually small-diameter cylindrical tubulars. The chords are cylindrical tubulars, rectangular tubulars, or wide-flange members. Pipelines are run underneath the walkway when the walkway is on the top of the cross section; pipelines are overhead when the walkway is on the bottom of the cross section. Figure 3-9 shows several geometries that have been used for catwalks.

Triangular cross section catwalks have also been constructed with the pedestrian walkway either at the top of the cross section or at the bottom. The triangular catwalk with the walkway at the bottom of the cross section is the most common; this type of catwalk uses either two Warren or two Pratt bridge trusses. These same trusses may be used when the walkway is at the top of the cross section; a modified Warren bridge truss has also been used. (See side views of trusses C and D in Figure 3-9.)

The catwalk leading to a flare jacket with flare tower is not required to support very heavy loads. Usually, the pipelines supported by such a catwalk are the main gas line, the gas line for the pilot flame, flame igniter line (flame-front generator), and electric conduits for aerial warning lights. Such catwalks are made with two Warren roof trusses for the sides and either a Warren or Pratt bridge truss for the bottom; the pipes lie inside the triangle on the bottom truss.

**Platforms, Catwalks and Heliports** 39

The catwalk from the production or treatment platform to the quarters platform should support pipelines for potable water, utility water (semipurified water for cleaning purposes), electrical conduits, and communications lines.

The catwalk between the drilling/well-protector platform and the treatment platform should support pipelines for the crude petroleum, utility water, fire safety water (seawater), potable water, bypass to flare pipeline, electrical conduits, and communications lines.

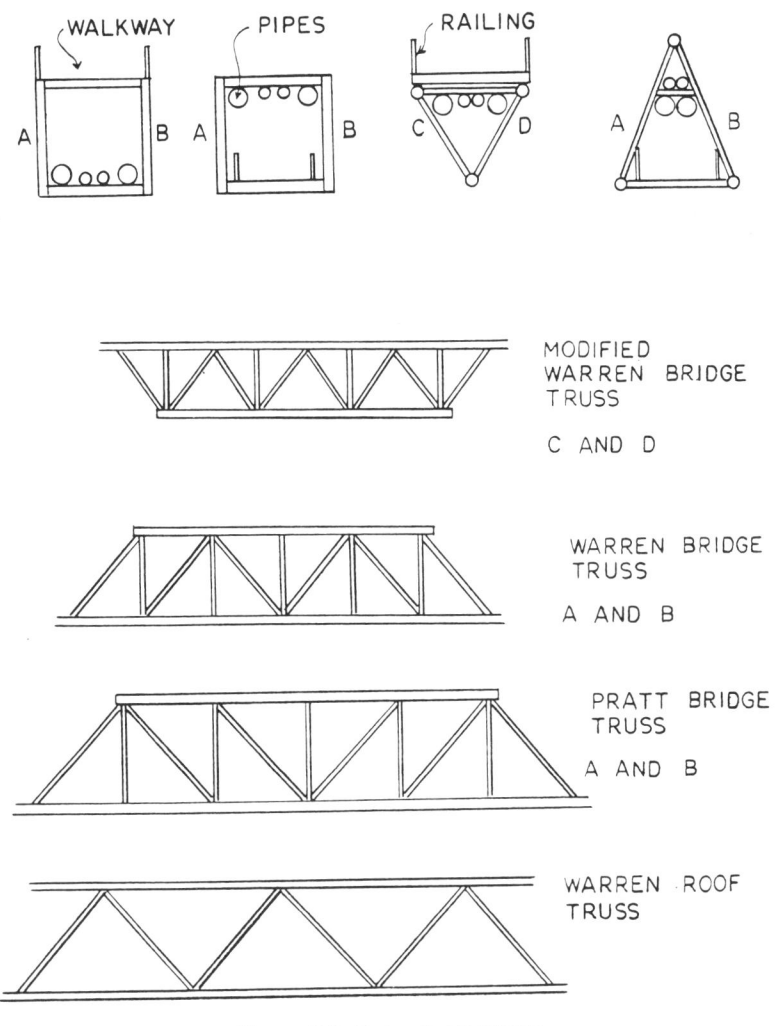

*Figure 3-9.* Catwalk structures.

## 40  Introduction to Offshore Structures

The walkway on a catwalk should be wide enough to permit the use of small fork trucks and pallets. There are always boxes, crates, bags, drums, barrels, or small machinery components to be moved. The decks of catwalks should be steel grating supported by grid systems of beams and stringers.

## Helicopters and Heliport Design

### Helicopters

When the distance to the offshore location is approximately 50 miles (80 km) or less, work crews are transported by boat. When the distance is more than 50 miles, helicopters are used.

There are several advantages to using helicopters for transportation. These include:

1. Considerable time-saving and, therefore, reduction in cost. A helicopter cuts travel time to about one-sixth of that by boat.
2. Transfers between boats and the offshore platform are sometimes impossible in high seas. Helicopter reliability and capability in bad weather are much better.
3. Boat-delivered crews sometimes arrive seasick and ill-prepared for work—not so for helicopter transported crews.
4. Supervisors or specialists can hop quickly from shore to platform and back, accomplishing their jobs more efficiently.
5. Emergency repair parts can be obtained more quickly; geological specimens can be more rapidly taken to shore for analysis.
6. Injured men can be flown to hospitals on shore.
7. Rapid evacuation of the platform is possible in the event of an emergency or severe storm.

As early as 1952, the Sikorsky S-55 helicopter was regularly used in offshore operations. The Bell 204B and the Sikorsky S-58 came into general offshore use in 1957. The Sikorsky S-62 was introduced in 1963, and the Sikorsky S-61, in 1965. Today, newer models of these helicopters and some by other manufacturers are used offshore. Table 3-1 lists the characteristics of many large helicopters.

Whereas the requirements of scheduled airliners must be met by the passengers, the helicopter must fit into the needs of the offshore oil operator, conforming to his requirements. The large helicopters used presently for offshore work are twin or multiengined machines capable of carrying 20 or more passengers in all weather conditions with airline dependability. In most cases these helicopters are readily convertible for cargo-carrying by folding up or removing the seats. In general, the cargo capacity of a large helicopter is in the range of 5000 pounds ($22.3 \times 10^3$N) carried either internally or in a cargo sling underneath the craft.

Platforms, Catwalks and Heliports    41

Table 3-1
Characteristics of Helicopters

| Make and model | Gross weight (lb) | Main rotor diameter (ft) | Overall length (ft) | Overall height (ft) | Tread of main gear (ft) | Wheel base (ft) | No. of seats Crew | No. of seats Pass. | Fuel capacity (gal) |
|---|---|---|---|---|---|---|---|---|---|
| Bell 206 B | 3,200 | 33.3 | 39.1 | 9.5 | 6.4 | skid | 1 | 4 | 76 |
| Bell 206 L | 4,050 | 37.0 | 42.7 | 11.7 | 7.7 | skid | 1 | 6 | 98 |
| Aerospatiale 350 | 4,300 | 34.0 | | | 6.6 | skid | 1 | 5 | 100 |
| Bell 222 | 7,850 | 39.8 | 47.5 | 12.0 | 9.1 | 12.2 | 1 | 7 | 188 |
| Sikorsky S-55 | 7,200 | 53.0 | 62.3 | 15.3 | 11.0 | 10.5 | 2 | 10 | 185 |
| Sikorsky S-62A | 7,500 | 53.0 | 62.3 | 16.0 | 12.0 | 18.0 | 2 | 10 | 188 |
| Bell 205A | 9,500 | 48.0 | 57.0 | 14.8 | 9.2 | skid | 2 | 13 | 215 |
| Bell 212 | 11,200 | 48.0 | 57.2 | 13.0 | 9.3 | skid | 2 | 13 | 215 |
| Bell 412 | 11,500 | 46.0 | 56.0 | 11.0 | 8.5 | skid | 2 | 13 | 217 |
| Sikorsky S-58 | 13,000 | 56.0 | 65.8 | 16.0 | 12.0 | 28.3 | 2 | 12 | 283 |
| Boeing Vertol 44 | 15,000 | 44.0 | 86.3 | 15.5 | 14.3 | 24.5 | 2 | 19 | 300 |
| Bell 214ST | 16,500 | 52.0 | 62.2 | 14.8 | 8.8 | skid | 2 | 18 | 412 |
| Boeing Vertol 107 | 19,000 | 50.0 | 83.5 | 16.8 | 14.5 | 25.0 | 2 | 25 | 360 |
| Sikorsky S-61L | 19,000 | 62.0 | 72.6 | 16.8 | 13.0 | 23.5 | 3 | 28 | 410 |

## Heliport Design

The heliport landing area must be large enough to handle loading and unloading operations. The surface must be clean, non-skid, well-drained, and strong enough to support impact landing loads. Although it is possible for helicopters to land and take off vertically, for economical operation with maximum payload, they must take off on a sloping ascent, preferably into the wind.

In the past, heliport sizes on offshore platforms have varied greatly in size. Some have been round, others square. Each heliport should be designed for the largest helicopter expected to land there. The basic heliport dimensions are determined from the overall length of that helicopter. Usually, the minimum dimension, that is, length of side on a square heliport, varies from 1.5 to 2.0 times the length of the largest helicopter expected to use that offshore facility. A circular heliport should have a diameter equal to the length of side on a square heliport. For example, offshore heliport sizes vary from 80 x 80 ft (24 x 24 m) to 160 x 160 ft (49 x 49 m).

According to one design approach, the heliport landing surface should be designed for a concentrated load equal to 75% of the gross weight of the largest helicopter, acting on any one square foot of surface. Another approach is to use an impact factor of two times the gross weight of the largest helicopter; this load must be sustained by an area of approximately 24 x 24 inches anywhere on the heliport surface. Fire-resistive materials must be used for offshore heliports. There should be some type of personnel safety shield slanting upward and outward all around the offshore heliport such as a strong metal mesh fence. This shield, or shelf, should measure 5 ft (1.5 m) wide horizontally.

A large equilateral triangular marker approximately 30 ft (9.2 m) on a side should be painted to encompass the center of the landing surface. One corner of the marker indicates magnetic north; the other two corners in the pattern are left unpainted. Inside the triangle there should be painted a large H, approximately 10 ft (3 m) tall by 5 ft (1.5 m) wide. The width of the triangular markings should be 2 ft (0.6 m); that of the H should be 18 inches (0.45 m).

There should be a wind indicator adjacent to the heliport to provide true wind direction. Sometimes yellow boundary lights are used to outline the landing surface. Floodlights are usually provided also.

# Chapter 4

# *An Overview of Engineering Procedures*

The design and construction of an offshore platform involves many related and interdependent endeavors. Table 4-1 lists the various stages in the overall project.

Several of these endeavors—two related to determining criteria, two related to analysis, design, sizing of members, and construction and installation— require additional description. These are briefly mentioned here and are presented in more detail in later chapters. Figure 4-1 shows the principal technologies involved in offshore platform design.

### Operational Criteria

This phase consists of determining the number of wells to be drilled, the type of drilling equipment and materials to be used and, if production activities are to be accomplished later, the specific requirements of that activity. The amount of deck space needed for the various operations must be determined, and the number of decks must be decided. The mode of oil transportation—whether by tanker, barge, or pipeline—must be determined, as well as the mode of oil storage. Also to be determined is the assurance that the platform configuration required to meet the operational criteria can be fabricated with the installation equipment available.

### Environmental Criteria

After the number of decks and the space requirement on each deck are determined, it is then necessary to determine the environment to which all of the equipment will be subjected. This involves determining the forces imposed on the platform by waves and wind. Many environmental factors must be assessed before the forces can be estimated: water depth, tide conditions, storm wave height, storm wind velocity, current, and sometimes earthquake

## Table 4-1
## Major Events in Building a Fixed Offshore Structure

A. Preliminary phase
   1. Recognition of need and setting of operational criteria
   2. Determination of environmental criteria
   3. Feasibility studies and cost estimates
   4. Financing arrangements and monetary allotments

B. Design phase
   1. Preliminary studies and special investigations
      a. Soils
      b. Size selection for derrick and transportation barges
      c. Conditions relative to corrosion, ice, earthquakes, product transportation, and crew transportation
   2. Design and preparation of engineering drawings
      a. Foundation design
      b. Structural design
      c. Preparation of drawings
   3. Document preparation
      a. Specifications
      b. Contracts
      c. Bid reply sheets
      d. Rental contracts for derrick and transportation barges
      e. Rental contracts for tug and work boats

C. Bidding phase
   1. Selecting bidders
   2. Sending and receiving bids
   3. Evaluation of bids
   4. Award of contracts

D. Construction phase
   1. Fabrication onshore
      a. Ordering and receiving materials
      b. Fabrication of specialty items
      c. Layout and patterns
      d. Welder qualification
      e. Cutting, fitting, and joining members into components
      f. Coating of components and other corrosion protection
   2. Loading for transportation
   3. Erection offshore
      a. Placement of underwater components
      b. Installation of pile foundation
      c. Setting of above water components and equipment
      d. Miscellaneous associated construction
   4. Acceptance

Engineering Procedures 45

## OFFSHORE PLATFORM DESIGN

### TECHNOLOGY

| OCEANOGRAPHY | FOUNDATION ENGINEERING | STRUCTURAL ENGINEERING | MARINE CIVIL ENGINEERING | NAVAL ARCHITECTURE |
|---|---|---|---|---|
| WIND <br> WAVE <br> CURRENT } FORCES <br> TIDE <br> ICE | SOIL CHARACTERISTICS | MATERIALS SELECTION AND CORROSION | INSTALLATION EQUIPMENT | FLOTATION & BUOYANCY |
| | VERTICAL PILE – SOIL CHARACTERISTICS | STRESS ANALYSIS | INSTALLATION METHODS | TOWING |
| | LATERAL PILE – SOIL CHARACTERISTICS | WELDING | NAVIGATION SAFETY INSTRUMENTATION | LAUNCHING |
| | SCOUR | STRUCTURAL ANALYSIS | | CONTROLLED FLOODING |
| | | DESIGN FOR FABRICATION AND INSTALLATION | | |
| | | APPURTENANCES | | |

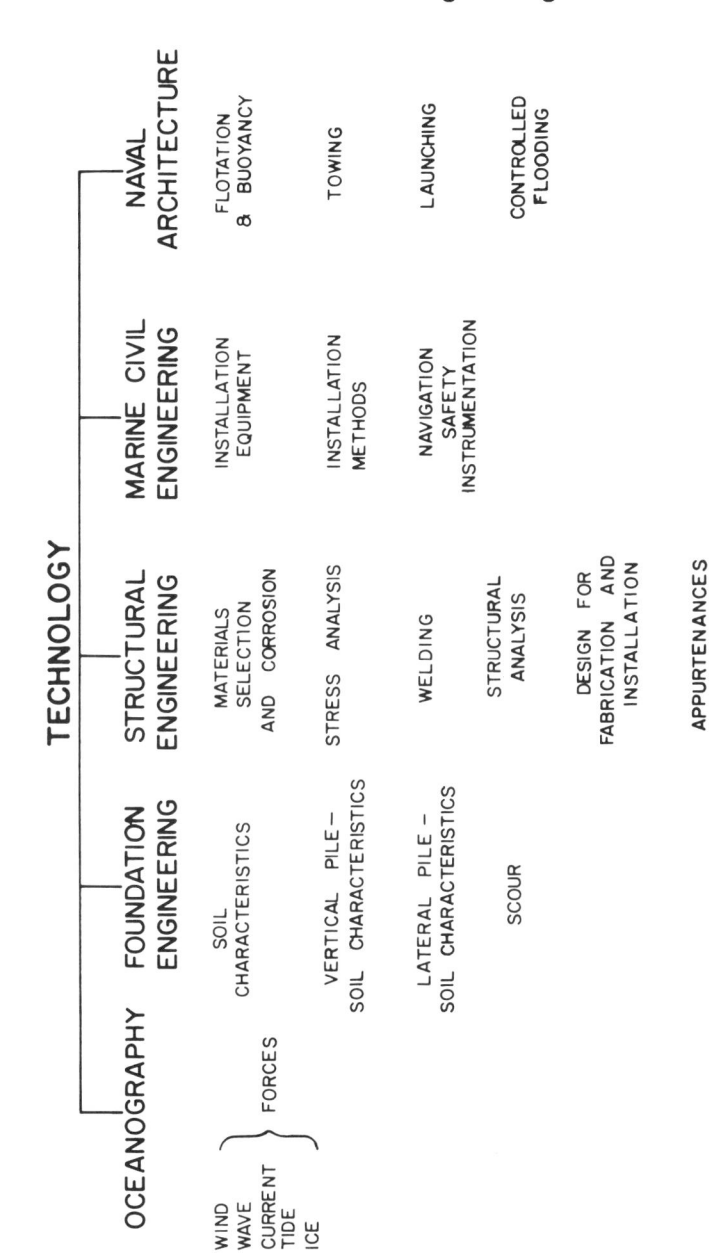

**Figure 4-1.** *Technologies involved in offshore platform design.*

and ice conditions. All of the environmental factors which impose loads on the platform must be investigated. Oceanographers and meteorologists are responsible for this investigation.

### Foundation Design

Before analysis and design of the foundation, it is necessary to determine the soil characteristics of the ocean floor where the platform is to be placed. The required information consists of the geological history of the area, soil boring data, and results of experimental pile driving. Geologists and specialists in soil mechanics are responsible for evaluating the collected data and translating the information into the capacity of the soil to resist operational and environmental forces transmitted to it through the structure.

### Structural Design

Following the determination of operational, environmental, and foundation characteristics, the next step is the analysis and design of the structure of the platform. The number of decks and the structural configuration of the deck substructure must be selected so as to support the operational loads and to provide foundational resistance to environmentally inposed forces. When the selection of deck configuration and platform structural type is complete, preliminary estimates of the sizes of the various members are made. The design is then redone in a second go-round. Revised computations of operational and environmental loads are made, the foundation requirements are again evaluated and, finally, the various structural member sizes are again determined. The overall process cycles among these major aspects until an adequate and safe design, meeting all the criteria, is evolved.

There are many details which must be designed after the main structural member sizes have been determined, including boat docks, stairs, hand railing, heliports, launch rails, lifting eyes, etc. As part of the structural design, there must be an analysis of the structure to ensure that it will withstand the loads imposed on it during fabrication and installation.

### Construction and Installation

After the platform has been designed, it must be fabricated and installed. Most of the fabrication occurs in a construction yard onshore. On-site installation is limited to launching and upending the jacket, driving pilings, placing the deck structure, and welding all of these into a single unit.

The components are prefabricated into the largest units that can be economically and quickly transported from the fabrication yard to the offshore site. Prefabrication allows for a minimum amount of construction time at sea so that operating losses due to rough weather are minimized.

## Engineering Procedures 47

All materials are ordered well in advance of the first day of construction. Construction can last from 4 to 12 months, depending on the complexity and size of the structure.

The jacket is usually assembled by constructing its narrow dimension frames lying flat on the ground. These are then rotated by cranes into a vertical position where cross bracing, guides, and other members are then added. Thus, when finished, the jacket is lying on one of its long sides. The two middle legs on the long side are usually parallel, and the jacket is constructed with these legs lying on the launch beams used to skid the jacket off the shore onto a barge.

Once the jacket and its deck sections are completed, the components are pulled or lifted onto barges and transported to the offshore site. At the site a winch and cable assembly pull the jacket off its barge into the water. The lower sections of the jacket legs are then allowed to flood so that the jacket rests in an upright position in the water. A large derrick barge lifts the jacket and places it in the designated spot for drilling. Piles are driven through the jacket legs and through the skirt pile guide tubes if skirt piles are used. The deck sections are then mounted in place on top of the piles and welded into place. Prefabricated modules containing living quarters, pump assemblies, and other equipment are brought out by barge and lifted into place on the deck substructure to complete the installation.

For the *Cognac* platform (see Chapter 5), the jacket was constructed in three sections. The base was fabricated in an upright position due to its size. The intermediate and upper jacket sections were fabricated in the same way as ordinary jackets to go in 200 or 300 ft (60 or 90m) of water. After the base was positioned and fastened by piling, the intermediate section was guided onto the top of the base and fastened. The process was repeated for the upper jacket section.

### Resources

Three publications which are excellent guides to the design and construction of offshore platforms are:

1. "Rules for the Design, Construction, and Inspection of Offshore Structures, Det Norske Veritas, Oslo, Norway, 1977.
2. "Planning, Designing, and Constructing Fixed Offshore Platforms," an American Petroleum Institute Recommended Practice, API-RP-2A, Tenth Edition, Dallas, Texas, U.S.A., March 1979.
3. "Requirements for Verifying the Structural Integrity of Outer Continental Shelf (OCS) Platforms," United States Geological Survey, Conservation Division, OCS Platform Verification Program, Reston, Virginia 22092, October 1979.

# Chapter 5

# *The World's Tallest Platform*—Cognac

## Announcement

In January 1976, Shell Oil Company, the oil industry's largest producer in the Gulf of Mexico, announced plans to build and install the world's tallest offshore self-contained template drilling and production platform to be located in the Gulf of Mexico on the continental slope (as distinct from the continental shelf) about 100 miles (160 km) southeast of New Orleans, Louisiana. The site is about 15 miles (24 km) southeast of the edge of the Mississippi River Delta.

## Description

The platform, *Cognac,* was installed between July 25, 1977, and September 25, 1978, in water with an average depth of 1020 ft (311 m). Plans call for drilling 62 wells using two drilling rigs simultaneously. (See Figure 5-1.) At an elevation of 14 ft (4.3 m) above the mean water line (MWL), the jacket measures 84 x 164 ft (26 x 50 m). The jacket base section measures 380 x 400 ft (117 x 122 m) at the mudline. There are eight main legs which extend the full height of the jacket, and two framing legs which extend from elevation -400 from MWL to the ocean bottom.

Because of the water depth, the jacket was divided into three sections. These were fabricated, transported to the offshore site, and launched as independent units. After the base section was secured to the bottom with piles, the intermediate and top sections were joined to the base section underwater using elaborate positioning systems. All of the legs are seven ft (2.1m) in diameter. The long connecting pins inserted in the legs that join the sections are six ft (1.8 m) in diameter. The base is held in position by 24 vertical skirt piles driven 450 ft (137m) into the soft clay bottom. (See Figure 5-2.) The piles are seven ft (2.1 m) in diameter and are grouted into skirt pile sleeves eight ft (2.4 m) in

## World's Tallest Platform—Cognac 49

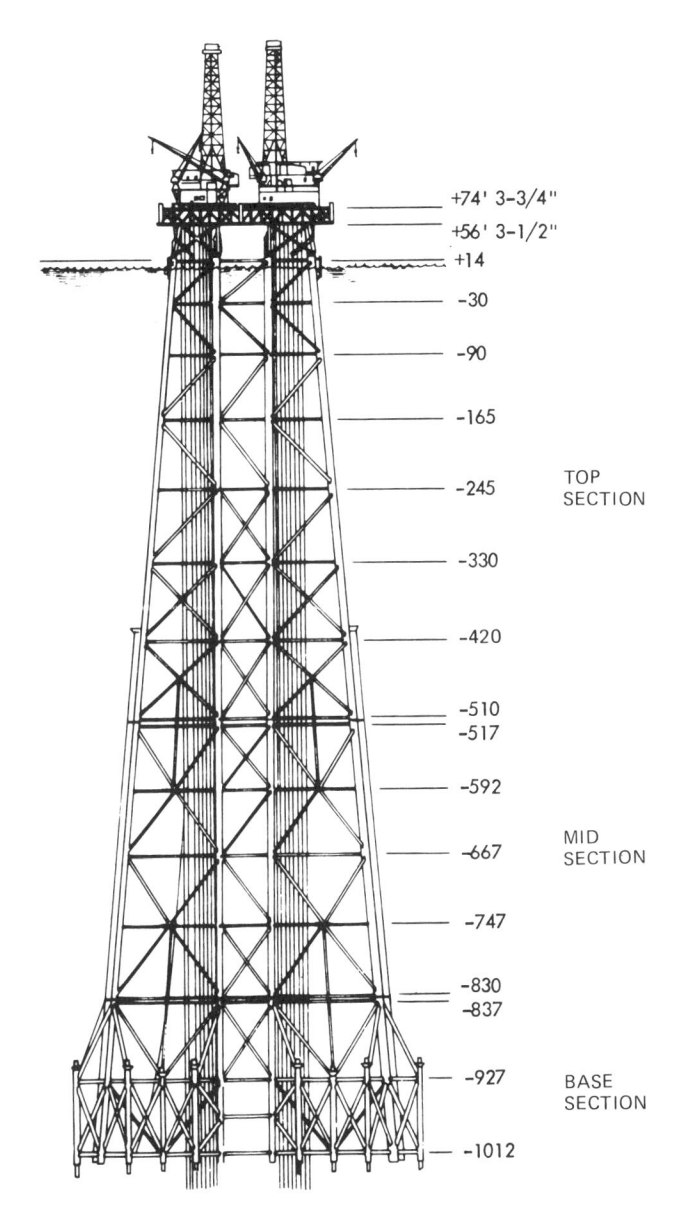

**Figure 5-1.** Assembled Cognac platform. (Copyright © 1979 Offshore Technology Conference.)

## 50  Introduction to Offshore Structures

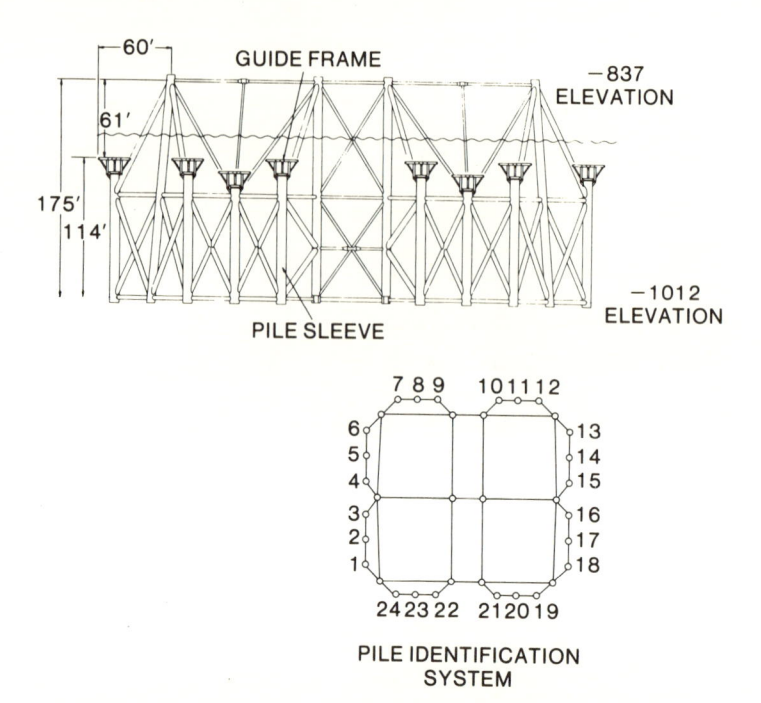

**Figure 5-2.** Location of the pile sleeves on the Cognac jacket base section. (Copyright © 1979 Offshore Technology Conference.)

diameter. The annular areas between the 1000-ft (305-m) long connecting pins and the inside diameters of the jacket legs were grouted to bond firmly the three jacket sections into one unit.

The maximum wall thickness of the skirt piles is 2.50 inches (6.35 cm) at the mudline. The minimum thickness is 1.5 inches (3.81 cm), and the average wall thickness is 2.0 inches (5.1 cm). Each pile is designed to carry 8600 kips (38 x $10^6$ N) of axial load and 650 kips (2.9 x $10^6$ N) of shear. The jacket braces are 5 ft (1.5 m) and 2 ft (0.6 m) in diameter. Because the jacket was to be placed in 1020 ft (311 m) of water with a relatively low weight, all skirt pile sleeves, main legs, and many structural members were stiffened with external ring stiffeners. The rings were designed to have a safety factor of 2.0 against expected hydrostatic pressure loads and 1.4 against emergency overloading. The factor of safety is that multiplier applied to the service load to determine the maximum load to be accomodated in design.

The skirt piles and connecting pins weigh 16,000 tons (14,512 tonnes). The three jacket sections weigh a total of 33,500 tons (30,386 tonnes). The conductor tubes for well drilling weigh 7000 tons (6349 tonnes). The deck structural weight is 2500 tons (2268 tonnes). Thus, the total weight of the steel used to build the platform is 59,000 tons (53,515 tonnes). The base section is 178 ft (54 m) high and weighs 14,000 tons (12,700 tonnes). The intermediate or mid-section is 320 ft (98 m) tall and weighs 8500 tons (7710 tonnes). The top section is 530 ft (162 m) tall and weighs 11,000 tons (9978 tonnes).

### Deck Structure

The decision to use two drilling rigs simultaneously dictated much of the deck configuration. A two-level deck with a total of 66,500 sq ft (6178 sq m) of usable work area was chosen. In plan view the dimensions are approximately 180 x 200 ft (55 x 61 m). The top level is 74 ft (22.5 m) above MWL; the lower level is 56 ft (17 m) above MWL.

The deck substructure, a tubular truss system, was installed in two lifts, one of 254 tons (230 tonnes), and one of 262 tons (238 tonnes) as shown in Figure 5-3. The lower level deck beams and plating, along with the upper deck truss structure, beams, and plating were fabricated in six segments to facilitate lifting into place. The segment weights are given in Figure 5-3. Lifting was accomplished with two derrick barges, one on each side of the jacket.

The deck section segments required a little over one day for setting in place. Twenty-two days were required for welding, sandblasting, and painting. Figure 5-4 shows the deck structure completed and ready for mounting the drilling rigs and modular equipment.

### Oceanographic Criteria

The environmental forces characterized by hurricane waves, wind, and current were the basic loads used to determine platform size and weight. The so-called normal or ordinary environmental forces were the controlling factors in the fatigue behavior analyses. The non-hurricane wave, wind, and current forces for the four summer months, June through September, primarily influenced the construction operations.

As Shell Oil Company and other companies in the *Cognac* venture had been recording environmental data in the adjacent South Pass offshore leasing area since about 1969, developing the design conditions for *Cognac* was a simple combination and prediction exercise. Table 5-1 lists the oceanographic criteria. The long-term data were analyzed to develop the frequency of occurence distributions of a range of significant wave heights and periods for use in the fatigue behavior studies.

Because the installation technique required the platform sections to be lowered into position with wire ropes, extensive current measurements were made at the *Cognac* site over a four-year period. The water depths of greatest concern were near the bottom, and at elevations corresponding to the mating

**52** Introduction to Offshore Structures

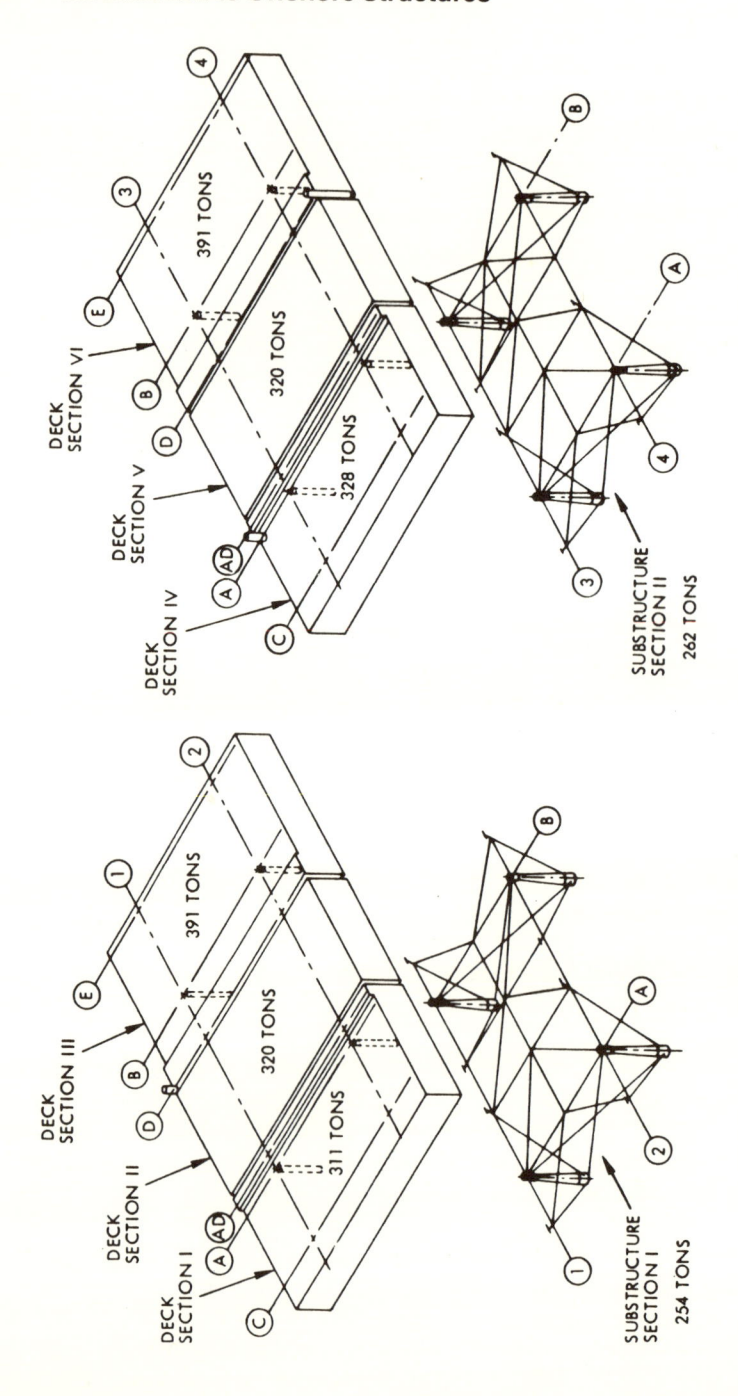

*Figure 5-3.* Cognac deck structure. *(Copyright © 1979 Offshore Technology Conference.)*

## World's Tallest Platform—Cognac

*Figure 5-4.* Cognac jacket and superstructure completed September 1977. (Copyright © 1979 Offshore Technology Conference.)

## Table 5-1
## Oceanographic Criteria

| Environmental load | Current considered | Current combined into design wave |
|---|---|---|
| Design wave | Height: 70 ft<br>Period: 11.5 sec | 78 ft<br>12 sec |
| Wind speed | 125 mph: structural members<br>150 mph: deck equipment | 0 |
| Current | 4 ft/sec at surface<br>0 ft/sec at 150 ft | 0 |
| Cd | 0.6 | 0.6 |
| Cm | 2.0 | 2.0 |

Note: 3.28 ft = 1 meter
1 mph = 26.8 m/sec
1 ft/sec = 0.305 m/sec

of the jacket sections. Records indicated that actual currents during 1977 and 1978 (while installation took place) were lower than was to be expected from information gathered earlier. Seldom was the current at the various points of interest greater than 0.75 ft/sec (0.23 m/sec).

### Structural Design

The deck layout is shown in Figure 5-5. This arrangement was dictated by the decision to use two drilling rigs. The functional requirements are summarized in Table 5-2. Figure 5-6 shows the arrangement of horizontal and diagonal bracing at various levels within the jacket.

For preliminary dynamic structural analysis, a multiplier of 1.15 was applied to the static overturning moment to estimate the dynamic overturning moment. Likewise, a base shear amplification factor of 1.5 was used. As the design progressed, three detailed solutions were undertaken using random wave time domain, wave spectrum frequency domain, and regular wave time domain methods. The structural model contained 2000 members and 700 joints. Figure 5-7 shows typical results from these analyses. At different levels in the structure, the dynamic amplification for static shear varied from about 1.2 near the top to 1.5 at the base. The dynamic amplification factor for static overturning moment varied very little with height from a value of 1.3 except at the top where it reached 1.5.

# World's Tallest Platform—Cognac 55

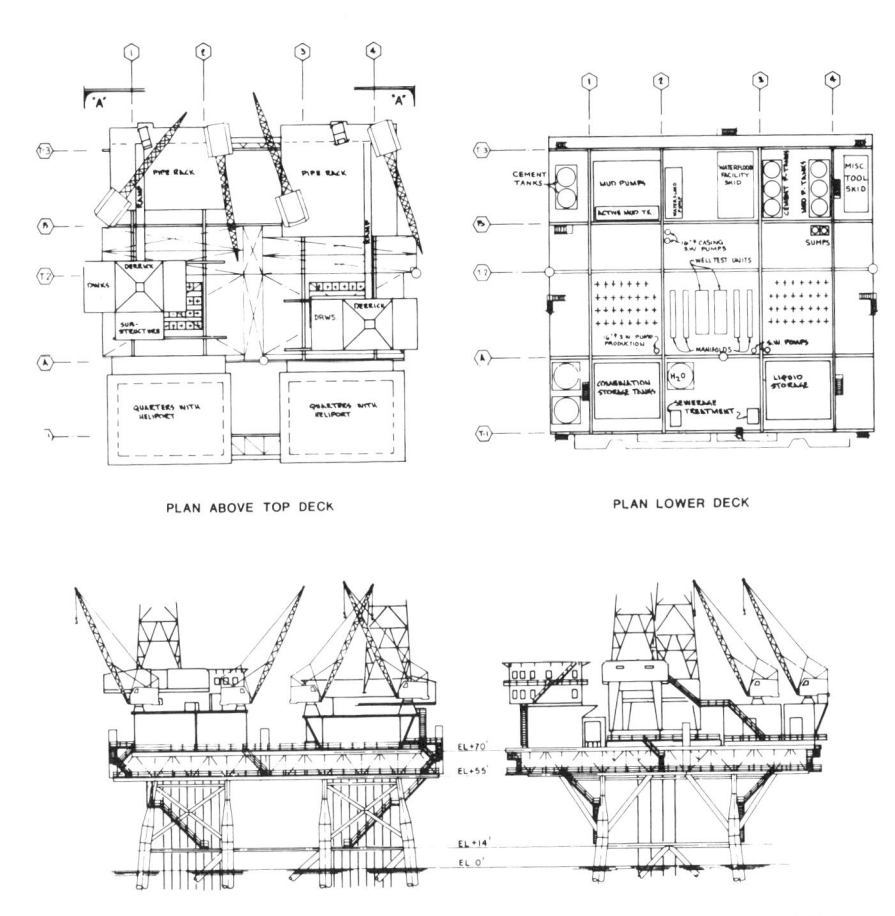

*Figure 5-5. Drilling rig layout for final platform concept. (Copyright © 1979 Offshore Technology Conference.)*

### Table 5-2
### Cognac Functional Requirements

Two drilling rigs operating simultaneously
Oil and gas production facilities
62 well conductors (60 straight & 2 curved at 6°/100 ft) (6°/30.5 m)
8 flowline risers
2 - 12-inch (0.305 m) oil pipeline risers
1 - 16-inch (0.406 m) gas pipeline riser
Weld-on sacrificial corrosion anodes

**56** Introduction to Offshore Structures

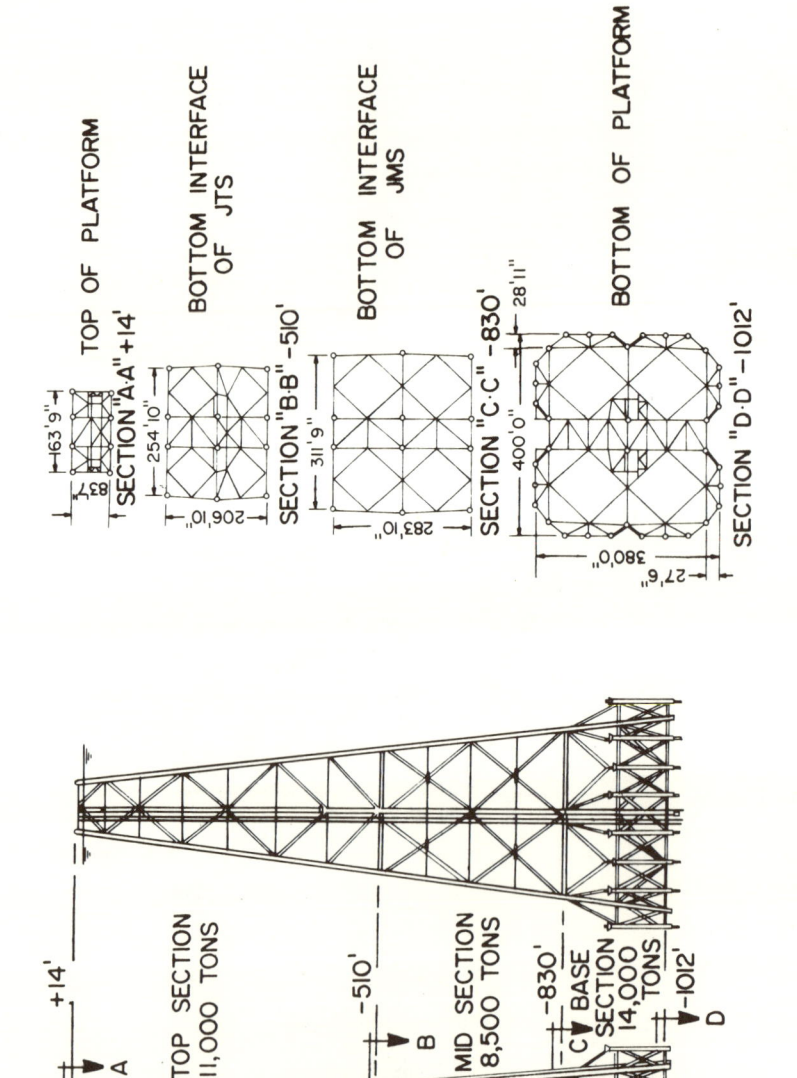

*Figure 5-6b.* Cognac platform plan dimensions. Note: JTS stands for jacket top section and JMS stands for jacket midsection. (Copyright © 1979 Offshore Technology Conference.)

*Figure 5-6a.* Cognac platform configuration. (Copyright © 1979 Offshore Technology Conference.)

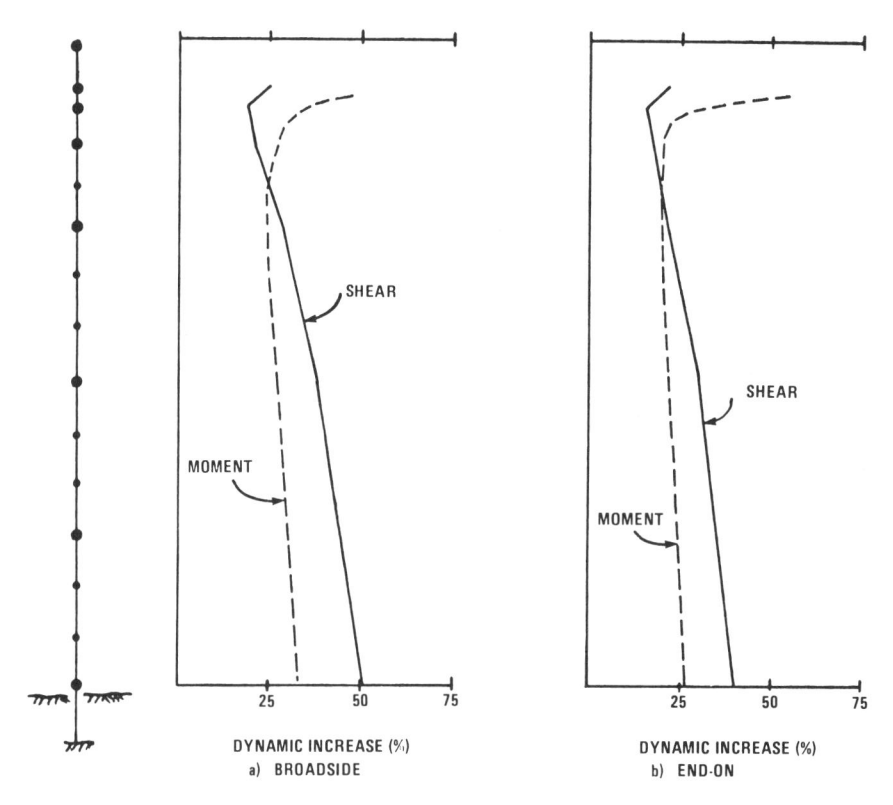

*Figure 5-7. Dynamic amplification. Notice in the simplified model at the left that the mudline is indicated just above the bottom of the model. (Copyright © 1979 Offshore Technology Conference.)*

## Fabrication

Because of size, fabrication of the jacket sections involved innovations in the customary construction methods. The same fabrication techniques were used for both the intermediate and the top jacket sections. Fabrication began with two narrow-dimension planar frames, or bents, being constructed lying flat on the ground. Each bent was rotated about one of its jacket legs into the vertical for installing the connecting braces to make a three-dimensional space frame. The intermediate section bents had to be separated into two pieces for handling; the top section bents had to have temporary vertical members welded in to carry the out-of-plane loading.

## 58    Introduction to Offshore Structures

**Figure 5-8.** Roll-up of midsection bent; 495 tons 316 feet into the air. (Copyright © 1979 Offshore Technology Conference.)

The first two bents were rotated into the vertical so as to lie on parallel skid beams as shown in Figure 5-8. The jacket section was completed by rolling up a bent on either side of the two on the skid beams. The outer bents are held in position by cranes until the connecting braces are fastened. Figure 5-9 shows the rolling up of a bent for the top jacket section. Figures 5-10 and 5-11 show the jacket base section being towed along a bayou to the open water and during launch at the offshore site in the Gulf of Mexico.

World's Tallest Platform—Cognac 59

*Figure 5-9.* Roll-up of top-section bent; 2200 tons. (Copyright © 1979 Offshore Technology Conference.)

*Figure 5-10.* Cognac base section being towed through Bayou Chene. (Copyright © 1979 Offshore Technology Conference.)

**Figure 5-11.** Cognac *base section being launched. (Copyright* © *1979 Offshore Technology Conference.)*

## Installation

Figures 5-12 and 5-13 show the installation procedure schematically. The two derrick barges were positioned within a 12-point mooring array. The launch barge was positioned between the derrick barges. Lowering and control lines were fastened to each jacket section before it was launched.

Once the sections were upright in the water, large winches mounted in structural frames on the decks of the two derrick barges lowered the jacket sections at rates of three to five ft/min. Acoustic beacons on the ocean bottom echoed signals to the control center on one of the derrick barges. The acoustic signals were analyzed with an onboard computer, and real-time positions of the barges and the jacket section were calculated as the operation proceeded.

Via the umbilical, or control line, data from the jacket section relative to water levels in legs, tilt, list, and azimuth were fed back to the control center computer. Within the control line were hydraulic and pneumatic hoses for opening and closing valves for flooding, deballasting, and grouting, and for activating release mechanisms for detaching, lowering, and control lines.

Figure 5-14 shows diagrammatically the alignment and docking system. Details of the docking pole are shown in Figure 5-15. The operation of the alignment system is too detailed for explanation in this brief discussion. The references listed at the end of the chapter give considerably more information.[1-8]

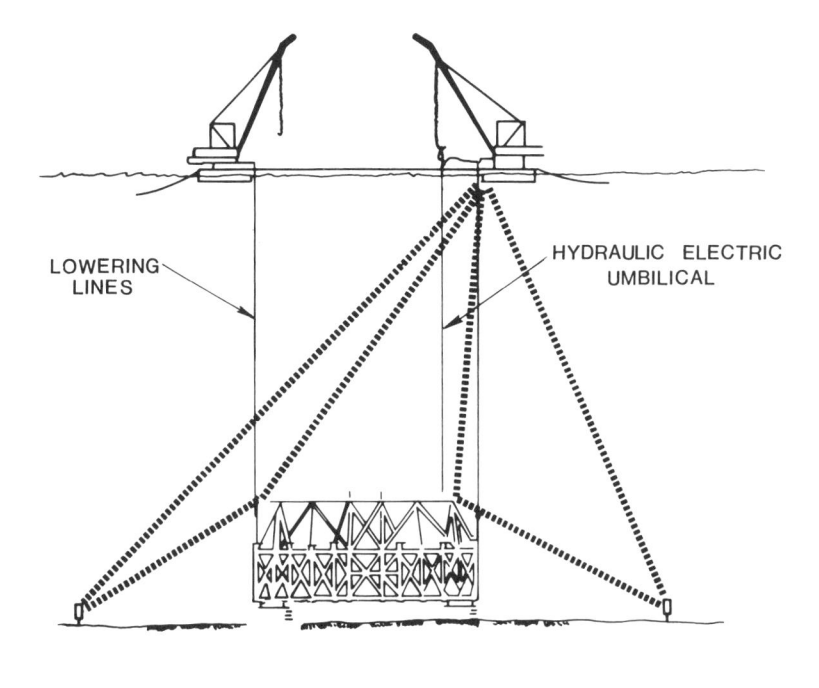

- LOWER BASE TO ±20 FEET OFF BOTTOM
- ORIENT PLATFORM WITH ACOUSTICS
- ADJUST MUDMATS
- LOWER & RELEASE LINES

*Figure 5-12.* Lowering *Cognac base section to the bottom. (Copyright © 1979 Offshore Technology Conference.)*

The unique lowering, positioning, and docking operations were accomplished without significant problems, since the platform was successfully installed within 30 ft of the desired target and only one-half degree off the desired orientation.

## Cost

Fourteen other companies joined Shell Oil Company in this venture. The platform is estimated to have cost between $200 and $250 million. This cost includes the specialized equipment for lowering; the controlling of two derrick barges simultaneously with an onboard computer; the various model tests and experiments; and the underwater hammer for pile driving.

## 62 Introduction to Offshore Structures

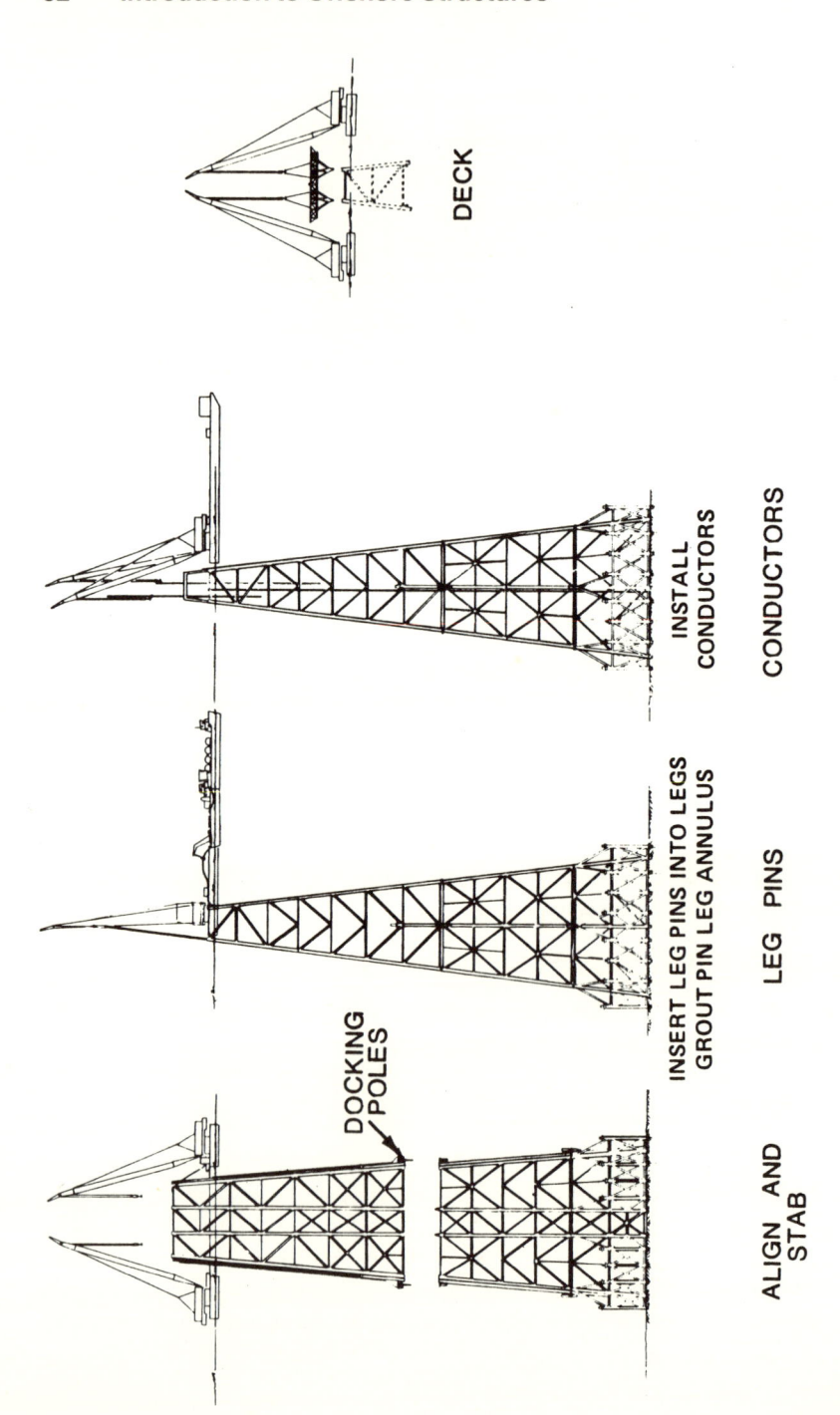

*Figure 5-13.* Cognac platform installation concept. (Copyright © 1979 Offshore Technology Conference.)

World's Tallest Platform—Cognac 63

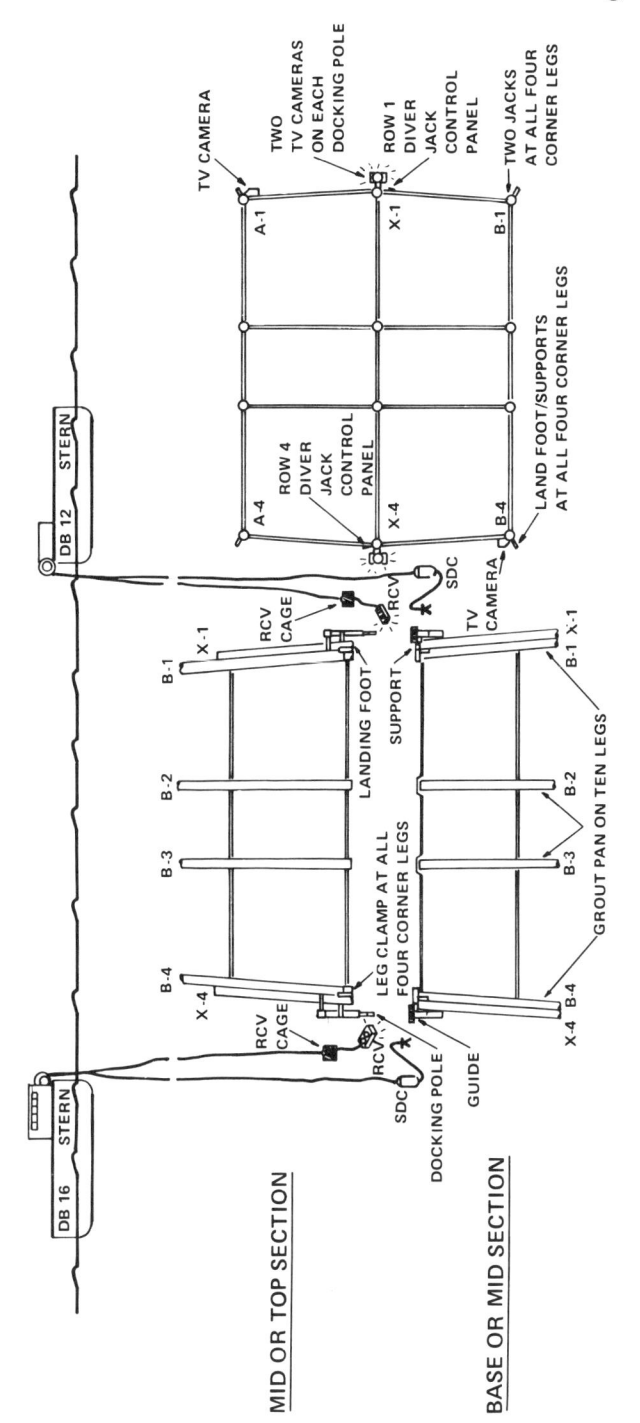

*Figure 5-14.* Cognac platform alignment and docking system. (Copyright © 1979 Offshore Technology Conference.)

## 64    Introduction to Offshore Structures

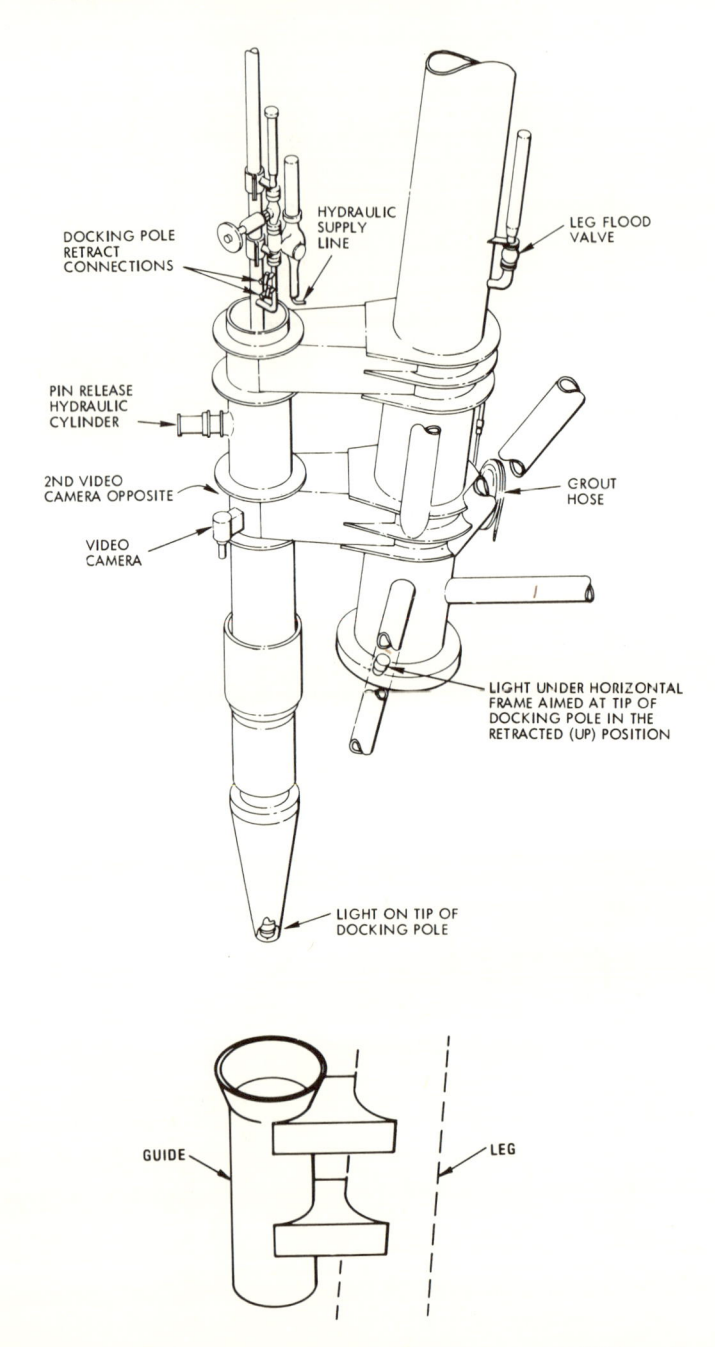

**Figure 5-15.** Cognac *docking pole details. (Copyright © 1979 Offshore Technology Conference.)*

## World's Tallest Platform—Cognac

### References

1. Sterling, G.H.; Casbarian, A.O.; Godfrey, D.G.; and Dodge, N.L.; "Construction of the Cognac Platform in 1025 Feet of Water, Gulf of Mexico," 1979 Offshore Technology Conference, Paper #3493.
2. Sterling, G.H.; Cox, B.E.; and Warrington, R.M.; "Design of the Cognac Platform for 1025 Feet of Water, Gulf of Mexico," 1979 Offshore Technology Conference, Paper #3494.
3. Godfrey, D.G. and Huete, D.A., "Measurement Techniques and Misalignment Analysis for the Three-Part Cognac Platform Installation," 1979 Offshore Technology Conference, Paper #3495.
4. Piter, E.S.; Sterling, G.H.; and Cox, B.E.; "Development of the Alignment System for the Three-Part Cognac Platform Jacket," 1979 Offshore Technology Conference, Paper #3496.
5. Mayfield, J.G.; Strohbeck, E.E.; Olivier, J.C.; and Wilkins, J.R.; "Installation of the Pile Foundation for the Cognac Platform," 1979 Offshore Technology Conference, Paper #3497.
6. Collipp, B.G. and Johnson, P.; "Marine Equipment and Procedures for Cognac Platform Installation," 1979 Offshore Technology Conference, Paper #3498.
7. Simpson, W.F.; "The Instrumentation System for the Cognac Platform," 1979 Offshore Technology Conference, Paper #3499.
8. Peterson, D.C.; "Cognac Positioning System," 1979 Offshore Technology Conference, Paper #3500.

# Chapter 6

# Design Loads and Forces

The oceanographic and meteorological data needed for the design of a platform are called environmental criteria. Such information includes storm wave heights and wave periods, storm wind speeds and gust conditions, tides, swells, ocean bottom scours or slides, currents, icing conditions, earthquakes, etc.

The most important design considerations for an offshore platform are the storm wind and storm wave loadings it will be subjected to during its service life. The theories concerning the modeling of ocean waves were developed in the nineteenth century. Practical wave force theories concerning actual offshore platforms were not developed until 1950, when the Morison equation was presented.[1]

For the Gulf of Mexico, the importance of correct wind and wave force prediction was emphasized by the damage caused by hurricanes Hilda and Betsy in the mid-1960s. Since then, much effort has been devoted within the offshore industry to develop models of wind and wave fields that allow accurate forecasting of wind speeds, wind forces, wave heights, and wave forces.

## Wind Forces

Studies of wind loading extend far back in the history of engineering. Accurate representation did not come about until the development of aerodynamic theory.

The force of the wind on a structure is a function of the wind velocity, the orientation of the structure, and the aerodynamic characteristics of the structure and its members. The forces exerted on a structure by the wind are expressed as

$F_D = C_D 1/2 p V_z^2 A$ (force parallel to wind)
$F_L = C_L 1/2 p V_z^2 A$ (force perpendicular to wind)

# Design Loads and Forces

where:

- $C_D$ = drag coefficient
- $C_L$ = lift coefficient
- $p$ = density of the air
- $V_z$ = wind velocity at height $z$
- $A$ = area perpendicular to wind velocity

Values for $C_D$ and $C_L$ may be found in Section 12 of *The Handbook of Ocean and Underwater Engineering* (McGraw-Hill Book Co., 1969). In preliminary calculations the wind force perpendicular to the wind direction is often neglected.

Because of the shear forces with the earth's surface, the wind velocity is not constant, but is zero at the surface and increases exponentially to a limiting maximum speed known as the gradient wind. The wind speed at any elevation above a water surface is represented as

$$V_z = V_{30} \left[\frac{z}{30}\right]^{1/7}$$

where:

- $V_{30}$ = wind speed at a height of 30 feet (the customary elevation for such measurements)
- $z$ = the desired elevation in ft
- $30$ = the reference height, ft

This equation is called the one-seventh power law. In metric units the equation is

$$V_z = V_{10} \left[\frac{z}{10}\right]^{1/7}$$

where:

- $V_{10}$ = the wind speed at a height of 10 meters
- $z$ = desired elevation, m
- $10$ = reference height, m

The power law has been widely used. A similar equation with the power varying from 0.143 to 0.3 has been recommended for coastal areas depending on whether $V_{30}$ is more or less than 60 mph (1620 m/sec).

## 68 Introduction to Offshore Structures

In the wind force expression the aerodynamic effects are accounted for in the drag coefficient term, $C_D$. This coefficient can range from 2.1 to 0.7, depending on the shape of the structure.

The wind effects on all parts of the structure should be calculated, especially on the cables, as they are susceptible to violent motion from vortex shedding at high wind speeds.

To determine wind forces on structures, it is not sufficient to know only the sustained wind speed, but the gust as well. The gust factor is that multiplier which must be used on the sustained wind speed to obtain the gust speed. The average gust factor $F_g$ is in the range of 1.35-1.45; the variation of gust factor with height is negligible.

The velocity of the wind used in design should be consistent with the risk assumptions for the structure. The wind force contributes about 25% of the total overturning moment on offshore structures in water over 150 ft (46 m) deep, and an even larger percent for structures in shallow water. The wind force is dependent upon the conditions of exposure and the shape of the structure.

Structural components loaded primarily by wind (as opposed to wind and wave) should be designed for the fastest-mile-velocity with a period of recurrence at a given site of 100 years. As a large variation of wind direction may occur during a storm, all orientations of the structure should be analyzed to assure a safe design. The U.S. Weather Bureau uses the term "one minute average" wind speed, which in other terminology has been called the sustained wind speed. This indicator is normally quoted for a height of 30 ft (10 m) above the earth's surface. The term "fastest-mile-velocity" is equivalent to the sustained wind speed multiplied by a gust factor. The fastest-mile-velocity should be used for designing pieces of equipment located on the platform decks and for tie-down design.

For computation of wind forces in conjunction with maximum wave forces, the 100-year sustained wind velocity should be used. Table 6-1 lists some design wind pressures that have been used in conjunction with a 100-year

### Table 6-1
### Design Wind Pressures on Structural Components
### For a 100-Year Sustained Wind Velocity of 125 mph

| Component | Pressure (psf)* |
|---|---|
| Flat surfaces such as wide flange beams, gusset plates, sides of buildings, etc. | 60 |
| Cylindrical structural members | 48 |
| Cylindrical deck equipment ($L = 4D$) | 30 |
| Tanks standing on end ($H \leq D$) | 25 |

*Note: 1 psf = 47.88 Pa

### Design Loads and Forces

**Table 6-2**
**Wind Shielding Factors for Structural Components in Series**

| Component | Shielding factor |
|---|---|
| Second in a series of trusses | 0.75 |
| Third or more in a series of trusses | 0.50 |
| Second in a series of beams | 0.50 |
| Third or more in a series of beams | 0.00 |
| Second in a series of tanks | 1.00 |
| Short objects behind tall objects | 0.00 |

sustained wind velocity of 125 mph. Allowance must be made for the shielding some components provide for others by multiplying the force expected if fully exposed by the shielding coefficient listed in Table 6-2.

### Wave Forces

Wave force calculations are approached from two different directions. For small platforms, or platforms in relatively shallow water, the design loading from wave forces is calculated as a static force applied to the structure. On taller structures, the natural period of vibration of the structure approaches the period of the ocean waves, and a more complicated dynamic analysis must be performed. Only static design is mentioned in this discussion.

Design wave forces should be based on the worst expected storm conditions with an average expected recurrence interval of 100 years. Storm surge and astronomical tides should be included in computing the storm water depth. The design wave for a static wave analysis is specified by the user, that is, the company or division of a company for whom the structure is being designed.

The horizontal force exerted by waves on a cylindrical member consists of a drag force related to the kinetic energy of the water, and an inertial force related to the acceleration of the water particles. The pressure of the water is multiplied by the volume, or the projected area of the structural member perpendicular to the direction of wave advance, to obtain the total force on the member. This force per unit length on a member is computed using the equation developed by Morison, O'Brien, Johnson, and Schaaf in 1950.[1] This equation represents the force on a cylindrical tube in the form of two components: inertial and drag. The equation is expressed as

$$F = F_I + F_D$$
$$F = C_m \frac{w\pi}{g} \frac{D^2}{4} \frac{du}{dt} + C_d \, 1/2 \frac{w}{g} D u |u|$$

where:

- $C_m$ = mass coefficient
- $C_d$ = drag coefficient
- $w$ = weight density of seawater, $N/m^3$
- $D$ = diameter of the cylinder, m
- $F$ = wave force per unit length acting perpendicular to the member axis, $N/m$
- $F_I$ = inertia force per unit length along the member, $N/m$
- $F_D$ = drag force per unit length along the member, $N/m$
- $u$ = horizontal water particle velocity, m/sec
- $|u|$ = absolute value of $u$, m/sec
- $\dfrac{du}{dt}$ = horizontal water particle acceleration m/sec$^2$
- $g$ = gravitational acceleration, m/sec$^2$

The values of $C_m$ and $C_d$ are determined through experimentation and modeling and are dependent on the size and the added mass of the cylinder in the water as well as the type of wave theory used to determine the particle velocities and accelerations. The dimensionless drag and inertial coefficients $C_d$ and $C_m$ most often used in the Morison expression are 0.7 and 2.0, respectively.

Water particle velocity and acceleration are functions of wave height, wave period, water depth, distance above bottom, and time. Several theories are available by which the water particle velocity and acceleration may be calculated. The most elementary wave theory was presented by Airy in 1845. The Airy theory is a linear solution to the differential equation for water particle behavior based on the assumption that the wave height is small in comparison to the wave length. This assumption is accurate within a certain range, but in storm conditions with steeper waves it is inaccurate.

A nonlinear theory derived by Stokes gives a more accurate representation of the wave profile and, as a result, is most often used to determine wave particle characteristics.[2]

Another widely used theory, known as the stream function theory, is a nonlinear solution similar to the Stokes' fifth order theory.[3] Both use summations of sine and cosine wave forms to develop a solution to the original differential equation. The stream function theory determines the coefficients to obtain a least-squares fit to the posed free-surface boundary conditions so that (although the assumptions made in this theory are the same as those presented by Stokes) a better solution is obtained. The theory to be used for a particular offshore design is determined by the policy under which the designing engineers are working.

## Design Loads and Forces

The size of the wave to be used in design (height and period) has been the greatest problem in offshore structure design. Methods of determining the wave height that would exist in a certain ocean area due to storm conditions have been investigated by many people over the past 50 years. Tables 6-3 and 6-4 list the wave characteristics resulting from different wind velocities blowing over various fetches and durations.

Investigations have established that the only accurate method for describing the ocean surface is by a probability function or a wave spectrum. Several theories have been developed to describe the ocean surface as a function of the wind speed, the area or fetch over which it is acting, and the amount of time it is blowing. The Peirson-Neumann, Peirson-Moskowitz, and Neumann spectrums are all mathematical models that determine a design wave height.

The wave spectrum method defines the ocean in terms of the energy distribution through the frequency range. As the wind starts to blow, the energy in the surface motion builds to a certain point at which there is a constant amount of energy on the surface in the wave action. This energy is distributed over a range of frequencies with a peak at one particular frequency. If the assumption is made that this spectrum is in the form of a Rayleigh distribution, probability theory may be used, and a significant wave height and period may be obtained. Furthermore, estimates of the amount of energy in the waves can be varied to represent the conditions in a storm that has a recurrence interval of 25 years at a given location or one that occurs every 100 years.

Present design criteria for offshore platforms generally use the 100-year storm interval to determine design wave heights and frequencies. Data for this magnitude of storm are determined through the recorded history of previous storms or hurricanes and through theoretical predictions. Common practice in the Gulf of Mexico today is to use a design wave height of 70-75 ft. This value varies somewhat, depending on the water depth, the distance to shore, and the shoaling effects brought about by the immediate geography. As an initial value, a wave period of 12-14 seconds may be used to determine preliminary wave force data. For different parts of the structure, different periods are used; use whichever wave period produces the maximum stress in the particular element being designed.

No allowance is made for shielding of one structural member by another in computing the total wave force on each member and for determining moments produced by wave forces on a member. All members of the structure and all appurtenances below the surface of the design wave should be included in the wave force calculations. Figure 6-1 shows the wave force components that act on a cylindrical pile. Notice that the drag and inertia components are 90° out of phase. For a small pile acted on by a high wave, the resultant force is primarily from the drag component. On the other hand, a large pile acted on by a low wave is subject to a resultant force primarily due to the inertia component.

## Table 6-3
## Wave Growth and Wind Duration

| Wind speed (knots) | 6 Hours L | 6 Hours C | 6 Hours H | 6 Hours T | 12 Hours L | 12 Hours C | 12 Hours H | 12 Hours T | 24 Hours L | 24 Hours C | 24 Hours H | 24 Hours T | 48 Hours L | 48 Hours C | 48 Hours H | 48 Hours T |
|---|---|---|---|---|---|---|---|---|---|---|---|---|---|---|---|---|
| 10 | 38.0 | 8.0 | 1.5 | 2.7 | 49.0 | 9.4 | 1.9 | 3.1 | 59.0 | 10.3 | 2.3 | 3.4 | 66.0 | 11.0 | 2.6 | 3.6 |
| 13 | 56.0 | 10.0 | 2.3 | 3.3 | 74.0 | 11.5 | 3.0 | 3.8 | 90.0 | 12.9 | 3.6 | 4.3 | 108.0 | 14.0 | 4.2 | 4.6 |
| 18 | 94.0 | 13.0 | 4.0 | 4.3 | 123.0 | 15.0 | 5.0 | 4.9 | 164.0 | 17.0 | 6.3 | 5.6 | 197.0 | 19.0 | 7.4 | 6.2 |
| 24 | 144.0 | 16.0 | 6.0 | 5.3 | 200.0 | 19.0 | 8.0 | 6.3 | 265.0 | 22.0 | 11.0 | 7.2 | 320.0 | 24.1 | 13.0 | 7.9 |
| 30 | 200.0 | 19.0 | 8.1 | 6.7 | 285.0 | 22.5 | 11.5 | 7.4 | 375.0 | 26.0 | 16.0 | 8.6 | 470.0 | 30.0 | 20.0 | 9.6 |
| 36 | 270.0 | 22.0 | 11.0 | 7.2 | 375.0 | 26.0 | 16.0 | 8.6 | 502.0 | 30.0 | 21.0 | 9.9 | 650.0 | 35.0 | 26.0 | 11.3 |
| 43 | 345.0 | 25.0 | 14.0 | 8.2 | 486.0 | 30.0 | 20.0 | 9.7 | 662.0 | 35.0 | 28.0 | 11.3 | 870.0 | 40.0 | 36.0 | 13.0 |
| 51 | 435.0 | 28.0 | 18.0 | 9.2 | 620.0 | 34.0 | 25.0 | 11.0 | 870.0 | 40.0 | 35.0 | 13.0 | 1160.0 | 46.0 | 45.0 | 15.0 |

**L** = Wavelength (ft)
**C** = Wave propagation velocity (knots)
**H** = Wave height (ft)
**T** = Wave period (sec)

Note: 1 ft. = 0.305 m
  1 knot = 1.84 km/hr

## Table 6-4
## Wave Growth and Wind Fetch

### Wind Fetch

| Wind speed (knots) | 10 Miles L | 10 Miles C | 10 Miles H | 10 Miles T | 50 Miles L | 50 Miles C | 50 Miles H | 50 Miles T | 100 Miles L | 100 Miles C | 100 Miles H | 100 Miles T | 500 Miles L | 500 Miles C | 500 Miles H | 500 Miles T |
|---|---|---|---|---|---|---|---|---|---|---|---|---|---|---|---|---|
| 10 | 28.4 | 7.1 | 1.2 | 2.3 | 47.6 | 9.4 | 1.9 | 3.1 | 56.3 | 10.0 | 2.3 | 3.3 | 73.4 | 11.6 | 2.7 | 3.8 |
| 13 | 39.0 | 8.3 | 2.0 | 2.8 | 70.0 | 11.3 | 3.0 | 3.7 | 86.0 | 12.5 | 3.6 | 4.1 | 115.3 | 14.6 | 4.2 | 4.8 |
| 18 | 56.1 | 11.0 | 2.4 | 3.3 | 108.2 | 14.5 | 4.4 | 4.6 | 134.3 | 16.2 | 5.4 | 5.1 | 196.7 | 18.8 | 7.6 | 6.2 |
| 24 | 77.0 | 12.0 | 3.4 | 3.9 | 153.4 | 17.0 | 6.2 | 5.4 | 200.0 | 20.0 | 8.0 | 6.3 | 315.4 | 24.0 | 12.3 | 7.9 |
| 30 | 95.4 | 13.5 | 4.3 | 4.4 | 202.3 | 19.0 | 8.2 | 6.2 | 265.0 | 22.0 | 10.6 | 7.2 | 447.3 | 28.4 | 17.2 | 9.4 |
| 36 | 124.1 | 15.0 | 5.4 | 4.9 | 251.0 | 21.2 | 10.3 | 7.0 | 329.0 | 24.5 | 13.3 | 8.1 | 565.8 | 31.9 | 23.0 | 10.6 |
| 43 | 148.8 | 16.4 | 6.6 | 5.4 | 307.5 | 23.4 | 12.8 | 7.7 | 406.3 | 27.0 | 16.7 | 8.9 | 736.4 | 36.4 | 29.0 | 12.0 |
| 51 | 177.0 | 18.0 | 8.0 | 5.9 | 370.5 | 26.0 | 15.5 | 8.5 | 501.8 | 30.0 | 20.3 | 9.9 | 917.6 | 40.6 | 36.5 | 13.4 |

L = Wavelength (ft)
C = Wave propagation velocity (knots)
H = Wave height (ft)
T = Wave period (sec)

Note: 1 ft. = 0.305 m
1 knot = 1.84 km/sec

## 74  Introduction to Offshore Structures

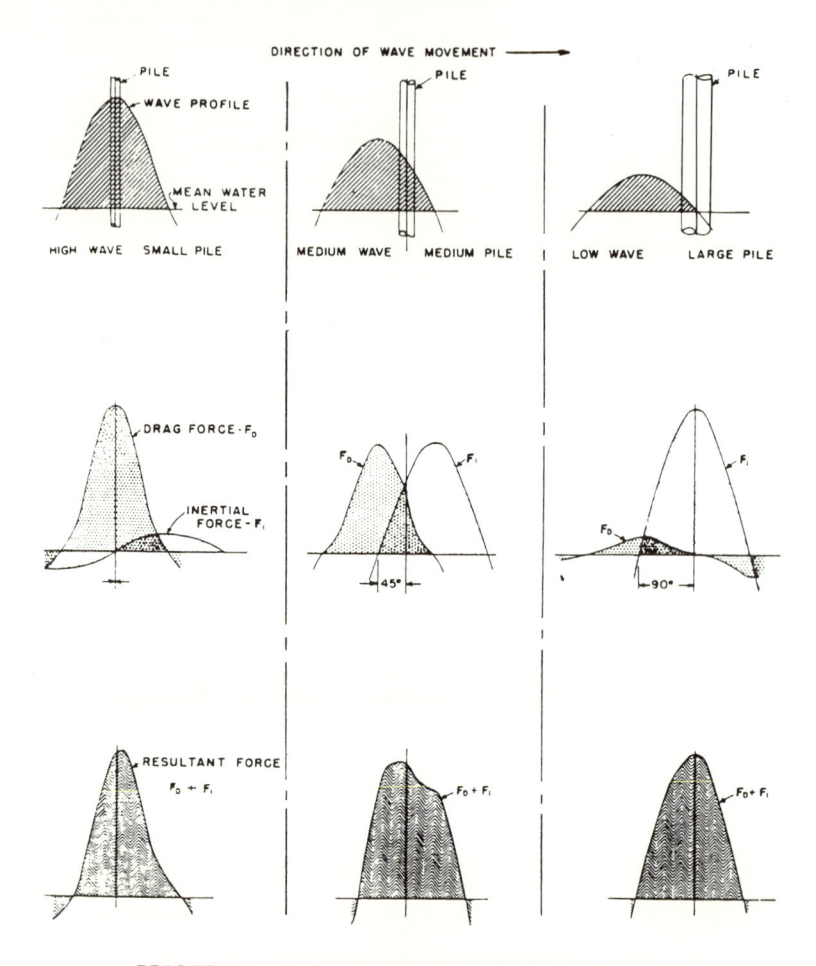

DRAG FORCE IS PROPORTIONAL TO THE VELOCITY SQUARED
TIMES THE PROJECTED AREA
INERTIAL FORCE IS PROPORTIONAL TO THE
ACCELERATION TIMES THE VOLUME

*Figure 6-1.* Typical wave forces on a cylinder. (Source: Randolph Blumberg paper, American Association of Oilwell Drilling Contractors School of Offshore Operations, Houston, Texas, 1972.)

The wave crest should be positioned relative to the structure at more than one position so that the effect of wave position on applied wave force may be found. The wave position that causes the maximum horizontal force on the overall structure will not necessarily be the wave position that most affects a particular member. Maximum stress in a particular member may occur at a wave position, wave height, or wave period different from that which causes

## Design Loads and Forces 75

the maximum force on the structure as a whole. Various directions of wave approach should be considered inasmuch as the structure may react more flexibly because of a wave from one direction than from another. For example, wave approach from a diagonal direction might produce maximum stress in a jacket leg or pile.

In computing the bending moment in individual structural members, either vertical or inclined, the full wave force as computed by the combined drag and inertial expression should be applied as a uniform load perpendicular to the member. As the beam ends are neither fixed nor freely supported, the maximum bending moment $M$ should be taken as

$$M = wL^2/10$$

where:

$w$ = average load per unit length
$L$ = member length.

Appurtenances and exposed structural components of offshore structures are more sensitive to local increases in wave pressure from the irregularity of real waves than is the structure as a whole. For design of such components, the following minimum localized design pressures (representing the envelope of measurements made in 100 ft of water) are recommended until they can be replaced by data commensurate with the water depth of the particular platform: at 25 ft (7.6 m) elevation, above mean water line (MWL), 300 psf (1.44 x $10^4$ Pa); at the MWL, 200 psf (9.6 x $10^3$ Pa); at 25 ft (7.6 m) elevation below MWL, 100 psf (4.8 x $10^3$ Pa); and at 50 ft (15.2 m) below MWL, 75 psf (3.6 x $10^3$ Pa).

### Current Force

The presence of current in the water produces three distinct effects. First, the current velocity should be added vectorially to the horizontal water particle velocity before computing the drag force. Because drag depends on the square of the horizontal particle velocity, and because the current velocity decreases slowly with depth, a comparatively small current can increase drag significantly. The second effect is a steepening of the wave profile from changing the wave celerity. This effect is very small and may be neglected. The third effect makes the structure itself generate waves which in turn create diffraction forces. However, these diffraction forces are negligible for realistic values of current acting on normal-sized members. In design, the maximum wave height is sometimes increased by 5-10% to account for the current effects, and the current per se is neglected.

## Dead Load

The dead load is the weight in air of the overall platform structure including piling, superstructure, jacket, stiffeners, piping and conductors, corrosion anodes, decking, railings, grout, paint, and other appurtenances. Sealed tubular members are to be considered either buoyant or flooded, whichever produces maximum stress at the point under consideration. Excluded from dead load are the following:

1. Drilling equipment weight—the weight of drilling equipment placed on the platform, including derrick, drawworks, mud pumps, mud tanks.
2. Production or treatment equipment weight—the weight of treatment equipment located on the platform, including separators, compressors, piping manifolds, storage tanks.
3. Weight of drilling supplies—the variable loads employed during drilling, such as: drilling mud, water, diesel fuel, casing.
4. Weight of treatment supplies—the variable supplies employed during production, such as: fluid in the separator, storage in the tanks.
5. Drilling load—any appropriate combination of derrick load, pipe storage (set-back), or rotary table load.

## Live Load

Those loads which act in addition to the equipment and structural weights taken statically are called live loads. In general, live loads include impacts from loads that are suddenly or dynamically applied, and dynamic amplification of cyclic loads which excite some component of the platform, or the entire platform, at a frequency near its natural frequency.

Live loads are often designated as a given factor times the applied static load. It is the responsibility of the platform user, that is, the company or division of a company for whom the structure is being designed, to stipulate the live load factors for the static design. In the absence of more specific requirements, the minimum live loads listed in Table 6-5 may be used—either the distributed load or the critically placed concentrated load—whichever produces maximum stress in the member.

Sometimes in design, when it is desired to produce the maximum stress in a member, it becomes necessary to account for the effect of a live load on a portion of the structure. Special consideration must be given to the effects of sidesway and unbalanced live loads whenever columns and/or trusses are battered.

The live load to be used in combination with hurricane forces should be the maximum load reasonably expected to result from the intended use of the structure, with due regard for omitting those loads which will not occur while the platform is evacuated; e.g., hook loads, crane loads, impact loads.

## Design Loads and Forces

**Table 6-5**
**Live Load Factors**

|  | Uniform load beams and decking (lbs/sq ft) | Concentrated line load on decking (lbs/linear ft)* | Concentrated load on beams (kips) |
|---|---|---|---|
| Walkways and stairs | 100 | 300 | 1 |
| Areas over 400 sq ft | 65 | | |
| Areas of unspecified light use | 250 | 750 | 60 |
| Areas where specified loads are supported directly by beams | | 500 | |

*Midway between and parallel to supporting beams.

Note: 1 lb/sq ft = 47.88 Pa
1 lb/ ft = 14.6 N/m

Unbalanced deck loads, which in combination with storm loads produce maximum stress in part of the substructure, need only be considered to the extent that they are consistent with the intended use of the structure. Both windward and leeward load unbalance should be investigated.

### Impact

For structural components carrying live loads which induce impact, the stipulated live load must be increased sufficiently to account for the impact effect. Unless otherwise specified, the increases listed in Table 6-6 may be used.

### Other Forces

Structures in localities subject to earthquakes and other extraordinary conditions must be designed with due regard to such conditions.

The risk of losing a structure because of a freak wave higher than the maximum design wave may be appreciably reduced by providing "air gap" clearance. Where raising the superstructure can be accomplished at nominal cost, and no loss of platform utility results (from inability of derrick barges to lift and place equipment packages), a deck clearance of 5-10 ft (1.5-3 m) is suggested.

Jacket members in contact with the ocean bottom should be designed for the distributed soil pressures acting on them.

## 78  Introduction to Offshore Structures

**Table 6-6**
**Impact Factors for Live Loads**

| Structural item | Load direction Vertical | Horizontal |
|---|---|---|
| Rated load of cranes* | 100% | 20% |
| Drilling hook loads | 0 | 0 |
| Supports of light machinery | 20% | 0 |
| Supports of reciprocating machinery | 50% | 50% |
| Boat landings | 200 kips | 200 kips |

*Horizontal crane forces should be considered as applied at the location of the lifted load in whichever position and direction will produce the maximum stress in combination with the vertical load plus impact.

Note: 1 kip = 4.45 kN

### Deck Floor Loads

The distributed floor loads range from 250 lbs/sq ft (11,975 N/sq m) in the drilling rig and derrick area to 1500 lbs/sq ft (71,851 N/sq m) under the mud, water, and fuel tanks. The distributed floor loading for areas like the pipe rack, power plant, and living accommodations is about 1000 lbs/sq ft (47,900 N/sq m).

### References

1. Morison, J.R.; O'Brien, M.P.; Johnson, J.W.; and Schaff, S.A.; "The Force Exerted by Surface Waves on Piles," *Trans. of AIME,* Vol. 189, 1950, pp. 149-154.
2. Hendrickson, J. and Skjelbreia, L., "Fifth Order Gravity Wave Theory," *Proceedings of 7th Coastal Engineering Conference, Vol. I,* 1961, Ch. 10, pp. 184-196.
3. Dean, R.G., "Stream Function Representation of Nonlinear Ocean Waves," *Journal of Geophysical Research,* Vol. 70, 1965, pp. 4561-4572.

# Chapter 7

# *Pile Foundations*

Pile foundations used offshore by the oil industry are designed for loading conditions vastly greater than those onshore. The oil industry has operated on the premise that the risk is financial and acceptable because maximum foundation loads occur during storms. (It is standard practice to remove personnel before a storm strikes.) Although the loading ranges of documented pile load tests continue to increase, the placement of platforms in deeper and deeper water with attendant larger loads has increased also so that the uncertainties involved in pile design remain.

Immediately after World War II when piled structures were placed in 10-20 ft (3-6 m) of water, the piles were considered independent of one another, and often little or no bracing other than the platform itself connected the piles. Offshore piles are characterized by their large ratios of lateral to vertical loading due to hurricane or storm forces. As platforms were placed in deeper water, the concept of a template (jacket) was developed to enable the piles to resist lateral loading in a unified way. With the development of tower template platforms, the behavior of pile groups became important.

The large lateral forces on the combined jacket and pile structural system from waves and wind require that the piles penetrate great distances into the soil of the ocean bottom. Penetrations on the order of 250-400 ft (76-122 m) are common. Since the piles usually extend up through the jacket legs to the water surface, the total pile length may easily exceed 600-800 ft (183-244 m). Figure 7-1 shows typical features of pile design and fabrication. The pile in Figure 7-1 has a diameter of 4 ft (1.2 m) and a length of 610 ft (186 m); it was designed for an offshore location where the water depth is 280 ft (85 m) and the bottom soil is soft-to-stiff clay.

This pile varies in wall thickness from 0.625 to 1.25 inches (16-32 mm). Note that the largest wall thickness occurs in the vicinity of highest bending stress from a short distance above the mudline to a considerable depth below the mudline. The cross section of the pile at its tip is also heavy wall to serve as a driving shoe.

## 80 Introduction to Offshore Structures

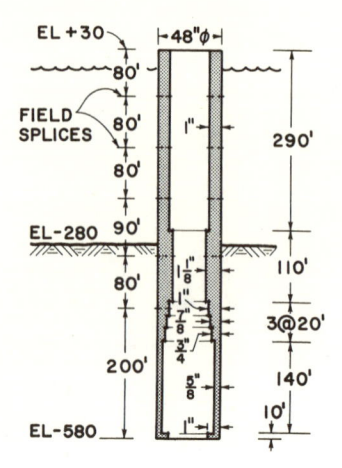

**Figure 7-1.** Example of design and assembly schedule for offshore pile.
(Source: McClelland, B., Focht, J.A., and Emrich, W.J., "Problems in Design and Installation of Heavily Loaded Pipe Piles," Proceedings of the First Civil Engineering in the Oceans Conference, 1968, p. 603.)

Piles are usually made up in lengths called segments, and the segments are welded end-to-end in the process of driving to achieve the desired overall length. New segments are called add-ons. Noted at the left in Figure 7-1 are the locations where field welds occur. These welds are located some distance from the points where wall thickness changes occur.

Early pile hammers had energy capacities for driving of 20,000 (27,200); 40,000 (54,400); and 60,000 foot-pounds (81,600 J). Hammers with energy capacities of 300,000 (408,000) and 550,000 foot-pounds (748,000 J) are common today. A hammer capable of operating in a deballasted chamber underwater with a capacity of 797,000 foot-pounds (1,084,000 J) was used in 1977-78 to drive the seven-foot diameter piles of the *Cognac* platform to an average depth of 450 ft (137 m) in the soft clay bottom of the Gulf of Mexico.

### Axial Loads

Axial pile capacity can be determined by static or dynamic analysis.

**Static Analysis**

This method is based on empirical data gathered from model studies and field load tests interpreted in accordance with accepted theories of soil mechanics. The axial load taken by a pile is transferred to the soil in two ways:

## Pile Foundations 81

through skin friction and end bearing. Thus, the ultimate pile capacity $Q$ at a given soil penetration $D$ is the sum of $Q_s$, the frictional resistance all along the outer surface of the shaft and $Q_p$, the end-bearing load capacity:

$$Q = Q_s + Q_p = fA_s + qA_p$$

where $A_s$ and $A_p$ represent the embedded surface area of the pile and its end area, and $f$ and $q$ represent the unit skin friction or soil-pile adhesion and the unit end-bearing.

Successful application of the static method depends on the selection of appropriate values of $f$ and $q$. These must take into account the combined effects of soil conditions, pile type and dimensions, method of pile installation, and manner of loading. The discussion of these parameters here is applicable for driven open-end steel pipe piles.

Figure 7-2 shows the basic concepts of the static method and lists the expressions used to calculate the design factors.

For piles in clay, the immediate or quick shear strength is the most appropriate *in situ* strength property for evaluating pile capacity. The end-bearing resistance of deep piles in clay may be estimated by the equation

$$Q_p = cN'_c A_p$$

where:

$N'_c$ = bearing capacity factor commonly assumed to have the value nine

$c$ = cohesive shear strength determined from the *in situ* strength profile

$A_p$ = end area of the pile.

In most cases for piles in clay soils the value of $Q_p$ is small in comparison to the total frictional resistance $Q_s$.

The value of the unit skin friction $f$ at any point along a pile in clay is equal to or less than the cohesive shear strength $c$ of the soil. In general, the value of $f$ must be established empirically from load tests. Some guidelines as to its value are; for normally consolidated or underconsolidated clay (firm-to-soft soil), the value of $f$ is approximately equal to the value of $c$. For overconsolidated clay (stiff-to-hard soil), the value of $f$ may be taken as 0.5 tons/sq ft (4.89 tonnes/sq m) for the first 50 ft (15.2 m) of penetration after which the value increases approximately at the rate of 0.005-0.008 tons/sq ft/ft (0.16-0.26 tonnes sq m m) of added penetration. Usually $f = 1/2\ c$.

The unit skin friction $f$ for a pile driven in sandy soil is expressed as

$$f = Kp_o \tan d$$

## 82   Introduction to Offshore Structures

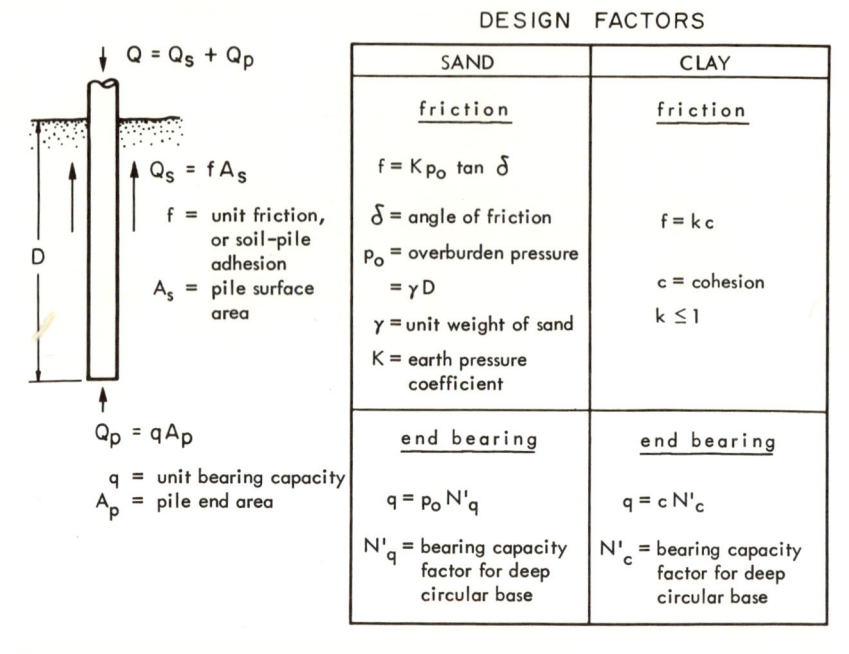

**Figure 7-2.** Basic concepts and design factors in static method for computing pile capacity. (Source: McClelland, B., Focht, J.A., and Emrich, W.J., "Problems in Design and Installation of Heavily Loaded Pipe Piles," Proceedings of the First Civil Engineering in the Oceans Conference, 1968, p. 605.)

Diagram shows: $Q = Q_s + Q_p$; $Q_s = f A_s$ where $f$ = unit friction, or soil-pile adhesion, $A_s$ = pile surface area; $Q_p = q A_p$ where $q$ = unit bearing capacity, $A_p$ = pile end area.

### DESIGN FACTORS

| | SAND | CLAY |
|---|---|---|
| **friction** | $f = K p_0 \tan \delta$ <br> $\delta$ = angle of friction <br> $p_0$ = overburden pressure <br> $= \gamma D$ <br> $\gamma$ = unit weight of sand <br> $K$ = earth pressure coefficient | $f = kc$ <br><br> $c$ = cohesion <br><br> $k \leq 1$ |
| **end bearing** | $q = p_0 N'_q$ <br> $N'_q$ = bearing capacity factor for deep circular base | $q = c N'_c$ <br> $N'_c$ = bearing capacity factor for deep circular base |

where:

- $p_0$ = the overburden pressure, a product of the unit weight of soil and the pile depth
- $K$ = a coefficient of lateral earth pressure
- $d$ = skin friction angle between the soil and the pile.

For open-end driven steel pipe piles, $K$ may be taken as 0.7 for compressive loads and 0.5 for tensile loads. For piles driven in undersize holes, $K$ ranges from 0.4 to 0.7. For piles driven in oversize holes, $K$ ranges from 0.1 to 0.4.

For medium-dense to dense sandy soils, representative values for the other parameters to be used with driven pipe piles are listed in Table 7-1. Table 7-1 gives limiting maximum values of $f$ and $q$ for piles with large penetrations into sandy soils. For preliminary calculations in the absence of accurate data for a specific pile, no increase in $K$ should be allowed for the possibility of the formation of a plug in an open-end pile. If the soil is loose, the parameters listed in Table 7-1 should be decreased somewhat.

## Table 7-1
## Representative Parameters for Driven Pipe Piles in Sandy Soils

| Soil type | Internal friction angle of soil $\phi$ (degrees) | Skin friction angle $d$ (degrees) | Tangent $(d)$ | Limiting skin friction[1] ($f$ max) | Bearing capacity factor ($N\dot{q}$) | Limiting end bearing[1] ($q$ max) |
|---|---|---|---|---|---|---|
| Clean sand | 30-35 | 30 | 0.5772 | 1.0[2] | 41 | 100[2] |
| Silty sand | 25-30 | 25 | 0.4663 | 0.85 | 22 | 50 |
| Sandy silt | 20-25 | 20 | 0.3640 | 0.7 | 12 | 30 |
| Silt | 15-20 | 15 | 0.2680 | 0.5 | 8 | 20 |

[1]Tons/sq ft (1 ton/sq ft = 9.76 tonnes/sq m)
[2]These limiting values are usually encountered at a depth of about 80 ft.

## 84   Introduction to Offshore Structures

For piles driven sufficiently in sandy or granular soils to develop a competent or proper plug, the unit end-bearing is expressed as a bearing capacity factor $N_q'$ times the overburden pressure $p_o$. For preliminary design, values of $N_q'$ and limiting end-bearing $q_{max}$ appropriate for medium-dense to dense granular soils are given in Table 7-1.

The total capacity of a pile in sandy soil increases in approximately a linear fashion with depth of penetration because both $f$ and $q$ are directly proportional to overburden pressure. Limited experimental data indicates that the unit end-bearing capacity $q$ is not proportional to pile penetration but tends to reach a maximum value and then remain almost constant with increasing depth.

For a pile with its tip embedded in a loose sandy stratum, the pile capacity determined from the parameters given in Table 7-1 should be decreased. The computed end-bearing capacity $Q_p$ of an open-end pile should be included as part of $Q$ only if the frictional resistance of the soil plug inside the pile is expected to be greater than $Q_p$. Jetting or drilling through the pile destroys the plug inside the pile. The potential end-bearing capacity can be positively restored only by constructing a cement grout plug for the pile tip.

### Dynamic Analysis

Dynamic formulas are based on the amount of energy necessary to drive a pile through a given soil stratum. This analysis is done through a one-dimensional wave equation that describes mathematically the travel of an impulse down a pile as it is driven. By representing the soil, pile, pile cap, cushion, and hammer system as a series of masses connected with springs and dampers, an iterative computer solution is obtained through which the pile load capacity may be calculated. Figure 7-3 shows the physical and analytical model of a typical pile. This type of analysis considers that the stress wave produced by the hammer blow travels along the length of the pile at the speed of sound. The entire pile length is not stressed simultaneously. A high-speed digital computer is required for the analysis.

### Safety Factors

Pile driving through underconsolidated clays is easy, whereas driving in dense sands can be very difficult. Driving resistance may increase markedly as the pile encounters different strata within the earth, especially cemented sand layers. Also, when driving in clays and a pause occurs to permit splicing on another pile length (or for some other reason), the soil "set-up" may become significant and cause increased driving resistance until the pile is again moving. These factors must be accounted for in determining the driving power required.

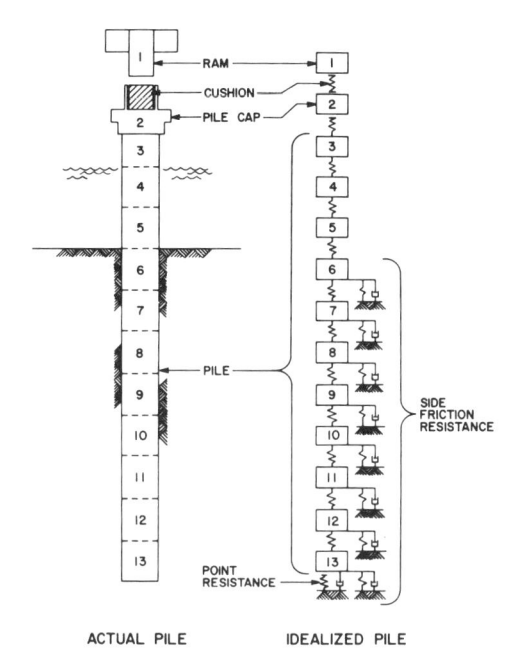

*Figure 7-3.* Hammer/pile/soil model. (Copyright © 1979 Offshore Technology Conference.)

The ultimate capacity of a pile driven in clay is two to five times the resistance developed at the time of driving. For sand, the ultimate pile capacity is approximately equal to the driving resistance. When driving is resumed after a delay, it is usually assumed that the driving resistance is equal to the ultimate static capacity of the pile at that penetration. Experience indicates that the increased driving resistance in clay is not significantly reduced until about 10 diameters of new penetration is achieved. Recovery from set-up occurs more quickly in sandy soil.

The pile wall thickness is often made greater than what is needed for the structural requirements of the platform design. This thicker wall on the pile is necessary to achieve more effective use of the hammer's energy, that is, to make the lower end advance when the hammer is driving on the upper end.

Pile load tests have shown that for two adjacent similar piles the capacities may differ by 10-20% as a result of natural minor variations in subsurface materials and installation procedures. Likewise, computed and actual ultimate capacities may differ by 10-20%—even when the computations are based on subsurface data and computation procedures of superior reliability. Greater differences are to be anticipated for data and procedures of average or

## 86    Introduction to Offshore Structures

less-than-average reliability. Situations that tend to create lesser degrees of reliability are:

1. Large and irregular variations in soil strengths
2. Great extrapolation of current empirical correlations
3. Installation techniques that alter the soil properties
4. Irregular soil stratification

For normal operating conditions (ordinary environmental loads, including the effects of frequently occurring storms) on economically important structures, a safety factor of 2.0 is recommended when there is an average level of confidence in both the subsurface data and the pile design criteria. A safety factor of 3.0 is recommended when there is less-than-average reliability on either subsurface data or design criteria.

For maximum possible design loads with a recurrence interval of 100 years, or for structures that may be reasonably exposed to a calculated risk of failure, a safety factor of 1.5 is recommended when there is an average level of confidence in both the subsurface data and the pile design criteria. A safety factor of 2.0 or higher is recommended when less-than-average reliability exists with respect to subsurface data or design criteria.

### Installing Undrivable Piles

Increasing pile penetration does not necessarily increase pile capacity. However, there are supplemental methods for increasing pile penetration which can bring about increased capacity over that achieved in just driving the original pile to refusal. These supplemental methods may be divided into four categories:

1. Driving a smaller size pile through an initially installed larger pile
2. Fastening a pile inside an oversize hole with grout
3. Driving a pile concentrically along the path of an undersize pilot hole
4. Driving a pile into a hole created by uncontrolled jetting or drilling

In the following sections these four methods are briefly discussed. The methods are shown in Figure 7-4.

### Insert pile

This name is given to a smaller-diameter pile driven to a greater depth through the original pile. (See Figure 7-4a.) Insert piles are useful in increasing the pile capacity when the original pile cannot be driven any deeper and yet is short of its intended penetration, or when the original pile fails to encounter

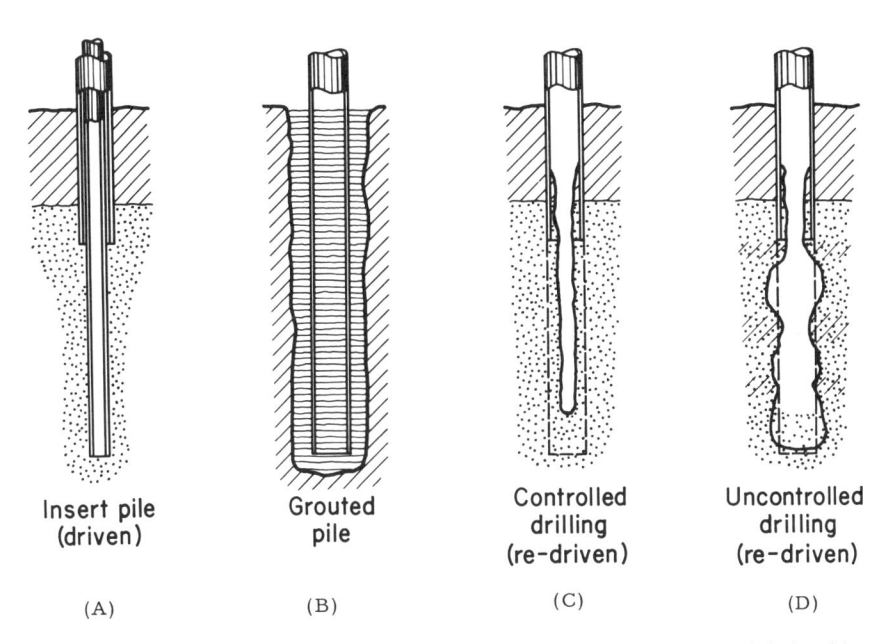

**Figure 7-4.** Installation procedures for piles that cannot be installed solely by driving. (Source: McClelland, B., Focht, J.A., and Emrich, W.J., "Problems in Design and Installation of Heavily Loaded Pipe Piles," Proceedings of the First Civil Engineering in the Oceans Conference, 1968, p. 628.)

the expected soil resistance when driven to its maximum design length. The insert pile is welded to the original pile at the top of the jacket leg, and the annular space between the piles is filled with grout.

## Grouted pile

Figure 7-4b shows a pile that has been placed concentrically in an oversize hole and fastened with grout. This method requires the drilling of an oversize hole to the proposed pile penetration depth. The grout fills the annular space between the outside wall of the pile and the surrounding soil material.

This procedure is the only means of installing piles with tension capacity in hard soils or soft rock. However, the use of grouted piles increases the level of design uncertainty. When some hard-soil formations are drilled, they soften when exposed to water; this softening causes the wall of the hole to become a plane of weakness, producing lower skin-friction resistance. Lateral yielding of the formation into the drilled hole represents another uncertainty; again skin friction is reduced. If drilling mud is used in the drilling, the inside surface of the hole may be left coated with mud. It is believed that this method also lowers the skin friction of the final assembly.

### Controlled Pilot Hole

As shown in Figure 7-4c, this method is useful when the original driven pile encounters hard driving or apparent refusal before reaching the desired penetration. A pilot hole of controlled, reduced diameter is drilled through the pile to some predetermined depth and then the original pile is driven further (redriven). Should the original pile again reach refusal, the process is repeated until the desired depth is achieved.

If the pilot hole is sufficiently small, and if its depth is stopped short of the design depth of the pile, it is probable that the skin friction developed by the pile will be equivalent to that of a pile installed entirely by driving. Pilot hole diameters generally range from one-half the pile diameter to about 4 inches (10 cm) less than the pile diameter. To ensure end-bearing capacity, the pile should be cleaned out, and the bottom of the inside of the pile filled with grout. The cleaning out should be done in a manner that will not result in upward yielding of the soil formation below the pile tip.

### Uncontrolled Drilling or Jetting

This method is shown in Figure 7-4d. It produces a hole of uncertain size and doubtful stability. The properties of the surrounding soil are so adversely affected that the capacity of the final pile installation is unpredictable. This method is seldom used as a supplement to pile driving for offshore structures.

### Lateral Pile Load Analysis

The generalized solution for a laterally loaded pile considers the horizontal load and moment with variations in magnitude, direction, and time. Different pile lengths, different rigidities, and battering of the pile must be taken into account. This problem is complex; many simplifying assumptions must be made for solution.[1,2]

The problems associated with lateral loads can be divided into two classes: first, the determination of soil stress-strain characteristics as they pertain to laterally loaded piles; and second, the mathematical determination of the pile deflection curve once the soil characteristics are known. A necessary prerequisite for solving the laterally loaded pile problem is to predict the lateral load resistance along the pile as a function of the deflection of the pile.[3] This soil reaction is a function of soil type, pile properties, loading, depth, and deflection at that depth. The ratio between the soil reaction or resistance at any point and the pile deflection at that point is defined as the elastic soil modulus. Although the soil modulus varies with depth, it is quite often assumed to be constant at any given depth as a mathematical necessity. Some computer methods allow soil modulus variation with depth to be assumed as either constant, linear, or a random variation in an actual analysis.

The usual approaches to handling the problems of laterally loaded piles are the limit analysis method, the point of fixity method, and *P-y* methods. The limit analysis method assumes that the limiting, or maximum, soil resistance acts against the pile when the ultimate load is placed against it, the soil is of constant strength with depth, and the pile deflects sufficiently to develop the full soil resistance all along the length considered. There is no method of computing the deflection of the pile, and there is no way to know whether or not the full soil resistance has been developed in this limit analysis method.

The point of fixity method makes the assumption that the pile is completely fixed against rotation at some point below the mudline. This point selection is made from experience, and the soil above this point is usually neglected. The point of fixity method is only a means of simplifying the procedure to obtain a solution—the structure does not act this way.

*P-y* analysis is the most common and widely accepted method and does apply when the soil is acting plastically. This method offers the most rational procedure for determining the actual behavior of a laterally loaded pile. In this method an elastic soil modulus, either constant or varying with deflection, is obtained for the soil. The selection of this soil modulus is the most difficult procedure in this method. The soil modulus is adjusted with deflection to account for the essentially nonlinear nature of soil resistance and to obtain a better picture of how the pile actually acts. A desirable feature of this elastic method is that the equation yields values of deflection, slope, moment, shear, and resistance for all points along the pile.

The critical factor for determining the size of the piles is the portion of the stress resulting from bending moment. Therefore, the bending moment in the piles should be predicted accurately. With the elastic method, it is possible to go from the load diagram through the shear, moment, and slope diagrams to the deflection diagram by a simple process of integration, provided that sufficient boundary conditions are known. It should also be noted that a simple process of differentiation can be used to go from the deflection to the load diagram and through the slope, moment, and shear diagrams.

The three most common methods of elastic analysis are beam on elastic foundation, nondimensional solutions, and the difference equation. The beam method is usually employed when the soil modulus is assumed to be constant with depth. This method utilizes the basic strength of materials approach. The difference equation is often used when the soil modulus has a random variation with depth and requires a digital computer. This method is valuable only if it is programmed such that the pile is divided into a large number of increments.[4,5]

Nondimensional solutions may be developed for any fixed form of variation of the soil modulus with depth and are the most commonly used method of analysis.[6] Dimensionless curves, such as the plot of lateral deflection coefficient versus depth coefficient for variable angular restraints of the pile and resisting force versus lateral deformation characteristics of the soil, must be known. These latter curves are known as *P-y* curves.[7]

### Battering of Piles

To increase pile efficiency against lateral loads, piles are usually installed in inclined or slanted positions with respect to a vertical reference line. Thus, the piles are said to have *batter*. Battering with a slope as small as 1:12 will sometimes decrease pile deflection as much as 40%. A slope of 1:12 means that in a vertical distance of 12 units the axis of the pile will have moved away from the vertical by a distance of one unit. The decrease in pile deflection due to batter rises to around 60% for the 1:8 slope typically used on the broadside or longside of offshore jackets.

### Group Piles

When piles are grouped closely, an element of soil will feel stress effects from two or more piles. These stress effects reduce the efficiency of the pile group when compared to the summation of the individual pile bearing capacities. The interaction must be properly taken into account or unpredicted settlement and reductions in capacity will result.[8,9]

### References

1. Matlock, H., "Correlations for Design of Laterally Loaded Piles in Soft Clay," 1970 OTC Preprints, Paper #1204.
2. Delenil, G.E.; Heyman, M.R.; Michel, D.J., "Behavior of Offshore Platform Three-Dimensional Batter Pile Foundation," *Proceedings of the Offshore Exploration Conference* (OECON), 1968, pp. 131-151.
3. Matlock, H. and Reese, L.C., "Foundation Analysis of Offshore Pile Supported Structures," Intnl. Society of Soil Mechanics and Foundation Engineering (paper), Paris, July 1961.
4. Gleser, S.M., "Lateral Load Tests on Vertical Fixed Head and Free Head Piles," Amer. Soc. of Testing Materials, Special Tech. Pub. #154, 1953, pp. 75-101.
5. Chan, J.H.C. and Matlock, H., "A Discrete Element Method for Transverse Vibrations of Beam Columns Resting on Linearly Elastic or Inelastic Supports," 1973 OTC Preprints, Paper #1341.
6. Matlock, H. and Reese, L.C., "Nondimensional Solutions for Laterally Loaded Piles with Soil Modulus Assumed Proportional to Depth," Bureau of Eng.

Research, Special Pub. #29, UT Austin, Sept. 1956. (*See also:* Matlock, H. and Reese, L.C., Generalized Solutions for Laterally Loaded Piles," *ASCE Trans.,* Vol. 127, 1962, Paper #3370.)

7. Myers, J.J. et al., "Lateral Loads on Piles," *Handbook of Ocean and Underwater Engineering,* McGraw-Hill Book Co., 1969, Chapter 8, pp. 8-115 through 8-125.

8. Focht, J.A., Jr. and Kock, K.J., "Rational Analysis of the Lateral Performance of Offshore Pile Groups," 1973 OTC Preprints, Paper #1896.

9. O'Neill, M.W.; Ghazzaly, O.I.; and Ho Boo Ha, "Analysis of Three-Dimensional Pile Groups with Nonlinear Soil Response and Pile-Soil-Pile Interaction," *1977 OTC Proceedings,* Paper #2838.

# Chapter 8

# Design for an Eight-Leg Jacket

The first offshore platforms were contructed of timber in the early 1900s. These platforms were designed to carry primarily vertical loads, because they were placed in the shallow protected waters of inland lakes, swamps, and bays. Only as the continued search for oil and gas brought about the location of platforms in the open seas did designers become concerned about lateral forces from waves, wind, and current. Allowing for these environmental forces on the structure greatly complicated the designs. At first, in the early 1950s, only cursory attention was paid to these environmental forces. It was not until hurricanes destroyed several offshore structures in the open waters of the Gulf of Mexico that such lateral force effects were adequately considered.

## General Description

There are three main components to a steel template platform: (1) jacket, (2) piles, (3) and topside facilities, often called the superstructure or simply the deck. The topside facilities may be divided into deck substructure and deck modules. The jacket is crucial to the entire project, and most of the design effort is spent on this component.

The topside facilities frequently comprise three decks: a drilling deck, a wellhead/production deck, and a cellar deck. These decks are supported on a gridwork of girders, trusses, and columns in two rows of four columns each. Beneath, and joining the ends of the deck columns, the piles project downward through the eight legs of the jacket into the ocean floor. Figure 8-1 shows the principal elements of an eight-leg platform. Figure 8-2 shows how the longitudinal and transverse planes through the columns are designated. For both the superstructure and the jacket, the column rows are designated as A and B in the longitudinal direction and 1, 2, 3, and 4 in the shorter transverse direction. The jacket consists of eight large-diameter tubular legs framed together by a large number of smaller tubular members called *braces*. The trusswork between each pair of outside jacket legs is usually of the Warren bridge type.

# Design for an Eight-Leg Jacket

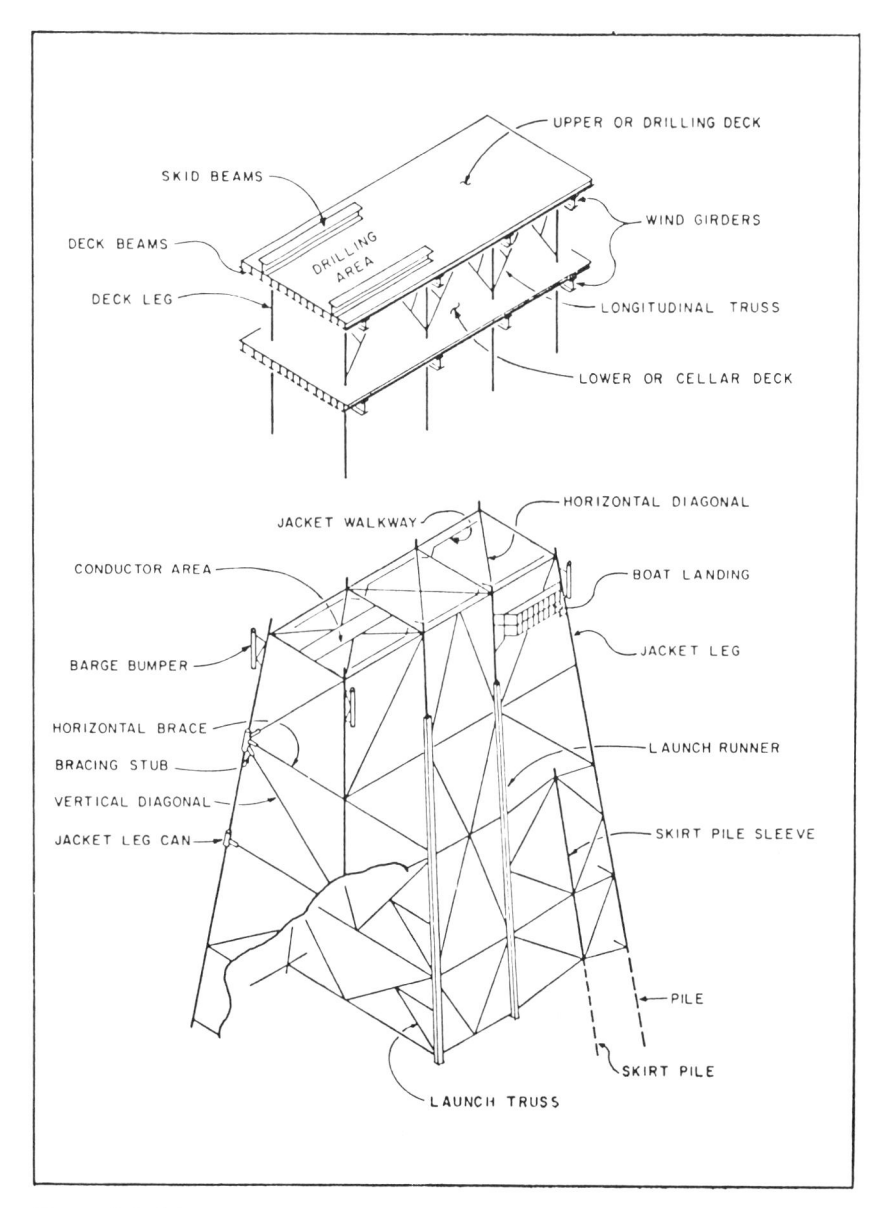

*Figure 8-1.* Principal elements of an eight-leg platform. *(Courtesy Brown & Root, Inc.)*

**94    Introduction to Offshore Structures**

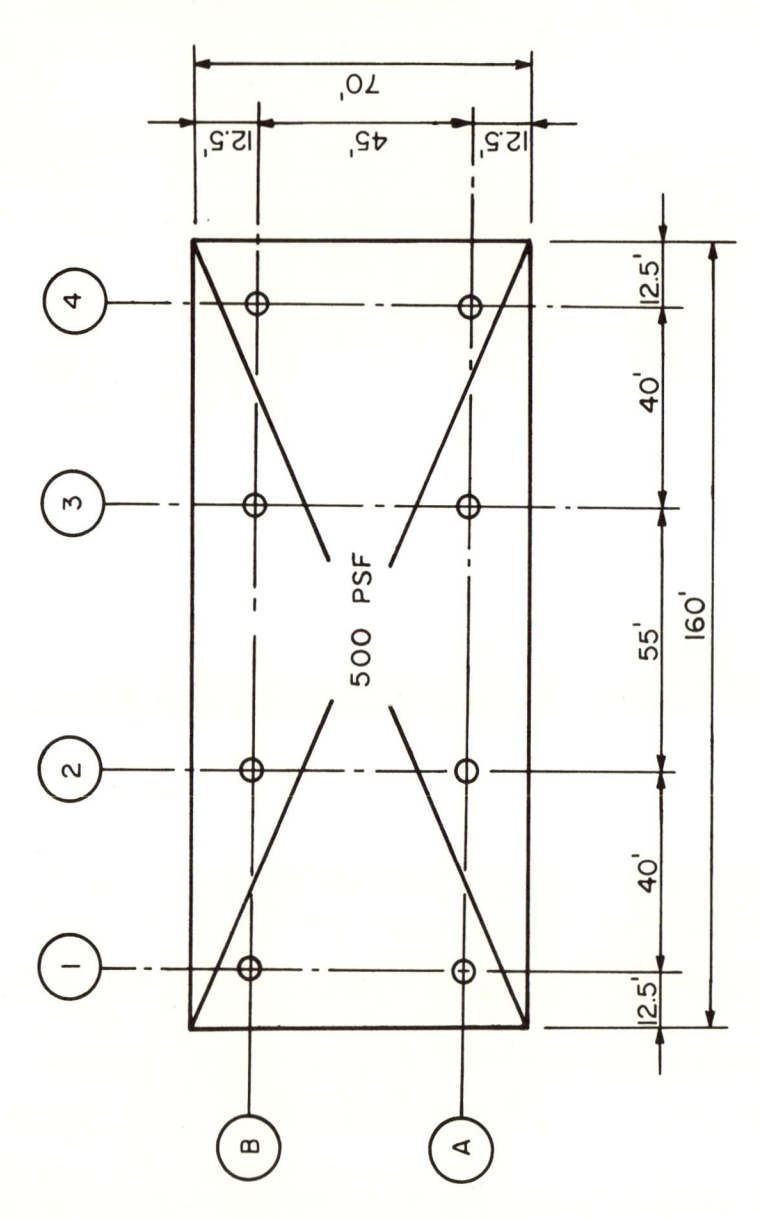

*Figure 8-2. Designation of longitudinal and transverse reference lines for an eight-leg jacket.*

## Design for an Eight-Leg Jacket

The leg spacings in a horizontal plane called seadeck level are usually stated at an elevation of 10-14 ft (3-4 m) above the mean water line (MWL). That is, the jacket extends from the mudline to 10-14 ft (3-4 m) above MWL. The reason the leg spacings are usually stated at this elevation above MWL is that the batter of the jacket, which usually begins just above the jacket walkway, begins in this range. The jacket walkway is at this elevation so that it will be above the normal everyday waves which pass through the jacket.

At the jacket top, the four legs along each side will be spaced approximately 36-45 ft (11-13.7 m) apart, although in some designs the center bay along the side may be as much as 60 ft (18.3 m). The center longitudinal spacing of 40-60 ft (12-18.3 m) is usually set by the availability of launch barges and the spacing of the launch runners on these barges.

In the transverse direction the leg spacing is approximately 45 ft (13.7 m). The transverse spacing is usually set by the dimensions of the drilling and/or production packages which are to be placed on the decks. The drilling deck and the wellhead/production deck extend out beyond the area outlined by the jacket legs. The length of this cantilever overhang is usually about 12-15 ft.

Sometimes jackets with eight legs will have larger-diameter legs at the corners. For example, the piles to be driven through the corner legs may be 60-inch (1.5-m) OD (outside diameter) and those for the in-between legs may be 48-inch (1.22-m) OD. Allowing one-inch annular clearance between the pile and the inside of the leg means that the legs will have internal diameters of about 62 inches (1.58 m) and 50 inches (1.27 m), respectively.

Jacket legs are not vertical. From the plan view rectangle size given at the 10-14 ft elevation, the legs flare out as they go down; they are said to have batter. Jacket legs are battered to provide a larger base for the jacket at the mudline and thus assist in resisting the environmentally induced overturning moments. The four legs along the side of the jacket usually have batter of one foot in seven or eight ft (one m in seven or eight m). In the transverse direction only the outer legs have batter, usually 1:10 or 1:12. The four legs designated A-2, A-3, B-2, and B-3 have batter only in the plane of the side of the jacket. This design provides parallel transverse trusses for skidding the jacket onto the transportation barge at load-out from the onshore construction yard.

After the governing topside facility geometry and loading are determined, the template structural geometry is established. Preliminary sizes are selected based on experience. Ocean-bottom soil conditions are analyzed to determine the number and size of piles needed. This analysis will also determine if skirt piles are needed. The design engineer must iterate the analysis of the jacket with that of the lateral load behavior of the piles to arrive at the final configurations of both the jacket and the piles.

Jackets are adapted to various water depths and different soil conditions in several ways: by making the corner piles and legs through which they pass larger than the in-between piles and legs or, if all the legs are the same size, by

## 96  Introduction to Offshore Structures

adding skirt piles in-between the jacket legs, or adding clusters of skirt piles about the corner legs. Because of the batter of the legs, there is increased separation between the legs at the mudline compared to the top. The main reason that skirt piles are used in lieu of larger-diameter jacket legs (with larger piles inside) or additional jacket legs (and piles) is the trade-off between the extra strength and support obtained versus the additional wave loading incurred. After the piling configuration has been selected, the configuration should be analyzed in tension loading for the maximum pull-out loads on the piles.

The skirt-pile sleeves are incorporated into the jacket structure between the two lowest levels of horizontal bracing. The sleeves must be sufficiently offset from the plane of the jacket side to permit the piles to pass parallel to the plane of the jacket side through pile guides. In deepwater designs the skirt piles are clustered around bottles (enlarged corner legs). These modifications increase the resistance the overall structure offers to the overturning moment caused by waves and wind. (See Figure 8-3.)

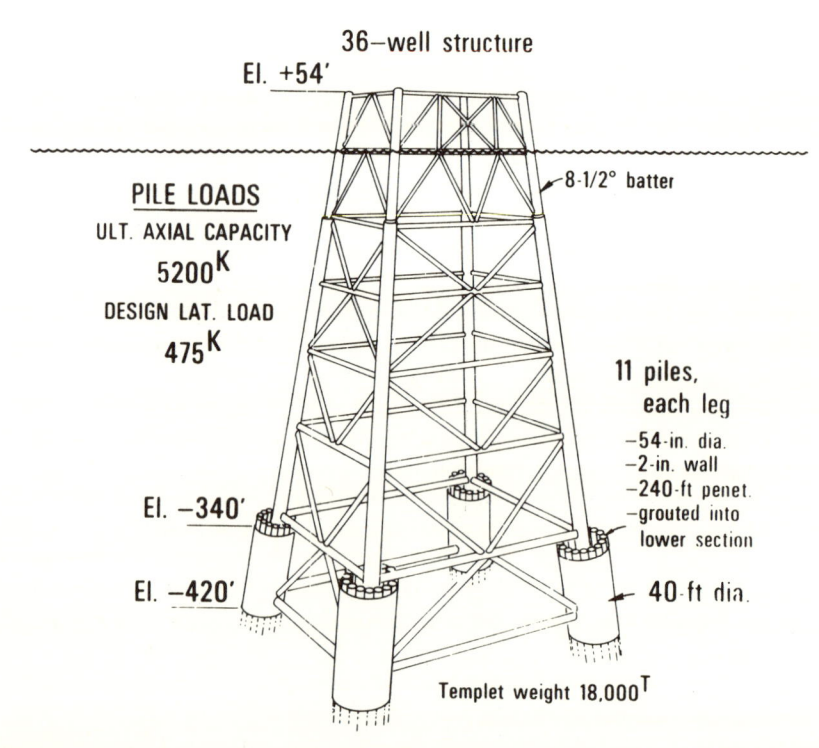

**Figure 8-3.** Jacket for a North Sea platform. Note the bottles at the lower ends of the legs for groups of piles. (Source: B. McClelland paper, American Association of Oilwell Drilling Contractors, School of Offshore Operations, Houston, Texas, 1972.)

## Design for an Eight-Leg Jacket

The deck substructure is usually made up of a group of parallel Warren trusses with Warren type cross-bracing to form the planar trusses into a space frame. The upper and lower chords of the planar trusses may be wide flange members or tubulars. Web members are usually tubular. The deck substructure supports the deck modules which are proportioned to fit into and on the deck substructure space frame. The deck substructure extends beyond the spacing of the jacket legs at the 10-14 ft (3-4 m) elevation by 10-15 ft in both horizontal directions. Thus, the plan size of the deck substructure may be from approximately 60 x 120 ft (18.3 x 36.6 m) to 75 x 170 ft (22.9 x 52 m). Plan size varies depending on the number of jacket legs and the functional requirements of the platform. Sometimes braces must be initially left out of the deck substructure to provide openings through which to lower and/or slide skid-mounted equipment into position. The braces are then field-welded into position.

The arrangement of deck modules into the available space on the deck substructure is carefully planned during the preliminary design. Modules are fabricated with complete piping on shore and tested for proper operation before load-out to the barge. Pipe joints are planned and positioned so that final connection of modules at sea can be accomplished by simply bolting two flanges or welding a short transition pipe into place.

Deck areas not allocated as module locations are framed and covered with fixed steel plates except for the wellhead area on the top and intermediate levels. This area is framed and covered with removable hatch plates wherever there is a drilling location. While the intermediate level deck usually has the same basic dimensions and framing features as the drilling deck, the cellar deck is frequently contained within the space outlined by the eight columns. This deck is framed and covered with steel grating. Commonly the height from the cellar deck to the intermediate deck is 10-12 ft (3-3.7 m). The height from the intermediate deck to the drilling deck is usually 18-20 ft (5.5-6.1 m).

While not part of the basic framework needed for structural strength, provisions for conductors and risers must be made, since these are necessary for the functional requirements of the platform. The number of wells to be drilled is decided at the very beginning of the project; for example, 18, 24, 30, or whatever the economics of the particular project indicates. The wells are drilled through conductor tubes positioned in an array which can be reached by sliding the drilling derrick from location to location across the deck. Conductors are vertical tubes of 30-36-inch (0.76-0.91-m) diameter driven through guide rings to a depth of perhaps 200 ft (61 m) below the mudline. The rectangular array of these tubes is usually setup on a six to eight-ft center-to-center spacing.

Risers are various vertical tubes located within the jacket framework for pumping seawater to the decks, for pipeline connections to other platforms or to shore, for heat exchange, and many other processing functions. Risers can range from 14-16 inches (0.36-0.41 m) in diameter to diameters approaching

## 98  Introduction to Offshore Structures

the size of conductor tubes. The number of risers on even a small, self-contained drilling/production platform may easily exceed a dozen.

Most designs call for two boat landings, one on each longitudinal side of the jacket between columns 2 and 3. Each boat landing should have platforms for embarking and disembarking at two different elevations. Access from the boat landings to the various decks should be adequate if not generous.

Barge bumpers are required on each jacket leg. These should extend for a considerable distance vertically to accommodate loading and unloading in a variety of sea conditions.

A minimum of one pedestal-mounted crane should be provided. An 80-ton capacity crane is commonly located about 25 ft outboard of the deck at transverse plane 2 if the well drilling area is between transverse planes 1 and 2.

### Environmental Conditions

Environmental conditions are naturally different from one offshore location to another. Specialists in oceanography, meteorology, and soil mechanics are needed to interpret the often scarce physical data and delineate the environmental conditions necessary for preliminary design. During the months while the preliminary economic and functional design studies are being conducted, extensive measurements are made of the environmental factors surrounding the offshore site so that more accurate criteria will be available for the detailed design.

If the jacket is to support drilling equipment at first, and later storage tanks or treatment facilities, the weights and environmental loads of both kinds of equipment must be determined in the initial design. The platform should be designed for the larger set of loads. Changes in wind loading are of particular importance.

The environmental data presented here is representative of the conditions generally found either in the Gulf of Mexico or in the southern regions of the North Sea and is not intended to be specific for any particular location.

### Gulf of Mexico

**Water depth.** The typical water depth may be taken as 300 ft (91.5 m) MGL (Mean Gulf Level).

**Design wave.** The maximum hurricane wave with a recurrence interval of 100 years has a height of 60 ft (18.3 m) and a period of 15 seconds. The height of the total tidal change is six ft (1.8 m). Thus, the design wave has a height of 66 ft (20 m) and a period of 15 seconds.

**Average wave.** Normal sea conditions are needed for fatigue load calculations. The significant wave height is frequently taken as the maximum wave height divided by 1.87, and the average wave height is about 0.6 of the

## Design for an Eight-Leg Jacket

significant wave height. Thus, the average wave height is around 20 ft (6.1 m). The wave period associated with the average wave height is 12.5 seconds.

**Ocean current.** At the water surface, the current is 3.6 ft/sec (1.1 m/sec). At a depth of 86 ft (26.2 m), the current is 75% of the surface value; at a depth of 190 ft (58 m), it is 50% of the surface value. At the mudline, the current is 0.5 ft/sec (0.15 m/sec). The current is assumed to be capable of approaching the platform from any direction.

**Wind speed.** The maximum sustained wind speed based on a 100-year storm recurrence interval is 154 ft/sec (47 m/sec). The three-second gust velocity is 216 ft/sec (66 m/sec). The smallest wind magnitude comes from the west. While the maximum sustained wind speed could come from any direction, the most probable direction is from the south or southeast. These magnitudes may be taken as the wind speeds at the 30-ft (10-m) reference height.

**Geotechnical (soils) data.** For the most part, the bottom of the Gulf of Mexico is sand, silty sand, silty clay, and clay. Because of the many rivers emptying into the Gulf, the geotechnical parameters differ from one location to another. For the so-called representative site in 300 ft (91 m) of water mentioned earlier, the soils data may be determined. From the ocean bottom to a depth of 30 ft (9.1 m), the soil consists of soft clay with an unconfined compressive strength of 0.1 tons/ft$^2$ (0.98 tonnes/m$^2$). The next 12-15 ft (3.7-4.6 m) is silty sand with an effective friction angle of 25° followed by 15 ft (4.6 m) of sand with an effective friction angle of 30°. Below this level there is silty clay for 10 ft (3 m), and then stiff clay for 85 ft (26 m); the latter has an unconfined compressive strength of about 1 ton/ft$^2$ (9.8 tonnes/m$^2$). The next layer is 30 ft (9.1 m) of consolidated sand at an effective friction angle of 35° with clay below.

### Danish Sector of the North Sea

**Water depth.** The average water depth is 40 m, or about 131 ft.

**Design wave.** The maximum storm wave with a recurrence interval of 100 years has a height of 79 ft (24 m) and a period of 15 seconds. The height of the total tidal change is 3.3 ft (1.0 m). Thus, the design wave has a height of 82 ft (25 m), and a period of 15 seconds.

**Average wave.** This figure is needed for calculation of the fatigue behavior of the tubular joints during the operational life of the platform. The height of the average wave is about 65% of the maximum storm wave, or 51 ft (15.6 m). Based on information from other North Sea platforms, there will be 12,000 of these average waves during a 20-year operational life. Assuming that waves pass by the platform at the rate of 10 per minute, an operational life of 20 years corresponds to a total of 1 x 10$^8$ waves.

**Ocean current.** At the water surface, the current is 4.0 ft/sec (1.2 m/sec). At the mudline, the current is 2.3 ft/sec (0.7 m/sec). The current is assumed to vary linearly with depth over the 131 ft (40 m) and may approach the platform with the same magnitude from any direction.

**Wind speed.** The one-hour average wind speed is 121 ft/sec (37 m/sec). This sustained wind speed can be from any direction, although the most probable direction is from the north or northwest. The three-second average gust speed is 171 ft/sec (52 m/sec). The smallest wind magnitude comes from the northeast. These magnitudes should be taken as the wind speeds at the 30-ft (10-m) height.

**Geotechnical (soils) data.** From the ocean bottom to a depth of 20 ft (6 m), the soil consists of silty sand. The effective friction angle is 28-30°. From a depth of 20 ft (6 m) to 49 ft (15 m), there is clay with an unconfined compressive strength of 0.465 tons/ft$^2$ (4.5 tonnes/m$^2$). From a depth of 49 ft (15 m) to 164 ft (50 m), there is consolidated sand with an effective friction angle of 35°.

## Other Environmental Conditions

In addition to the principal environmental conditions mentioned in the two preceding sections, there are several other factors that may influence design criteria depending on local situations.

**Ice.** If the platform is to be located in a cold region, obviously not the Gulf of Mexico, there could be added load on various topside facilities due to atmospheric ice formation. Solid ice build-up of three inches (76 mm) is often assumed in design calculations for critical topside facilities in Cook Inlet, Alaska, platforms.

**Vortex shedding.** Underwater tubular members may be subject to vortex shedding as the current and waves flow past. Topside facilities, particularly cables and struts, may be subject to vortex shedding from storm winds.

**Bottom scour.** Because the top surface of the ocean bottom is usually composed of silted sand, there is the possibility of scour around the bottoms of the jacket legs. This possibility must be determined and prevented through proper design. If the first few feet of soil around each pile were removed by scour, it could seriously alter the ability of the soil-structure system to develop adequate lateral soil resistance.

**Calculating wave force.** Using the Stokes fifth order wave theory, the design wave of 82 ft (25 m) with a 15-second period has a profile as shown in Figure 8-4.

Design for an Eight-Leg Jacket 101

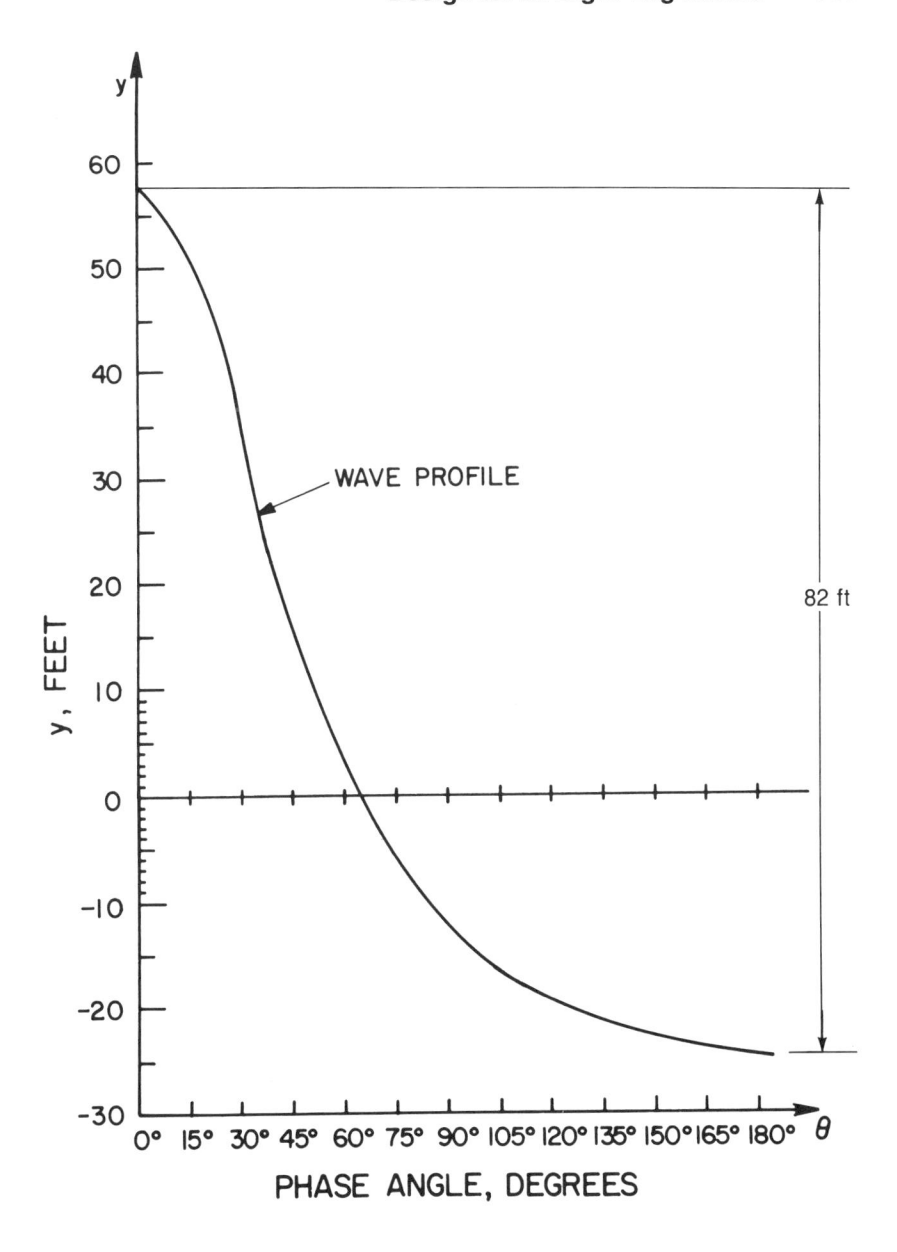

*Figure 8-4.* Stokes fifth-order design wave profile.

## 102 Introduction to Offshore Structures

The equation for this profile is $y = 33.9 \cos \theta + 14.2 \cos 2\theta + 6.1 \cos 3\theta + 2.4 \cos 4\theta + 1.19 \cos 5\theta$ where:

$$\theta = \frac{2\pi}{L}(x - Ct)$$

$C$ = wave celerity = wave length/wave period. The wave length is 991 ft (302 m). The wave celerity is 66 ft/sec (20 m/sec). Using the method described in the paper "Fifth Order Gravity Wave Theory" by Skjelbreia and Hendrickson mentioned in Chapter 6, the horizontal water particle velocity can be determined.

The horizontal water particle velocity diminishes with depth from the water surface. The distribution of force, as calculated from the Morison equation, is shown diagrammatically in Figure 8-5.

If the wave force is calculated by computer, the correct phase separation (90°) is kept between the drag and the inertia components. The horizontal drag pressures for different distances ahead of the crest and different water depths can be calculated as represented for a 55-ft, 13.5-second wave in Figure 8-6. Figure 8-7 shows the horizontal inertial pressures for the same 55-ft, 13.5-second wave. For preliminary design, it is desirable to apply the drag and inertial pressures shown in Figures 8-6 and 8-7 to member sizes other than that specified. This application can be done with conversion factors as shown in Table 8-1. In the case of inertial pressure, the member diameter is included in the conversion factor; inertial forces are expressed in pounds per square foot of projected area, and hence vary directly with diameter. If the forces on the submerged jacket member are calculated by computer, the information (from diagrams like Figures 8-6 and 8-7) for the particular diameter member is placed in the computer. The computer advances the wave through the structure in increments of phase angle and adds the force components each time.

If calculations are done by hand, the phase difference between the components is neglected, and the maximum pressure distribution of each component is added directly. This procedure is conservative; even so, the detail of the computations is quite lengthy. Usually, the length or height of the submerged member is broken into increments of several feet and pressures are averaged over that incremental length.

Wind forces on the topside facilities are calculated by hand for the various approach angles being considered and are added to the wave and current forces to give the total horizontal forces acting on the structure. The equations for calculating wind forces were given in Chapter 6.

### Design

It is not possible here to discuss in detail all the different design features for an eight-leg jacket; instead, some of the more important aspects will be

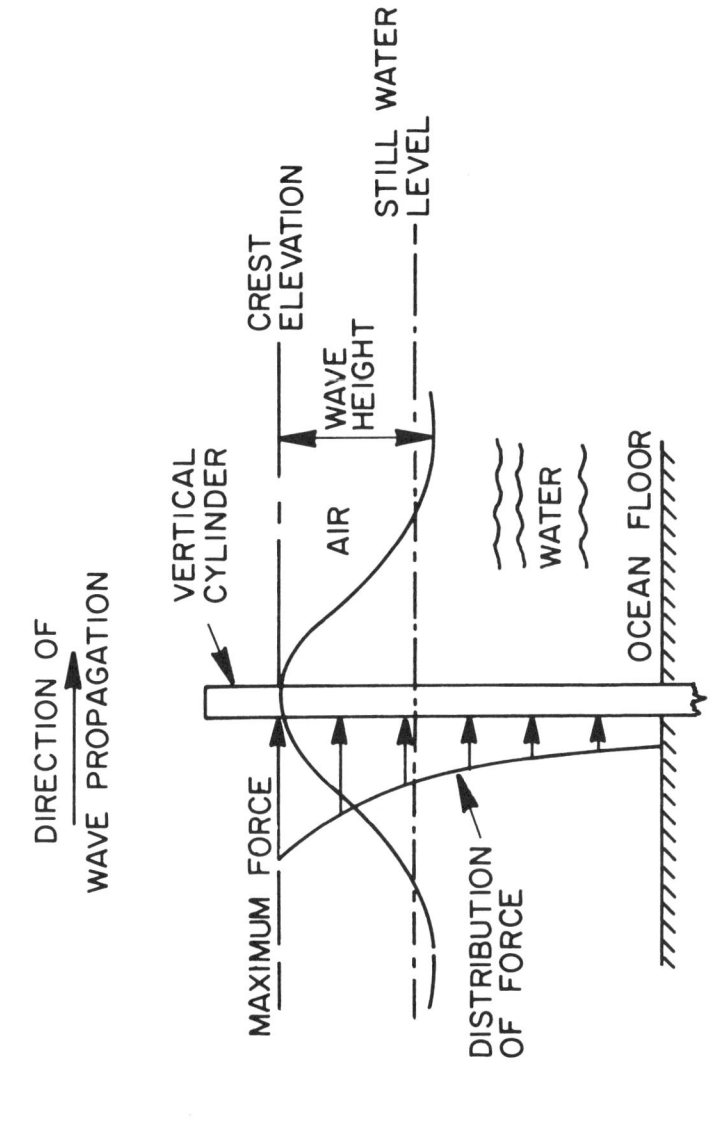

**Figure 8-5.** Sketch of wave and wave force on a vertical cylinder. (Courtesy Rockwell International Corporation.)

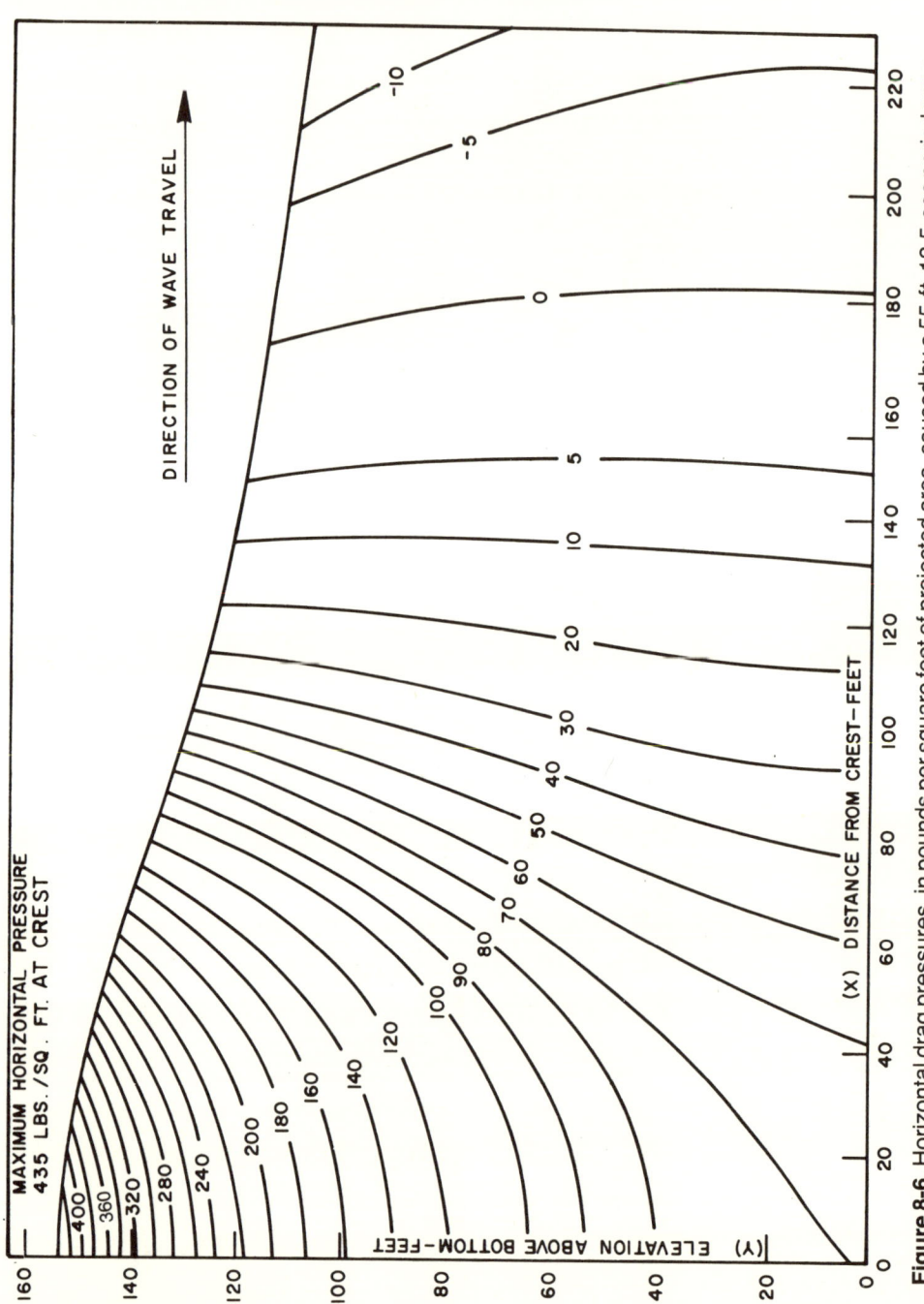

**Figure 8-6.** Horizontal drag pressures, in pounds per square foot of projected area, caused by a 55-ft, 13.5-sec period wave on a 12-inch OD cylindrical pile in 117-ft still water depth.

## Design for an Eight-Leg Jacket 105

**Figure 8-7.** Horizontal inertial pressures in pounds per square ft of projected area, caused by a 55-ft, 13.5-sec period wave on a 12-inch OD cylindrical pile in 117-ft still water depth.

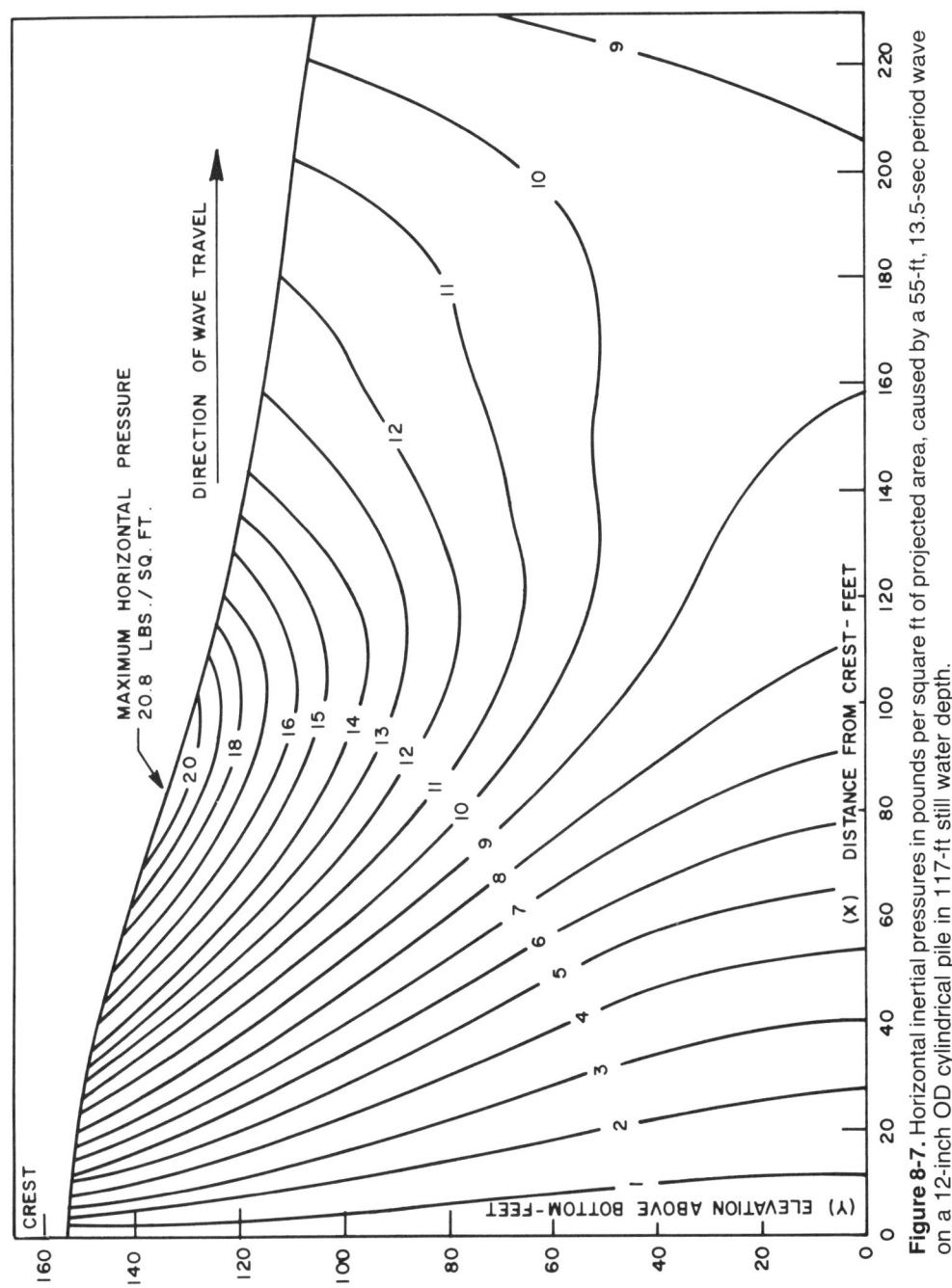

**Table 8-1**
**Factors for Converting Horizontal Drag and Inertial Pressures on 12-Inch Diameter Pile to Pressures on Piles of Other Diameters**

| Pile diameter (ft) | Multiplying Factor Drag | Inertial |
|---|---|---|
| 1 | 1.00 | 1.00 |
| 2 | 1.10 | 1.88 |
| 3 | 1.14 | 2.90 |
| 4 | 1.18 | 4.25 |
| 5 | 1.22 | 5.50 |
| 6 | 1.24 | 6.95 |
| 7 | 1.26 | 8.30 |
| 8 | 1.28 | 9.50 |
| 9 | 1.30 | 10.50 |
| 10 | 1.32 | 12.25 |
| 11 | 1.34 | 13.50 |
| 12 | 1.36 | 14.90 |

Note: 3.28 ft = 1 m

mentioned to highlight the overall process. Most of the preliminary design activity centers around problems related to the selection of the basic configuration and the initial sizing of members to satisfy a given set of operational requirements.

Experience plays a big part in design, so does creativity. Detailed design analysis enables the designer to determine the adequacy or conservatism he has included in creatively delineating the new design based on experience with earlier successful configurations. Thus, the essential ingredients in design are a knowledge of the past, an understanding of cost, an ability to assimilate and synthesize, and the proper application of analysis to corroborate or correct choices and decisions. The result of a good design is a structure which performs satisfactorily in service, requires no more than reasonable maintenance, and minimizes capital investment.

## Jacket Design

The jacket surrounds the piles and holds the pile extensions in position all the way from the mudline to the deck substructure. It supports and protects the well conductors, pumps, sumps, risers, etc., hence the name "jacket." The jacket legs serve as guides for the driving of piles and, therefore, it is properly called a template. The jacket is the basic structural element of the platform. It provides support for boat landings, mooring bitts, barge bumpers, corrosion protection systems, navigational aids, and many other platform components.

Table 8-2 shows the distribution of jacket weight among its various components. Each jacket is somewhat different to meet the functional requirements of its location; therefore, the figures in Table 8-2 are only representative and not actual at any particular site.

**Determining leg size and batter.** Virtually all of the decisions about the design of the jacket depend on the leg diameter. Choosing the proper leg diameter is all-important. Soil conditions and foundation requirements often tend to control the leg size. Sometimes the initial selection for leg diameter is a

Table 8-2
Distribution of Jacket Weight—
Typical Eight-Leg Drilling/Production Platform
in Approximately 300 ft of Water

| Component | Component weight (kips) | Total weight (%) | Subtotal weight (kips) | Total weight (%) |
|---|---|---|---|---|
| Legs | | | | |
|   Joint cans | 394 | 14.6 | | |
|   In-between tubulars and others | 686 | 25.4 | 1080 | 40.0 |
| Braces | | | | |
|   Diagonals in vertical planes | 516 | 19.1 | | |
|   Horizontals (including joint cans on braces) | 362 | 13.4 | | |
|   Diagonals in horizontal planes (including joint cans on braces) | 221 | 8.2 | 1099 | 40.7 |
| Other Framing | | | | |
|   Conductor framing | 78 | 2.9 | | |
|   Launch trusses and runners | 181 | 6.7 | | |
|   Miscellaneous framing | 5 | 0.2 | 264 | 9.8 |
| Appurtenances | | | | |
|   Boat landings | 62 | 2.3 | | |
|   Barge bumpers | 65 | 2.4 | | |
|   Corrosion anodes | 49 | 1.8 | | |
|   Walkways | 35 | 1.3 | | |
|   Mudmats | 11 | 0.4 | | |
|   Lifting eyes | 5 | 0.2 | | |
|   Closure plates | 5 | 0.2 | | |
|   Flooding system | 16 | 0.6 | | |
|   Miscellaneous | 9 | 0.3 | 257 | 9.5 |
| Total | | | 2700 | 100.0 |

Note: 3.28 ft = 1 m
    1 kip = .45 tonnes

**108   Introduction to Offshore Structures**

small change from a size known to have worked well previously in a similar situation. If sufficient preliminary design work has been done on the deck substructure, the initial selection for pile diameter may be the diameter required for the deck support columns. A good rule to keep in mind is that minimizing the projected area of the members near the water surface (in the high-wave zone) minimizes the load on the structure and reduces the foundation requirements.

All tubular products are susceptible to out-of-roundness. Since the pile may be somewhat out-of-round and even slightly crooked, the jacket leg must have a large enough inner diameter to accommodate reasonable variations. For the design of tubular braces, jacket legs, or any tubular members in general, weight, buoyancy, and hydrostatic pressure should be considered.

The connections between the tubular jacket legs and the smaller diameter brace tubes of the jacket are called tubular joints, or simply joints. This subject is treated in detail in Chapter 10. To provide sufficient strength to prevent collapse or the pulling out/punching through of the jacket leg (chord) under the sometimes large forces in the braces, the wall thickness of the chord is made heavier in the vicinity of the joint than what is required for the leg between joints. Generally, the smaller the jacket leg diameter, the thinner the thickness necessary for the heavy-wall chord section at the joint. The following empirical equation is useful in obtaining an initial estimate of the chord wall thickness $t$. Knowing the jacket leg radius $R$, the brace diameter $d$, the force in the tubular brace $F_b$, and the yield strength of the material $F_y$,

$$t = \left(\frac{0.9 F_b}{\pi d F_y}\right)^{0.59} (R)^{0.41}$$

In this equation $R$, $t$, and $d$ are in inches, $F_b$ is in kips, and $F_y$ is in ksi units.

The values for jacket leg batter given in the general design description are common. There are times when other values are more appropriate. The following is a list of items that change when the batter is increased:

1. The axial loads in the piles decrease.
2. More of the horizontal load on the pile head (mudline) is accounted for in the axial force of the pile.
3. The projected area of the pile in the horizontal plane increases.
4. The wave load on the jacket increases.
5. The jacket weight increases.
6. The add-on lengths of piling must be shorter.
7. Pile driving efficiency decreases.

Batter tends to become steeper as the water depth of the platform increases. Selection of the most economical batter is not easy; essentially it involves

### Design for an Eight-Leg Jacket 109

doing entire preliminary designs with two or more values for batter and comparing the results. Optimizing batter selection involves compromising among such factors as soil capacity, pile driveability, jacket and pile steel selection, whether or not to use skirt piles, fabrication costs, and offshore installation costs.

Maximum pile loads are obtained from the overall structural analysis. These pile loads are in the form of maximum reactions developed in the simulated foundation elements. These simulated elements are equivalent elastic springs developed from lateral load versus deformation ($P$-$y$) analyses (with knowledge of the soil properties into which the piles are to be driven). The simulated springs are used in the structural analysis computer program instead of the true inelastic pile-soil reactions. Figure 8-8 shows a diagram of the springs. Pile capacities and bending moments along the lengths of the piles may be computed by hand. The method is outlined in Chapter 8 of the *Handbook of Ocean and Underwater Engineering* (McGraw-Hill Book Co., 1969). With the bending moment distribution along the pile known, the wall

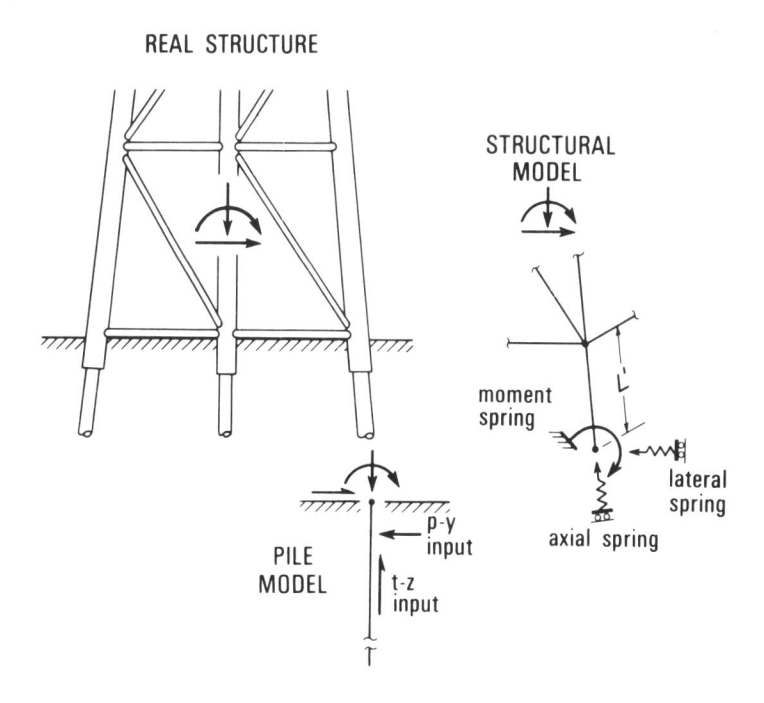

*Figure 8-8. Spring model to replace pile for structural frame analysis. (Source: B. McClelland paper, American Association of Oilwell Drilling Contractors, School of Offshore Operations, Houston, Texas, 1972.)*

## 110  Introduction to Offshore Structures

thicknesses of the different pile sections can be determined. Engineering judgment determines the lengths of the individual pile sections. It is undesirable to have a field splice occur in the vicinity of the maximum moment.

**Framing plans.** The legs of the jacket are interconnected and rigidly held by three kinds of braces: diagonals in vertical planes, horizontals, and diagonals in horizontal planes. Planes of horizontal bracing are spaced 40-60 ft (12-18 m) apart in elevation. Smaller spacings of about 40 ft (12 m) are often used near the water surface and, as the jacket plan size increases with water depth, the spacing of planes of horizontal bracing is increased.

The bracing system performs these general functions:

1. Assists in the transmission of the horizontal loads to the foundation.
2. Provides structural integrity during fabrication and installation.
3. Resists wrenching motion of the installed jacket-pile system.
4. Supports the corrosion anodes and well conductors; carries the wave forces generated by these elements to the foundation.

The experience of the designer to a large extent determines the framing plan. In the general design description it was mentioned that trusses of the Warren bridge type are often used. Several of the more commonly used framing plans are shown in Figure 8-9. The trusses in the vertical direction in each bay of the plan depicted as type 1 are of the Warren bridge type. The truss arrangements in the other framing plans shown in Figure 8-9 are common but do not have names. Most of the vertical bays use elements resembling the Pratt or Howe type truss bracing. The transverse plane vertical truss of the type 4 framing plan is a common bridge variety called a K truss.

Because of the way in which the diagonal braces intersect the jacket legs, some framing plans require joint cans of greater length than other plans. Optimizing jacket framing based on structural weight is at best a moot exercise. One preliminary study of framing schemes 1, 2, and 4 revealed that exclusive of launch trusses, the variation in weight among the three studies was only 6%. This study considered the total weight of joint cans, legs, and braces for three identical structures grouted to the piles in a location with ocean bottom soil of moderate strength. The choice of framing should offer the best horizontal and torsional resistance to the particular wave and current forces involved.

**Sizing brace members.** Tubular braces are beam-columns; however, they are usually subjected principally to axial forces. The diameter of such a member should be chosen so that its slenderness ratio, defined as the effective column length $kL$ divided by the cross-sectional radius of gyration $r$, lies in the range from 60 to 90. The variation of slenderness ratio from 30 to 100 is called

## Design for an Eight-Leg Jacket 111

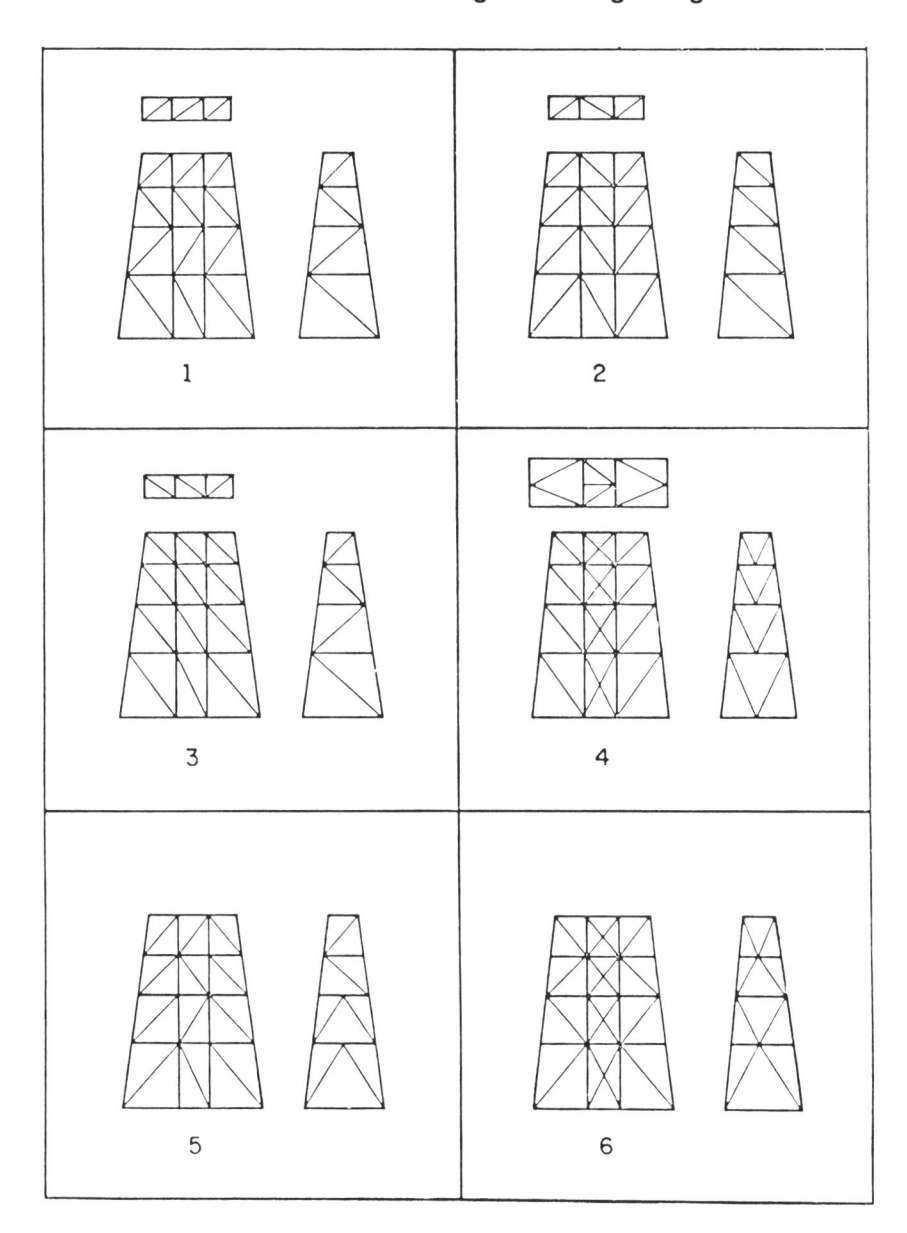

**Figure 8-9.** Commonly used jacket framing plans. (Courtesy Brown & Root, Inc.)

the intermediate column range. In the $kL/r$ range of 60-90 the column strength depends on the tangent modulus of the material and on $k$, where $k$ is the effective length coefficient that varies depending on the end conditions of the column.

For a pin-end column, $k$ is unity; for a column with both ends fixed, $k$ is 0.5; for a column with one end fixed and one end pinned, $k$ is about 0.7; and for a column with one end fixed and one end free from all restraint, $k$ is 2.0. Long columns (those with $kL/r$ greater than 100) are very sensitive to variations in $k$. On the other hand, in the $kL/r$ range from 60 to 90 the critical column stress is relatively insensitive to changes in $k$. A $k$ value of 0.8-0.85 is often used in designing tubular braces for jackets.

As a practical matter for small-diameter braces up to 18 inches in diameter, the wall thickness corresponding to standard pipe should be used as a starting point in design. For diameters approaching 30 inches, take the brace thickness initially to be 0.5 inch. For diameters 30-36 inches, start the design with a 5/8-inch wall thickness.

The transition between what is considered a thin- and thick-walled tube occurs for a ratio of tube diameter $D$, to tube wall thickness $t$, in the range of 15-20. Such thick tubes are seldom used as jacket braces. A tube with a $D/t$ around 30 will float. When $D/t$ reaches 90, local buckling problems are encountered. At such large $D/t$ ratios, one must investigate the stress difficulties caused by the hydrostatic pressure of the water surrounding the brace in service.

Long columns with $kL/r$ exceeding 100 fail according to the Euler elastic column equation, which is independent of material yield strength. Therefore, as the jacket brace length increases approaching a $kL/r$ of 90-100, the advantage of using a higher yield strength material diminishes.

The lower portion of the intermediate column range, $kL/r$ from 30 to 60, could be used for tubular braces. For a given brace length, increasing the diameter increases $r$, which in turn lowers the $kL/r$ ratio. It is difficult to say specifically why this range is not utilized but, in general, the jacket leg diameter is chosen first, and this choice restricts the diameter of the brace, since most braces are less than 70-80% of the diameter of the leg. The force on the brace developed by the passing wave increases as the brace diameter increases. Consequently, it is desirable to use smaller diameters (larger $kL/r$ ratios). The use of a lower $kL/r$ ratio for the same axial brace force implies a larger diameter and a thinner wall, increasing the $D/t$ ratio. This $D/t$ ratio enhances the possibilities of local buckling and hydrostatic pressure problems.

**Launch bracing.** While launching is treated in Chapter 11, it is necessary here to discuss the additional trusswork added to the jacket to accommodate the forces induced during this operation. The jacket is transferred from the

### Design for an Eight-Leg Jacket 113

onshore construction yard to the offshore drilling site lying flat on its side on a transportation barge. The two parallel planes of trusswork, called bents, in transverse planes 2 and 3 are fitted with the launch bracing, and continuous runners are added to legs A2 and A3, or B2 and B3, whichever are selected as lower chords for the launch trusses.

The jacket is usually launched upper end first from one end of the barge. The skid beams on the barge terminate in rocker arms and, as the jacket moves longitudinally on the barge by means of cables and winches and passes over the pivots, it rotates in a vertical plane and slides into the water.

The support forces exerted by the skid beams of the barge on the jacket should be evaluated for the full travel of the jacket. At the moment the jacket pivots on the rocker arms, all of the weight of the jacket is concentrated on them. Figure 8-10b shows this position.

Because the jacket overhangs the rocker arms fore and aft at the instant of launch, it is necessary to stiffen the two interior bents. The launch bracing shown in Figure 8-10b is of the Baltimore bridge truss type. The jacket also overhangs the two skid beams from side to side. (See Figure 8-10a.) Because of the overhang and the dynamically induced forces occurring at launch, it may be necessary to design many of the horizontal braces and diagonal braces in the horizontal planes of the jacket to resist launching forces rather than those forces developed during normal operation after installation. The seemingly large amount of bracing to the left side of Figure 8-10a is to position and hold the well conductors.

The launch bracing amounts to about 7% of the jacket weight. Therefore, the simulation of the rocker arms and the jacket position at launch must be done carefully to obtain a realistic evaluation of the distribution of support reactions on the structure. Significant economy in the amount of launch framing may be realized by more accurately determining the distribution of support forces to apply to the runners on the jacket legs.

**Skirt piles.** Skirt piles are added in-between the piles that go through the jacket legs or are grouped about the periphery of a jacket leg. Figures 8-1 and 8-11 depict skirt piles added in-between jacket legs. Figure 8-3 shows an example of skirt piles driven through guide tubes or sleeves in metal cylinders surrounding the bottoms of the corner jacket legs.

Skirt piles are used when it is necessary to increase the capacity of the structure to overcome overturning moment. Sometimes they are used to lessen the pile penetrations required at a site with unusual soil characteristics and thereby achieve more easy driveability of the piles.

A skirt pile develops its resistance by being grouted to the inside of the skirt-pile sleeve. Thus, the sleeves must be sufficiently long to provide enough bond strength to equal the ultimate capacity the pile can develop in the soil. The allowable bond stress for cement grout is ordinarily 20 psi (1.4 Kg/cm$^2$). For the 100-year storm condition, this figure may be increased by one-third.

**114** Introduction to Offshore Structures

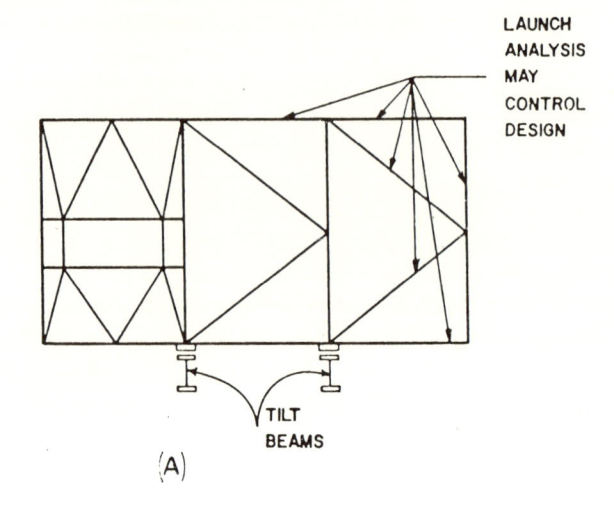

*Figure 8-10a.* Cross-sectional view of jacket on tilt beams. (Courtesy Brown & Root, Inc.)

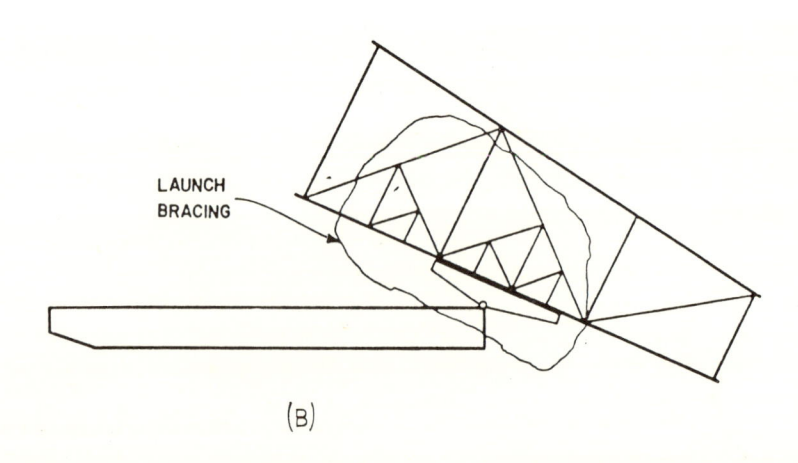

*Figure 8-10b.* Jacket launch bracing. (Courtesy Brown & Root, Inc.)

## Design for an Eight-Leg Jacket 115

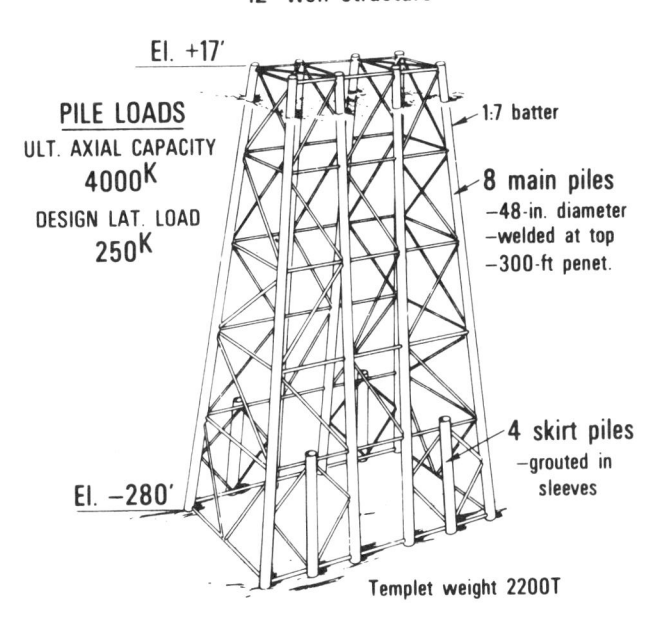

*Figure 8-11. Template type jacket for drilling platform to be located offshore of Louisiana in the Gulf of Mexico. Notice the sleeves for the four skirt piles. (Source: B. McClelland paper, American Association of Oilwell Drilling Contractors, School of Offshore Operations, Houston, Texas, 1972).*

Bonding between the skirt pile and the sleeve may be increased by welding circumferential or spiral rings on the inside of the sleeve and on the outside of the final add-on pile length that passes into the sleeve top.

### Foundation Design

**Pile penetration and mudline wall thickness.** The length of pile embedded in the soil strata of the ocean bottom is termed *pile penetration*. As discussed in Chapter 7, the amount of pile penetration required is that needed to develop the compression and tension (pull-out) capacities necessary to support a particular offshore platform.

The American Petroleum Institute recommended practice API-RP-2A[1] gives a method for determining pile capacity for axial bearing loads in either clay or sandy soils. The pile skin-friction resistance is a function of the undrained soil shear strength times the side surface area of the pile for clayey soils and the product of a lateral earth pressure coefficient, the effective overburden pressure, and the tangent of the soil friction angle for sandy soils. Other approaches have been used to predict pile capacities in clay, namely the method of Vijayvergiya and Focht[2] and Tomlinson's limited adhesion method.[3] Vijayvergiya and Focht's method assumes the unit skin-frictional capacity to be an empirical coefficient lambda times the sum of the effective vertical soil stress and the undrained soil shear strength. Tomlinson's limited adhesion method is believed to yield the most conservative answers (longest pile penetrations) of the three methods mentioned. The API-RP-2A method is generally recommended.

The designed pile capacity and penetration should not exceed the limits of experience under similar conditions with essentially the same pile diameter and installation equipment. Even so, if design penetration cannot be achieved, possible remedial alternatives should be investigated and defined before pile installation begins.

The pile foundation must resist large static and dynamic lateral loads. Soil disturbance near the mudline and later scouring action can significantly reduce the lateral load capacity of a pile. Scour is the removal of the first few feet of ocean bottom soil from around the pile due to the action of bottom currents and wave effects. It is a form of erosion and can be a natural geologic process or a result of the presence of the structural elements of the platform.

The need to develop sufficient lateral load resistance often dictates the thickness of the tubular pile walls at the mudline. The soil-pile interaction is a nonlinear system. Under lateral loading, clayey soils behave plastically, and it is necessary to relate pile-soil lateral deformation to soil resistance to predict behavior. This information developed from laboratory soil samples is called "$P$-$y$ data," where $P$ represents the lateral resistance of the soil, and $y$ represents the lateral deformation of the soil.[4] The soils specialist uses iterative procedures to develop a family of $P$-$y$ curves for different depths into the soil at a specific offshore location.

A procedure for using $P$-$y$ curves to determine lateral load pile capacity is outlined by the API. Vijayvergiya and Focht should be consulted for greater detail regarding $P$-$y$ curves for clay.[2] Although sand does not behave as plastically as clay, nevertheless, a method of constructing $P$-$y$ curves for sand has been devised. This method is discussed by Reese, Cox, and Koop.[5]

The rigid framework of the jacket exercises rotational restraint on the pile head (at the interface between the jacket and the pile). Thus, large bending moments are produced at the pile head by the lateral soil deformations caused by the lateral loads. It is this bending moment coupled with the pile axial load that establishes the wall thickness at the mudline and for a considerable depth beneath it. Pile sections with increased wall thickness are used for add-on lengths at the proper time so that they may come to rest near the mudline at the conclusion of pile driving. The location of this increased wall thickness is carefully determined ahead of time when the sequence of add-ons is devised.

## Design for an Eight-Leg Jacket 117

In addition to using increased wall thicknesses, sometimes high-strength steel is used for the tubular pile sections near the mudline to give increased pile strength.

**Pile driving.** Driving the tubular piles to design penetration can be the most time-consuming and costly operation in the installation of an offshore platform. The most desirable procedure for installing the piling is by driving only. Procedures which require jetting and driving, or drilling undersized holes and driving, can significantly increase pile installation costs.

Although piles have the same outside diameter, they are not of the same wall thickness throughout their length. The largest wall thickness should be located in the vicinity of the largest bending moment in the pile—either at or near the mudline. The impact stresses developed during the driving of the piles may dictate a greater wall thickness than that required to accommodate the maximum operational loading on the platform.

During the design stage a pile driveability study should be made to facilitate the selection of the proper hammer-pile combination to ensure that the piles can be installed to design penetration without jetting or undersize drilling.

In the past limited use has been made of vibratory pile drivers. Impact hammers have been far more common. The types are:

1. Drop hammers
2. Steam hammers which also operate on compressed air
   a. Single-acting
   b. Double-acting
   c. Differential-acting
3. Hydraulic-oil hammers
4. Diesel hammers

Some models of compressed air impact hammers can operate underwater. Based on hammer-pile selection experience, the minimum pile area in square inches should be not less than half the rated energy in foot-kips of the largest hammer to be used when driving the piling.

Dynamic-driving formulas, such as the one-dimensional wave equation, have been widely used to predict pile driveability. Studies of pile driveability indicate these factors are important in the analysis: pile makeup, hammer energy, hammer efficiency, pile-cap stiffness, cushioning blocks, soil-damping coefficients, soil-quake factors, and the soil resistance at the time of driving.

Maximum pile penetration and attendant maximum ultimate capacity generally means a hammer-pile combination with a thicker pile wall, stiffer pile cap, higher hammer energy and efficiency, lower soil-damping, higher soil-quake factor, and lower soil resistance at the time of driving.

Figure 8-12 shows an idealized deep-penetration pile suitable for the Gulf of Mexico. The pile is 700 ft (213 m) long and was embedded 390 ft (119 m) in

**118 Introduction to Offshore Structures**

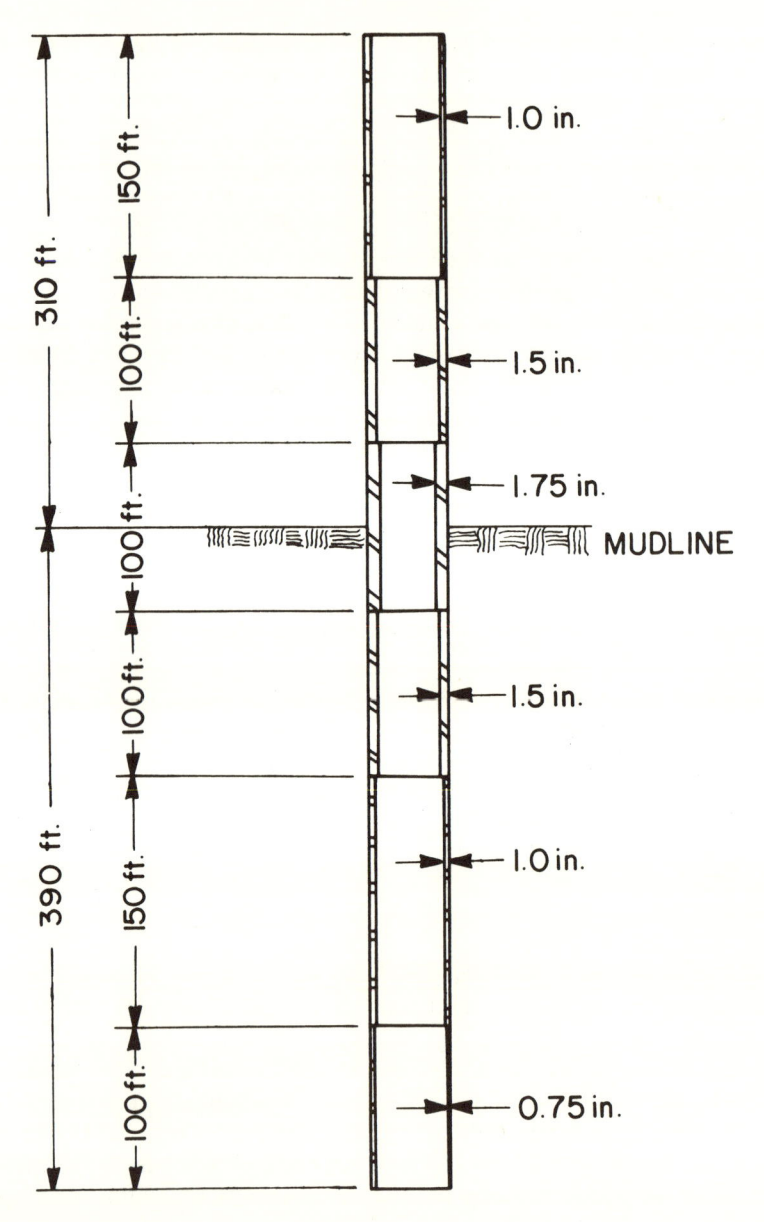

**Figure 8-12.** Steel pipe pile with constant outside diameter of 36 inches shown at full penetration.

normally consolidated, insensitive clay with ultimate soil resistance estimated at 12,000 kips (53 x $10^6$N).

**Mudmats.** Another design problem brought about by the fluidized nature of the initial soil layers encountered during jacket installation is that of providing adequate bearing area at the bottom of the jacket so that it will remain upright and stable while the first piles are being driven. These additional bearing areas are called *mudmats*. They are usually large in area and are proportioned for ocean floor penetrations of no more than a few feet.

Mudmats are normally made of heavy timbers fastened as beams across the corners of the jacket immediately beneath the lowest level of horizontal bracing. Alternatively, they may be made of light steel plate and structural shapes. The mats are placed adjacent to the jacket legs to minimize bending moments in the horizontal bracing of the jacket.

The required area of mudmats is determined by the effective weight of the jacket, and the bearing capacity of the ocean floor soil-surface layers. It is desirable to limit the average vertical soil pressure to not more than about 400 lbs/ft$^2$ (19,200 Pa).

## Deck Substructure Design

**Deck flooring.** Where the deck space is not covered with a module, the floor area is covered with steel plates, usually about one-half-inch thick. The thickness of the deck plating depends on the spacing of the floor beams and on the loads anticipated to be placed on the deck. Other factors—such as the amount of welding required, accessibility through the floor for maintenance operations on the level beneath, size of objects to be supported, floor deflection limitations, and availability of desired beam sizes—complicate the problem. Choice of deck thickness with beam sizing and spacing thus becomes a design compromise among many factors.

The supported weight on the deck floor must be transferred to the jacket columns via the systems of floor beams, girders, and deck substructure trusses. Consider the layout shown in Figure 8-2. The floor girders most likely span between the trusses in reference planes 1 and 2 and 3 and 4 because these are the shorter distances. The beams in these areas then span between girders (in a transverse direction). For the area between reference planes 2 and 3, the floor girders span between the trusses in reference planes A and B. The beams then span between girders and run in a longitudinal direction.

The deck beams and girders are proportioned for a uniformly distributed load, called the design load, which is usually furnished by the customer. After the beams and girders have been sized, they should be checked for the condition where the uniform load is replaced in certain locations by a concentrated load so positioned as to produce maximum moment. The object in this analysis is to determine the influence of this change on the moment distribution and deflection of the beam. Sometimes the absence of a concentrated load at certain locations along the beam or girder can be as detrimental as its presence.

The length of the cantilevered portion of the deck framing is governed by the design stress level. These beams and girders should not develop stress levels greater than those in the framing between the primary trusses. As the deck plates are seam-welded to the beams, effective portions of the plates may be added to the section moduli of the beams for computing flexural resistance. They must share the load, since they are welded together.

The framing of beams to girders and girders to trusses can be done in either of two ways. The beams may frame into the girder webs, and the girders into the sides of the top chords of the trusses; or the smaller pieces may simply rest on top of the larger structural members. Framing the smaller pieces into the sides of the larger pieces reduces the overall height of the superstructure by several feet. This framing has a beneficial effect from the standpoint of wind forces. It is also argued that such framing introduces partial restraint into the ends of the smaller of the two members making a given connection. However, this kind of framing is more expensive. The selection of the proper system of floor beams and plates is not necessarily the lowest weight solution or the least expensive solution; it is a compromise considering the number of cuts, ease of fitting, number, length and difficulty of welds, and the ease of operational maintenance. The cost of maintaining the system after functional operation of the platform begins is important.

**Beams and Girders.** Consider the deck substructure shown in Figure 8-13. There are two deck levels. The lower level is high enough from the water so that the crest of the storm wave clears the trusswork by a specified amount (the air gap) as it passes beneath the superstructure. The upper deck is placed at a sufficient elevation above the lower deck to provide operating clearance for the equipment on the lower level. However, the elevation of the upper deck must not be so high as to preclude placing equipment packages on this deck with the derrick barge during construction.

As shown in Figure 8-13, the arrangement is for drilling 18 wells—all at one end of the platform. The drilling rig is to be skidded from one well-conductor location to the next on the two beams shown using large hydraulic jacks. This layout is generally representative of many drilling/production platforms.

A summary of the weight analysis of two configurations of substructure for this platform is shown in Table 8-3. The dead and live loads from the equipment on the decks must be conveyed to the jacket and piles through the deck substructure. Also, the wind load brought about by the presence of the decks and equipment must be conveyed to the jacket and piles.

Design for an Eight-Leg Jacket    121

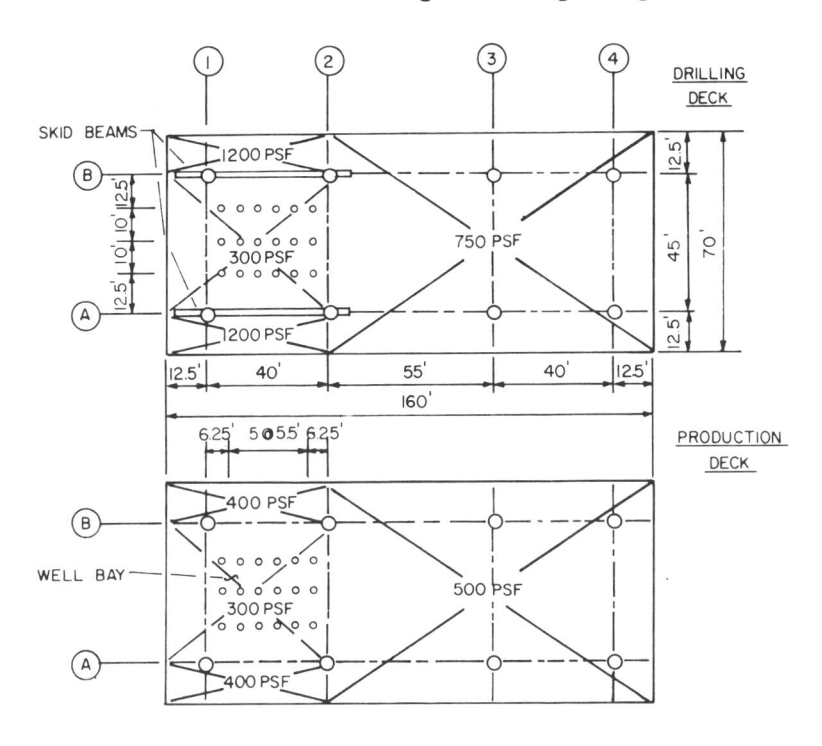

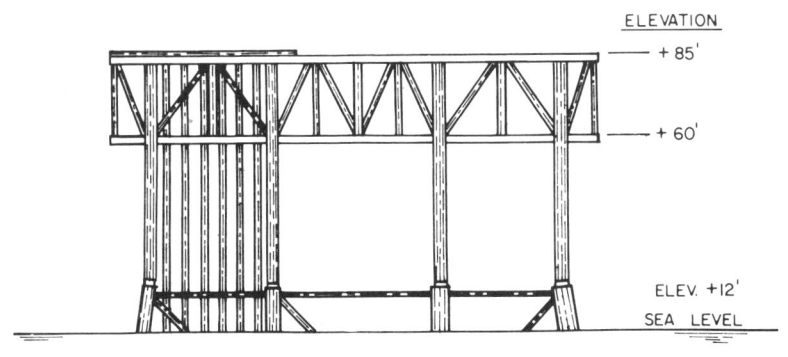

*Figure 8-13.* Plan and elevation views of deck substructure.

### Table 8-3
### Weight Comparison in Kips of
### Two Configurations of Deck Substructure—
### Eight-Leg Drilling/Production Platform

|  | Plate girder Substructure | Percent of total | Tubular truss substructure | Percent of total |
|---|---|---|---|---|
| Decking |  |  |  |  |
| Drilling deck |  |  |  |  |
|   Timber | 48.8 | 2.6 |  |  |
|   Plate | 137.0 | 7.4 | 158.9 | 11.0 |
| Production deck |  |  |  |  |
|   Plate | 100.0 | 5.4 | 115.7 | 7.8 |
|   Grating | 10.1 | 0.5 | 2.3 | 0.16 |
|   Total | 295.9 | 15.9 | 276.9 | 18.8 |
| Deck beams |  |  |  |  |
|   Drilling deck | 341.7 | 18.3 | 386.7 | 26.3 |
|   Production deck | 306.5 | 16.5 | 125.3 | 8.5 |
|   Total | 648.2 | 34.8 | 512.0 | 34.8 |
| Plate girders | 425.0 | 22.8 |  |  |
| Tubular trusses |  |  | 325.0 | 22.1 |
| Columns | 333.0 | 17.9 | 234.0 | 15.9 |
| Appurtenances |  |  |  |  |
|   Vent stack | 13.0 | 0.7 | 13.0 | 0.9 |
|   Stairs | 20.0 | 1.0 | 26.0 | 1.8 |
|   Handrails | 8.7 | 0.5 | 9.0 | 0.6 |
|   Lifting eyes | 5.0 | 0.3 | 4.3 | 0.3 |
|   Drains | 9.6 | 0.5 | 13.9 | 0.9 |
|   Sub-cellar | 22.7 | 1.3 |  |  |
|   Firewall | 29.8 | 1.6 | 25.3 | 1.7 |
|   Stiffeners (including ring stiffeners on tubulars) | 50.1 | 2.7 | 31.1 | 2.2 |
| Overall total | 1861 | 100 | 1471 | 100 |

Note: 1 kip = 4.45 kN

The columns, trusses, and girders of the deck substructure should be analyzed for the stresses induced during erection as well as for the in-place operational loads. The stresses induced during erection are of considerable magnitude, and sometimes temporary braces must be added to limit otherwise excessive stresses and deflections. For example, if lifting a portion of the deck substructure from the transport barge to the jacket top should cause the substructure columns to deflect outward beyond certain limits, their stabbing cones may miss connection with the mating pile extensions protruding from the jacket legs.

# Design for an Eight-Leg Jacket

**Trusses and Columns.** In the drilling area of the upper deck the skid beams and their supporting girders or trusses must be adequate in strength to handle all of the drilling rig positions.

Deck substructure trusses are initially sized using planar pin-joint analysis. Planar rigid-joint computer analysis is sometimes used for making refinements. Before the deck substructure design is completed, the entire three-dimensional girder, truss, and column system should be modeled to some extent with the jacket analysis to investigate possible regions of oversight or unacceptability.

Careful checking must be done of web-crippling needs in areas where concentrated deck loads are transferred to the flanges of girders or truss chords. Where uncertainty exists regarding deck loads, web stiffeners should be added as a matter of conservative design.

The columns of the deck substructure may fall into the category of long columns if their diameters are too small or the deck heights above the water are too great. Significant reductions in slenderness ratio of these legs may be achieved by the addition of a few diagonal braces between the lower chords of the trusses and the columns.

In some platform designs there are deck substructure legs that terminate in the framework of the jacket and do not stab directly into pile extensions. (See Figure 8-3.) In such situations the deck substructure and the trusswork of the upper levels of the jacket must be analyzed together. Variations in the vertical stiffness of the jacket will cause variations in the loading of the deck support columns.

Where diagonal braces of deck substructure trusses are in tension, and their ends terminate on the flanges of chord members, it should be noted that the strength of the connection depends on the through-the-thickness properties of the chord material.

## Materials

### Steel Selection

Selection of the proper steel to use in a particular offshore application is complicated and should receive special attention. Stress corrosion and corrosion fatigue resistances should be adequate for the long-term marine exposure or definite plans must be made to provide corrosion protection. From the fabricating standpoint, the quality of the steel must be carefully controlled; it must be possible to form the material into the various shapes and join these by welding into a useful structure. Thus, the ease with which the steel can be formed, machined, and joined is of primary importance. Because of the unique high loading on offshore structures, such mechanical properties as yield and ultimate strength are especially important. Table 8-4 summarizes the material characteristics which are important for underwater structures.

### Table 8-4
### Important Characteristics of Materials
### To Be Used In Underwater Structures

| Design | Fabrication | Service |
| --- | --- | --- |
| Yield strength | Joining | Corrosion resistance |
| Ultimate strength | Forming | Stress corrosion resistance |
| Young's modulus | Machining | Corrosion fatigue resistance |
| Density | Quality control | Repair |
| Poisson's ratio |  | Maintenance |
| Fatigue resistance |  | Creep resistance |
| Brittle fracture resistance |  |  |

It is common practice to use at least two grades of steel in an offshore platform: ordinary low-carbon structural steel and high-strength, low-alloy steel. Frequently, when the structure is to be subject to cold-climate conditions and located in deep water, a still higher-strength steel will be employed for critical components, i.e., those developing high stresses in service. Carbon-manganese heat-treated structural steel is an example of the steel used for critical components. The high-strength, low-alloy steels derive their increased strengths from the application of different amounts of alloying elements. The carbon-manganese steels are available in a normalized or quenched and tempered condition and depend on carbon content to develop their strength through the heat-treating process.

Ordinary low-carbon structural steels usually have a minimum yield strength around 36 ksi (249 N/mm$^2$) and a tensile strength of 58-75 ksi (400-520 N/mm$^2$). High-strength, low-alloy structural steels have minimum yield strengths in the range of 45-60 ksi (315-410 N/mm$^2$). Their ultimate or tensile strengths range from 60 to 80 ksi (410 to 550 N/mm$^2$). Depending on heat treatment, the mechanical properties of the high-strength carbon-manganese steels vary, but generally the yield strength is 50-80 ksi (345-550 N/mm$^2$) and the ultimate strength is 75-100 ksi (520-690 N/mm$^2$).

The higher tensile strength steels offer several advantages where predominantly static loads occur[6,7,8]:

1. Reduced thicknesses and thus easier welding
2. Less weight because of the reduced thicknesses
3. Reduction in fabrication costs
4. Lower transportation costs
5. Less welding filler material required
6. Reduction of residual stresses
7. Reduction in tendency for brittle fracture

The premium steels have some general disadvantages also. Sometimes the higher-yield strength is obtained at the expense of a reduced ductility. Ductility is important to develop the full reserve capacity of components like tubular joints. The higher-strength steels are not as weldable as low-carbon steel, and careful attention must be given to selection of welding electrodes and qualification of welders. Availability of the premium steel in the proper sizes may be limited and cost may be consequently higher.

Limiting the carbon content and maintaining close control over the chemistry during manufacture results in a fully killed (deoxidized) fine-grain steel with good impact properties and good weldability. The special high-strength steels are used for such highly stressed components as the tubular joint nodes. The intermediate strength steels are used for the lengths of tubulars spanning between nodes. Appurtenances such as railings, stairs, walkways, deck plating, stiffeners, etc., are made of ordinary structural steel.

The characteristics of ductile and brittle fractures are given in Table 8-5.

## Laminations and Lamellar Tearing

### Lamination

Studies of observed material failures in early Gulf of Mexico platforms showed that most of the failures occurred in or near the welded joints. This fact led to investigation of steel properties in the vicinity of the welds and heat-affected zones (HAZ) of the material.[9]

Platforms constructed in the early 1950s were made of low-carbon structural steel. As carbon content was not well controlled, strength and weldability varied. In general, the steels used had poor through-the-thickness properties. Notch brittleness of steel was not recognized as a pertinent parameter and welding quality varied.

The investigations of steel properties in failed components of platforms revealed that the material contained laminations. A lamination is a planar type discontinuity in steel plate resulting from flattening and elongating of

**Table 8-5
Characteristics of Fractures**

| Type of fracture | Fracture appearance | Form of failure | Extent of plastic deformation |
|---|---|---|---|
| Ductile | Fibrous | Shear | Large |
| Brittle | Granular | Cleavage | Essentially none |

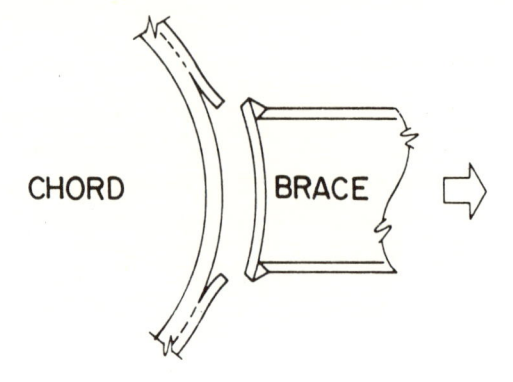

*Figure 8-14. Joint failure at a lamination.*

inclusions or voids during the rolling process.[10] Foreign particles such as slag and bits of furnace refractories may be trapped in the upper-central core of the steel ingot on solidification. Shrinkage of steel on solidifying may cause a phenomenon called *piping* to occur in ingots not having hot tops. Piping is a term given to the formation of cavities vertically along the upper core of the ingot. Air penetrates into these cavities and oxidizes the metal. The oxide-coated cavities or foreign particle inclusions become very thin layers inside the plates as the processing of the ingot occurs. These thin layers are observable at an exposed cross section, but often they occur within the geometry of plates, not extending to any edge. A hot top is an insulated riser on top of the ingot. Solidification of the metal is delayed in the hot top, and the liquid is available to flow into any regions within the ingot which otherwise would form cavities. By cutting off a sufficient part of the ingot top before processing begins, the occurrence of laminations can be prevented.

Today, it is well recognized that laminated steel plate cannot be used where short transverse (through-the-thickness) properties are important. Figure 8-14 shows the effect of a lamination in the chord of a tubular joint. The presence of the lamination reduces the effective thickness of the chord when subject to tension in the brace.

**Lamellar Tearing**

This phenomenon occurs in highly restrained joints within welded connections. Under high restraint, localized strains due to weld metal shrinkage can be several times higher than yield point strains. The strains resulting from applied loads are not of primary concern in lamellar tearing. The localized strains responsible for lamellar tearing are those associated with contraction of individual welds. Figure 8-15 shows a typical lamellar tear.

# Design for an Eight-Leg Jacket 127

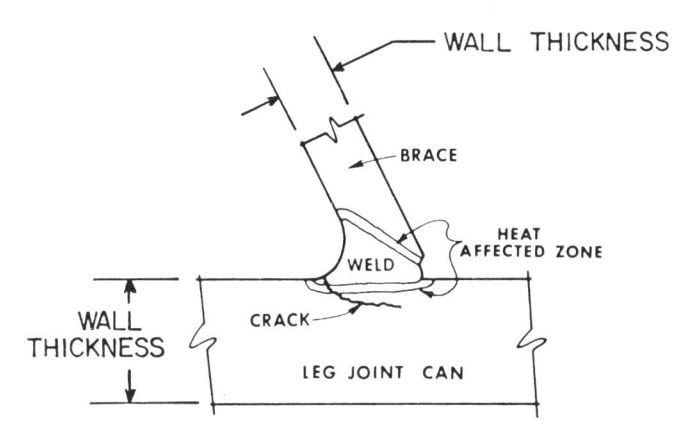

*Figure 8-15. Lamellar tear.*

Lamellar tearing is cracking within the original plate material adjacent to a weld. The cracks are caused by through-the-thickness strains produced by weld metal shrinkage. Sometimes these cracks originate in HAZ.

Lamellar tearing is a complex subject, and the average designer can do little to preclude its occurrence. The designer can specify steels with high short transverse strength properties. In conjunction with the welding engineer or metallurgist the designer may plan the weld sequences to minimize contraction strains, restraint, and distortion. Several references give helpful information.[11-15]

## Weldability

In evaluating the ease of jacket fabrication the relative weldability of the materials must be considered. Weldability is the capacity of a metal to be welded under specified fabrication conditions into a suitable designed structure which will then perform satisfactorily in the intended service.[16] The principal variables in weldability (besides chemical composition of the parent metal) are composition of welding electrodes and thermal input.

Plain carbon steel with greater than 0.35% C quenches readily to a hard microstructure which makes the material unweldable without special precautions.[17] In low-alloy steels carbon is held to about 0.20%; this limitation controls the maximum hardness obtainable. Elements such as Mn, Si, Cr, Mo, V, Ni, and Cu are added to raise the yield strength or to obtain other desirable properties such as improved notch toughness and resistance to atmospheric corrosion.

## 128    Introduction to Offshore Structures

"Carbon equivalent" is a numerical expression for equating the effect of carbon and manganese (and other alloys) on the mechanical properties and hardenability in terms of carbon content. Carbon equivalency may also be used as an indicator of weldability. Carbon equivalent is not a fundamental characteristic of the steel; there is no critical value at which welding problems begin. Other factors influence the welding characteristics of a steel such as preheat, weld heat input, joint geometry and accessibility for the welder, metal thickness, type of electrode, and the welding procedures. According to Carter[17] and Clark,[18] the carbon equivalent requirements of *Lloyd's Register* read (in part):

"The steel should be capable of being fabricated and welded under shipyard conditions. When the carbon equivalent calculated for the ladle analysis and using the formula given below is in excess of 0.45% approved low hydrogen higher tensile electrodes and preheating are to be used. When the carbon equivalent is not more than 0.45% approved low hydrogen higher tensile electrodes are to be used but preheating will not generally be required except under conditions of high restraint or low ambient temperature. When the carbon equivalent is not more than 0.41%, any type of approved higher tensile electrodes may be used and preheating will not generally be required except as above.

$$C.E. = C + \frac{Mn}{6} + \frac{Cr + Mo + V}{5} + \frac{Ni + Cu}{15}$$

"This formula is only applicable to steels which are basically of the carbon-manganese type containing minor quantities of grain refining elements, for example, niobium (columbium), vanadium or aluminum, and the proposed use of low alloy steels will be subject to special consideration."

According to Peterson,[16] limiting the carbon equivalent to 0.40%* will generally ensure that the steel will be weldable under field conditions. Carter[17] recommends holding to the 0.45% figure given in *Lloyd's Register* to ensure that the steel can be welded satisfactorily under field conditions without the use of excessive preheat. According to Sembritzki,[6] the International Institute of Welding proposed that to omit preheating, the carbon equivalent for uncomplicated welded joints should be limited to 0.41%.

### Stress Relieving

The microstructure of the weld zone and HAZ are influenced by heat input and cooling rate. Residual stresses from welding are high, often comparable to the yield stress. The danger of lamellar tearing is reduced by preheating the parent materials adjacent to the joint and by systematic sequencing of the

---

* Personal Communication from Peterson.

individual weld passes. Electrical heating has the advantage of more accurate control beforehand and during the welding plus control during the cooling-down phase.

Where the hardening effect within HAZ due to welding is high, stress relief heat treatment should be conducted after welding. Stress relieving consists of heating the welded component to 1000-1100°F (538-593°C), holding for a sufficiently long time to relieve the locked-up stresses, and then slowly cooling.

Today, there are many differing opinions as to the actual reduction of residual stresses achieved through stress relieving. There are indications that in the first few cycles of loading the residual stresses in structural steel undergo some relaxation as a result of local yielding occurring at points where the sum of the residual stress and the applied stress exceed the yield strength of the material. It is the author's opinion that appropriate metallurgical tests should be run at the start of a new project using the grades of steel and thicknesses envisioned to determine whether or not stress relieving is necessary in the particular case.

## Welding Specifications

Careful attention should be given to identifying proper welding procedures, operator qualification tests, and inspection methods to specify adequately the welding of a jacket.

Three steps are necessary to ensure satisfactory welding for a particular job: (1) establishing of good welding procedure, (2) use of prequalified welders, and (3) employment of competent inspectors for both shop welding and field fabrication. The principal factors that go into establishing good welding procedures are the properties of base metal; the selection of proper electrodes, current and voltage; the speed and number of passes; and the welding position and accessibility for welding. Reasonable intersection angles for tubular members should be used, and sufficient accessibility must be afforded for actual welding.

A good design on paper from the standpoint of welding cannot be assured in practice without knowledgeable inspectors to look for defects. The most important defects arising from improper welding technique are undercutting, lack of fusion and penetration, slag inclusion, and porosity. A good inspector can see that the proper preheat and postheat are used and can observe if the proper rate of cooling is being permitted. In addition, proper joint preparation and fit-up before welding can be observed and are important to the final quality of the connection.

Besides continual visual inspection, satisfactory weld quality is usually maintained through the frequent use of other inspection methods such as dye penetrant, ultrasonic, magnetic particle, and radiography.

## References

1. "Recommended Practice for Planning, Designing, and Constructing Fixed Offshore Platforms," API-RP2A, Dallas, TX, 10th Edition, Mar. 1979.
2. Focht, J.A. and Vijayvergiya, V.N., "A New Way to Predict Capacity of Piles in Clay," 1972 OTC Preprints, Paper #1718.
3. Tomlinson, M.J., *Foundation Design and Construction,* 2nd ed., John Wiley & Sons, Inc., 1969, pp. 387-392.
4. Matlock, H., "Correlations for Design of Laterally Loaded Piles in Soft Clay," 1970 OTC Preprints, Paper #1204.
5. Cox, W.R.; Koop, F.D.; and Reese, L.C., "Analysis of Laterally Loaded Piles in Sand," 1974 OTC Preprints, Paper #2080.
6. Sembritzki, E.; "Fabrication and Testing of Offshore Structures," Symposium of the International Scientific Association on the Safety of Offshore Structures, Copenhagen, September 8-9, 1977.
7. "Steels for Offshore Structures," *Welding Design and Fabrication,* November 1975, pp. 55-57. (Reprint available from Industrial Publishing Co., a division of Pittway Corporation.)
8. Tuthill, A.H. and Schillmoller, C.M.; "Guidelines for Selection of Marine Materials," Ocean Science and Ocean Engineering Conference of the Marine Technology Society, Washington, D.C., June 14-17, 1965. (Reprint available from the International Nickel Company, Inc.)
9. Peterson, M.L.; "Evaluation and Selection of Steel for Welded Offshore Drilling and Production Structures," 1969 Offshore Technology Conference Preprints, Paper 1075.
10. "Commentary on Highly Restrained Welded Connections," *Engineering Journal,* American Institute of Steel Construction, Vol. 10, No. 3, 1973, pp. 61-73.
11. Thornton, C.H.; "Quality Control in Design And Supervision Can Eliminate Lamellar Tearing," *Engineering Journal,* American Institute of Steel Construction, Vol. 10, No. 4, 1973, pp. 112-116.
12. "Lamellar Tearing of Welded Connections," *Civil Engineering,* December 1973, pp. 34-35.
13. Jubb, J.E.; "Lamellar Tearing," *Welding Research Council Bulletin,* No. 168, December 1971.
14. Farrar, J.C.; Dolby, R.E.; Baker, R.G.; Lamellar Tearing in Welded Structural Steels," *Welding Research* (supplement to the Welding Journal) Vol. 34, No. 7, July 1969, pp. 247S-282S.
15. Thomson, A.D.; Christopher, P.R.; Bird, J.; "Short Transverse Properties of Certain High-Strength Steels," *Transactions of ASME, Journal of Engineering for Industry,* 1968, pp. 1-9.
16. Peterson, M.L.; "Steel Selection for Offshore Structures," *Journal of Petroleum Technology,* March 1975, pp. 274-282.

17. Carter, R.M.; Marshall, P.W.; Swanson, T.M.; and Thomas, P.D.; "Materials Problems in Offshore Platforms," 1969 Offshore Technology Conference Preprints, Vol. I, Paper 1043.
18. Clark, J.E.; "High-Yield Ship Steels," Northeast Coast Institute of Engineers and Shipbuilders, April 1967.

# Chapter 9

# Computer Methods for Static and Dynamic Analyses

Prior to the early 1960s hand calculation was the only method available for analysis and design of offshore structures. To analyze a structure manually, the engineer had to make a number of simplifications and assumptions. He was dependent on geometric symmetry: the structure was divided into a series of identical trusses, plates, columns, etc., in which each component was analyzed as a separate two-dimensional problem. Simplified support conditions were assumed. Anticipated loads were increased to cover the uncertainties arising from the approximate nature of the analysis.

Even with all the simplifications, it still required weeks for an engineer to perform the analysis of an offshore platform. Realistically, such a structure could by analyzed manually for only a limited number of load conditions, and consideration of alternate designs was quite often impractical. With the computer, a large number of design parameters may be investigated. Consequently, the design engineer may obtain extensive analytical experience in a relatively short period of time. Many platform loading conditions can be quickly studied with a computer.

The first engineering-oriented computer programs were special purpose programs used to supplement or extend manual calculations. By 1965, use of the computer as an engineering problem solver was well-established; oil companies and engineering/construction firms were busy writing extensive programs for computer analysis of jacket space frames.

Initially, the commercially available computer programs for solving space frame problems were difficult to use because of the tediousness of preparing the data for input. Various companies either prepared their own large-scale static/elastic structural analysis systems or, by adding preprocessors and postprocessors to commercially available packages, devised programs particularly suited to the analysis of offshore platforms.

Today, computer service companies and computer programs are available to handle almost every aspect of structural analysis. For oceanography, there are programs to predict wave heights, determine wave properties, and compute combined wave and current forces on a structure. In soil mechanics and foundations programs are available to analyze both laterally and axially loaded piles and to predict the driving characteristics of piles in various soils. Three-dimensional framed structures can be analyzed statically with the well-known STRUDL program. Several companies have extended their in-house static programs to include calculation of hot-spot stresses (stresses at points of extreme stress concentration in welded tubular joints), and calculation of ratios of nominal stress compared to design code allowable stress for every structural member.

## Steps Involved in Static Analysis

In static/elastic analysis computer programs:

1. *The designer must define the structure in terms of physical dimensions, member sizes, and material properties.* Geometry of the structure is defined by joint locations and line members in a three-dimensional Cartesian coordinate system. Frequently, the initial values for tube diameters and wall thicknesses come from experience (what has worked satisfactorily in a similar application). Overall dimensions develop from deck layouts to meet operational requirements.
2. *The designer must input the soil conditions as interpreted for him by a soils specialist.* Some computer programs will construct lateral soil resistance versus lateral soil deformation or pile deflection curves (so-called *(P-y)* curves) from general soils data; in other programs the *(P-y)* curves themselves must be part of the input.
3. *The various loads must be entered into the program by the designer.* In some programs an Airy or Stokes type wave force is generated by the program from input such as maximum wave height, wave period, and mean water depth. Current force is entered separately or in some programs may be included in the wave-generated force by either specifying the current as input or calling for an appropriately larger maximum wave height.

    The vertical live loads and the equipment dead loads are determined from the deck layouts. It is not necessary to input the structural member dead loads as the program generates these from tube diameter, wall thickness, and joint coordinates. Also, this generation accounts for the buoyancy of flooded members because the list of flooded members is part of the input. Wind force and any other environmental

loads characteristic of the particular offshore location must be entered by the designer.
4. *The deterministic computer analysis passes the design wave through the structure from two directions: broadside and endwise.* Usually, the wave is passed through the structure at several other azimuth angles to determine the direction that produces the largest structural reactions. The computer advances the wave through the structure at specified increments, calculating total shear and overturning moment on the structure at the mudline of the jacket for each increment until the maximum values of base shear and overturning moment are found.
5. *The program will group various loads as prescribed by the designer to comprise loading conditions.* Many loading conditions will ordinarily be investigated. For each loading condition, the computer analysis gives (a) total base shear and overturning moment on the structure, (b) the member end forces and moments, (c) joint rotations and deflections, (d) external support reactions, and (e) an equilibrium check at each joint. The designer can usually specify which of these are to be printed as output.
6. *Stresses (axial, bending, combined axial and bending, torsion and shear) at the ends of all the structural members are next calculated by the computer.* It also compares these stresses to the allowable stresses as defined by the American Institute of Steel Construction (AISC) Specification or other applicable specification and prints tables of the comparison, usually expressed as a ratio of calculated value to AISC allowable value. The designer can specify which of several options he desires in the detail of the output.
7. *The soil-structure interaction has a strong affect on the structural response of a pile-supported fixed offshore platform.* In early computer efforts piles were broken into short-beam elements, and the lateral behavior of the soil for each element was modeled with linear springs. This method was cumbersome and is no longer used. Present-day soil-structure interaction programs use simpler pile modeling. The pile is replaced by a substructure consisting of lateral springs in two directions, an axial spring, and a moment spring—all inferred from the soils investigations. (See Figure 8-8.) The piling interacts with the surrounding soil in an inelastic manner, and the computer program linearizes this response to generate the equivalent elastic spring coefficients.

In some static analysis computer programs, the determination of the spring constants must be done as an auxiliary problem, and only the equivalent springs are put into the main program as input. The proper soil-structure interaction is achieved by iteration within the computer program. In the spring matrix simulating the pile it is important to evaluate accurately the terms accounting for the bending moment introduced into the pile head (at the mudline) caused by the shear displacement of the pile head from lateral soil deformation.

### Typical Static/Elastic Analysis Program

A typical computer program performs a linear elastic analysis of a two- or three-dimensional framed structure by the stiffness method. The program will combine applied loads, structural loads, and boundary support conditions in whatever manner prescribed by the designer, and thus handle either one or many loading conditions. Appropriate wave theory is included in the program. With wave theory and Morison's equation, the program will generate the wave forces acting on the members.

Morison's equation is applicable for tubular member jacket structures because the tube diameters are small in relation to the wavelengths of the passing waves. The Froude-Krilov hypothesis, which states that a submerged body does not affect the pressure field of the surrounding water-wave system, is used; such force components are frequently neglected. However, these force components cannot be neglected for structures like large submerged oil storage tanks where wave scattering becomes more important than viscous effects.

## Computer Options

The program will compute member-end forces, reactions at boundary joints, joint equilibrium checks at selected joints, and center-line moments at mid-span of each member.

## Output Options

The program will typically print or have the option of printing:

1. Structural geometry information
2. Applied joint loads
3. Applied member fixed-end forces
4. Computed member-end forces
5. Computed displacements of joints
6. Computed member-end stresses
7. Computed reactions at boundary joints
8. Imbalanced joint forces which exceed specified tolerances (This option requires the designer to state the force tolerance and the moment tolerance to be used in the equilibrium checks.)

## 136   Introduction to Offshore Structures

**Designer's Procedure**

**Given information.** At the beginning of the problem, the designer must have the following information:

1. Water depth
2. Location and bottom topography
3. Soils data
4. Type and functional use of the platform
   a. Deck loads
   b. Deck design
   c. Number of wells to be drilled
   d. Geometric constraints such as requiring two of the jacket legs to be parallel and provided with launch trusses
   e. Time constraints, if any
   f. Any special information that the operating company has that affects the design

**Review of given information.** The designer should review the accuracy of the following information:

1. Environmental criteria (This information may be from past studies if the platform is going into an existing field, or from new studies recently initiated if in a new field.)
   a. Wave height
   b. Wave period
   c. Tide and/or swell
   d. Current
   e. Wind speed
   f. Earthquake possibility
   g. Bottom stability, soil slides, sand waves
2. Functional requirements
   a. Whether or not standard deck loads are to be used
   b. Ancillary items like boat landing size, barge bumpers, railings, etc. (These components are included in each platform but not necessarily designed each time.)

**Initial hand calculations.** The designer calculates the following items and enters them as extra loads:

1. Simulations of wave force on appurtances and weights of appurtances
2. Wind force
3. Possible soil slide force
4. Deck loads

## Static and Dynamic Analyses

**Initial estimates by the designer.** The following items which meet the needs of the functional and environmental loadings comprise the platform configuration:

1. Jacket leg size and wall thickness
2. Bracing size, wall thickness, and member length to satisfy KL/r of column design
3. Number of piles, their size, and wall thickness
4. Number of skirt piles (if needed), their size, and wall thickness

**First Computer Run**

1. The program checks the jacket geometry for errors in coding to assure that all members are connected properly. The computer produces a diagram which includes every member and joint lying within a designer-specified distance either side of a specific plane defined by identifying three structural joints.
2. This initial run of the program generates the wave forces on the structure and subsequently the overturning moment due to the wave forces alone.

**Additional hand calculations.** The designer manually applies the calculated overturning moment to the preliminary structure and selects the pile batter. Moments are taken about a horizontal axis passing through the center of the structure at the mudline. Usually, the pile flexibility at the mudline is neglected at this stage in the computations. Total base shear is computed, accounting for the horizontal forces in the heads of the piles due to batter.

**Equivalent springs representing piles.** The designer either runs an auxiliary computer program to obtain the equivalent linearized springs or enters the information about the piles into the main program so that it can calculate the springs on the next pass.

**Main computer run.** The designer now activates the entire static analysis computer program to:

1. Pass the design wave through the structure at different azimuth angles to determine the base shear and overturning moment.
2. Calculate the stresses in each member.
3. Convert any renumbered joints of the reordered matrix bandwidth back to the designer's designations.
4. Compute the ratios of applied stress to specification allowable stress for each member. These are called stress ratios or unity checks.

5. Restart the program with modified member sizes and wall thicknesses, and recompute until the ratios of applied stress to allowable stress are less than unity for every member.
6. List for selected members the applied stresses and the ratios of applied stress to allowable stress to search for anomalies.
7. List all of the members that fall in certain ranges of applied stress to allowable stress to search for anomalies.

**Refinement.** Having accomplished the preliminary computer runs mentioned previously, the designer evaluates:

1. Configuration feasibility
2. Usefulness of initial sizes and wall thicknesses
3. Number of piles
4. Deflections of piles, deflections of one level of the structure relative to its adjacent levels
5. Practicability of structural joints
6. Whether or not the pile deflections at the mudline, the forces, and moments are consistent with the initially chosen piles

The designer repeats the previous steps as necessary until the design appears satisfactory.

## Dynamic Analysis

This section deals primarily with the dynamic analysis of the jacket and its soil-structure interaction. Dynamic analysis must be done by computer methods; the same is true of nonlinear analysis. Some people associate the two methods, although linearized dynamic analysis is in its own right a major advancement over static analysis—even computerized static analysis.

Dynamic analysis is becoming increasingly important for the following reasons:

1. For the larger and more costly structures, even small design refinements gained through dynamic analysis translate into huge construction cost savings.
2. As the offshore structural system becomes more complex, ordinary static analysis methods tend to result in higher risks in terms of property and life.
3. The harsh environmental conditions encountered at many deepwater sites simply cannot be adequately modeled by static analysis methods.
4. The capabilities of present-day dynamic programs offer increased opportunity to study different design configurations within the same

## Static and Dynamic Analyses 139

design time period so that overall reductions in construction schedules result by producing hardware according to the least time-consuming alternative.

The primary design load on an offshore platform is normally a dynamic load. In other words, the load is a function of time. The most common design load is the wind and wave loading that occurs during the 100-year design storm, particularly hurricanes for structures located in the Gulf of Mexico. Structures off the West Coast are subjected to earthquakes. In artic regions structures may be subjected to impact and oscillatory loads from ice.

Prior to 1960 most offshore structures were located in the Gulf of Mexico in water depths of 100 ft (30 m) or less. The dynamic loads resulted from storm conditions, and the loads were from waves with long periods compared to the vibration periods of the typical structure. Although the dynamic nature of the design loads was recognized, it was generally accepted that because the structures were stiff and damping was present, a static stress analysis was sufficient.

Between 1960 and 1966, the design philosophy changed. Structures were planned for locations off the West Coast where earthquake loading was known to be important, and in both the Gulf of Mexico and offshore of California structures were planned for much deeper water. It was apparent that in sufficiently deep water the vibrational period of the structure would be close enough to the period of storm waves or seismic disturbances that a static analysis would no longer be sufficient. Professor R.W. Clough at the University of California, Berkeley, was instrumental in helping to setup the initial methods for dynamic analysis.[1,2,3]

Today, the larger engineering companies designing offshore structures have available computer programs (usually developed in-house) for dynamic analysis. These programs generally compute in some time-history fashion the joint displacements and member-end forces for all the elements of the platform. Computer programs are also available to predict the behavior of a jacket during launching from a barge and during the upending operation. The better design programs are proprietary, although some computer service companies also offer dynamic analysis programs.

### Wave Forces

It is assumed that all applied forces act at the joints of the structure. The forces acting on a member at some point other than an end must be replaced by a statically equivalent set of end forces. The forces to be used are negative to the reactions at the beam ends that the applied force would cause if the beam were fixed at both ends. Each joint force is a function of time, and each has six components, including moments as components of the generalized force.

# 140 Introduction to Offshore Structures

The calculation of wave forces and the determination of fluid-structure interaction may be handled either by the deterministic or the stochastic approach. The deterministic approach uses regular waves; the stochastic approach uses the effects of random waves. The Stokes wave theory is usually used to describe waves in deep water, although it does not always provide the best fit to experimental wave data. It is used because the waves propagate without shape deformation and are periodic in space and in time. In the Stokes theory the wave amplitude of each term in the wave profile expression is not linearly related to the wave height as is true of the simpler one-term Airy theory. For this reason, the Airy theory is also widely used for deepwater wave calculations.

The wave spectrum, also called the wave spectral density function or the wave energy spectrum, is used in stochastic analysis to compute the structural response spectrum. The wave spectrum most often used is the Pierson-Moskowitz spectrum.[4] Based on actual wave measurements of extreme ocean states in the North Sea, another wave spectrum more sharply peaked than the Pierson-Moskowitz spectrum has been proposed by Rye, Byrd, and Torum.[5]

## Equations of Motion

Each member of the structure must be modeled mathematically. The finite element method of analysis is used and structural members are frequently divided into several elements for better representation of the complex behavior. Linear analysis is usually used. This analysis requires simplifying the drag force term in Morison's equation.

**Linear analysis.** When linearizing the drag force term in Morison's equation, the equations of motion in matrix form can be expressed as

$$M_v \ddot{u} + C\dot{u} + Ku = P(\dot{v}, \ddot{v})$$

where the object is to solve for $u$. In this family of equations:

$M_v$ = diagonal matrix of virtual mass

$C$ = matrix for structural and viscous damping

$K$ = square linear structural stiffness matrix

$P(\dot{v}, \ddot{v})$ = the load vector where $\dot{v}$ and $\ddot{v}$ are the water velocity and water acceleration

$\ddot{u}$ = structural acceleration

$\dot{u}$ = velocity

$u$ = displacement

## Static and Dynamic Analyses 141

**Comments on Equation of Motion Terms**

1. For each element, the virtual mass consists of the structural lumped mass and the added mass due to it being submerged.

2. The damping coefficient $C$ is larger than just the structural damping constant, since it includes the fluid-structure interaction which is essentially viscous damping. The damping coefficient ranges from 2 to 5% of critical damping. Maddox considers the total damping to be 2%.[6] For small damping, it has been shown by several authors that within broad limits the mathematical form of the damping force has very little effect on the motion; it is only the magnitude that is important.[7,8,9,10] Consequently, the simplest representation of the damping force is used, namely viscous damping. That is, the damping force acting at the joints is proportional to joint velocities.

3. In general, the load vector $P$ is different from one element to the next as $v$ and $\dot{v}$ change; thus, the data representing hydrodynamic and hydrostatic forces are voluminous.

4. The structural stiffness matrix is not used in its original, full form but is reduced extensively by restricting certain degrees of freedom of the system (deleting the rotational degrees of freedom).

The equations of motion are approximate; each term is linearized to simplify the solution. Linearization introduces little error as far as limited checking of the results with real structural behavior reveals. The computer solution of the equations of motion determines the first $n$ normal modes of vibration for the whole structure and their natural frequencies. The Rayleigh-Ritz iterative method of solution is employed. In this technique one first makes an estimate of the normal mode shapes. The Rayleigh-Ritz method then provides a means of computing approximately the natural frequencies and gives better estimates of the normal mode shapes. The amount of data to be handled is so large as to become impractical unless preprocessor and postprocessor programs are available to assist with the input and the output.

**Nonlinear analysis.** Presenting the equations of motion in nonlinear form is beyond the scope of this treatment. The three most important sources of nonlinear effects are the response of the pile-soil system, large displacements and plastic strains within certain highly stressed members of the structure, and drag force modification in the wave force equation. Bathe, Ozdemir, and Wilson present the equations of motion.[11] The equations are solved by iteration in time increments. Next, the stresses in the structural members are found with a general structural analysis program. Paulling mentions solving the equations of motion.[12] However, it is believed that there is no overall program available for public use for the nonlinear analysis of a flexible deepwater jacket structure. Special programs exist for small portions of the overall analysis.

Nonlinear behavior of material properties for steels in offshore platforms is ordinarily not considered. In comparison to other matters this subject is of lesser importance. Kaul and Penzien, and Selna and Cho discuss local yielding of the material.[13, 14]

## Comments on Solution Techniques

**Deterministic analysis.** The linearized equations of motion can be solved in two ways: modal analysis and time-history analysis. The wave and wind forces produce relatively slow motion of the structure. Because of this motion, many time-integration steps must be used for a time-history analysis. For solving linear systems of equations, modal analysis is preferable, since it allows the dependent equations to be decoupled. A good introduction to modal analysis is given by Biggs.[15]

Modal analysis assumes that each natural or normal mode of vibration of the structure may be expressed as a product of a shape function and an amplitude function. With orthogonality conditions and iterative techniques, the natural frequencies of the different mode shapes are found. Then the complete motion of the system is obtained by superimposing the independent motions of the several natural modes. The Duhamel integral greatly facilitates finding the solutions.[3]

Iterative methods are also used in nonlinear deterministic analysis, but discussion of these matters is beyond the present treatment.

Deterministic analysis is the standard approach taken in solving most offshore structural design problems. An outline of the steps in deterministic analysis is given in Figure 9-1.

**Stochastic analysis.** In offshore structural design stochastic analysis is generally used whenever the wave effects are predominant (when the cross-sectional dimensions of the structural elements are significant compared to the wave period).

A stochastic process is a random or chance process. It does not lend itself to explicit time description. The height of waves in a rough sea and the intensity of an earthquake are examples. It is not possible to predict the value of either at some future time; such phenomena are nondeterministic or random.

The response of a structural system to a random disturbing force is likewise random. Many random phenomena exhibit certain average patterns of behavior (their behavior can be described in terms of averages). Such data are referred to as statistics. When the averages take on recognizeable limits as the

## Static and Dynamic Analyses 143

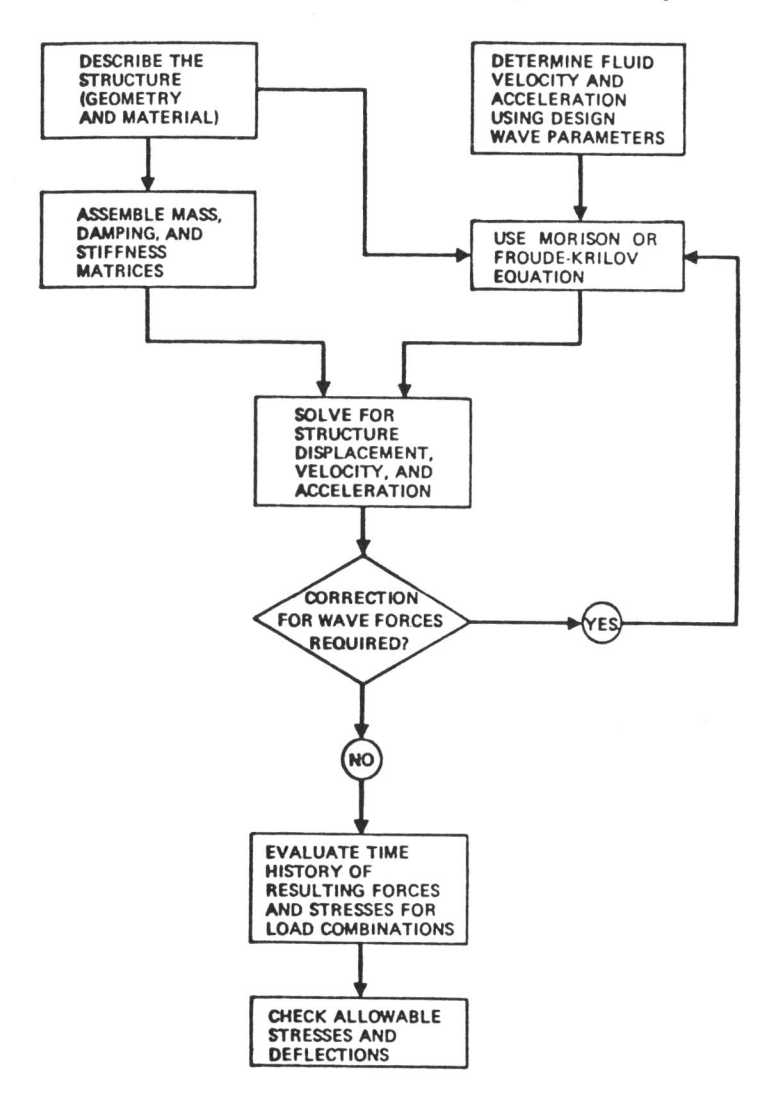

**Figure 9-1.** *Deterministic Method of Solution. (Courtesy of the Shock and Vibration Information Center, Naval Research Laboratory.)*

## 144  Introduction to Offshore Structures

number of phenomena samples become large, the random process is said to have statistical regularity. When the mean values of many samples of phenomena tend to be independent of time, the random process is said to be stationary.

Ocean waves are random processes that can be assumed stationary over short periods of time. Therefore, well-established statistical methods can be used to determine the occurrence probability of the maximum wave peaks.

Malhotra and Penzien and Chakrabarti discuss using stochastic analysis on offshore structures.[16,17] Further discussion of this matter is beyond the scope of the present treatment. An outline of the steps in stochastic analysis is given in Figure 9-2.

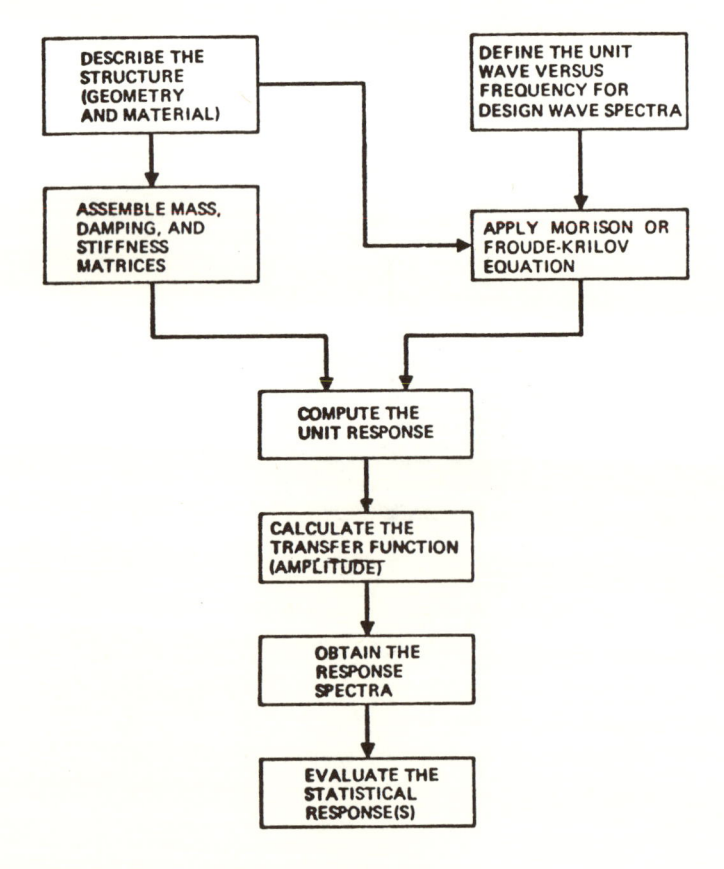

**Figure 9-2.** Stochastic Method of Solution. (Courtesy of the Shock and Vibration Information Center, Naval Research Laboratory.)

### Static and Dynamic Analyses 145

### Typical Offshore Structure Dynamic Analysis Program

The typical dynamic analysis program computes the elastic response of an offshore platform subjected to dynamic loadings such as wave loadings and earthquakes. The structure need not be in water for the program to be applicable.

### Results

As primary output, the program provides maximum values of member-end forces. Several other quantities are computed in the process:

1. Time-history of member-end forces
2. Time-history of joint displacements
3. Maximum values of joint displacements
4. Time-history of interstory shears
5. Time-history of overturning moments
6. Time-history of base shears
7. Time-history of axial pile loads

To compute the mass of the structure for the equations of motion, the dead weight of the structure and the equipment, supplies, and all other weights on or attached to the platform are required. This list includes the virtual mass due to being in the water, the mass of grout in those members designated to be grouted, and the water mass enclosed in flooded members. The static deflections from the weight and buoyancy loads are computed and can be added to the dynamic deflections if specified by the designer. Thus, the time histories of both joint displacements and member-end forces can include the effects of dead load.

### Designer's Procedure

**Background information.** As mentioned in the comments on solution techniques, there are two ways a dynamic analysis program may be used for design against earthquakes and/or extreme storm waves:

1. Time-history analysis of an earthquake ground motion (acceleration) or a wave loading.
2. Modal analysis, where a response spectrum of natural modes can be made the input, and the various frequencies of the vibration can be determined individually. The choice is left to the designer as to the proper combination of the modes in determining the overall response

(force, stress, deflections, etc.) of the platform. In some programs the designer is offered a choice of analysis methods:
   a. Root mean square
   b. The most significant mode plus the root mean square of the others
   c. Maximum value (the sum of the absolute values of the various modal responses)
3. The dynamic analysis program reduces the real structure to an idealized model of generalized coordinates. The user must pick the generalized coordinates by proper selection of load patterns. One natural frequency and its corresponding normalized mode shape is computed for each user-specified load pattern.

**Procedure**

1. The designer must first do a static/elastic analysis with a program such as that described earlier.
2. Spring constants must be linearized and linear *(P-y)* curves must be constructed to represent the soil response.
3. The designation of joints and members as given by the designer may be renumbered by the program to reduce the matrix bandwidth during computation. However, all output can be specified in the designer's original numbering scheme.
4. Classification of weights is different in the dynamic program from that in the static analysis program. In the static program distributed loads on the members may be used whereas in the typical dynamic program all loads must be joint loads.
5. To check the accuracy of input geometric coding, a dynamic load distribution can be applied as if it were a static load, and the program checks to see that appropriate forces are developed in the piles at the mudline.
6. Various earthquakes or waves are passed by the structure for the time-history analysis, or a smoothed response spectrum is employed to load the structure for modal analysis.
7. For the overall structure, the program determines the axial pile loads, base shear forces, and overturning moments at the mudline.
8. The program also determines member-end forces and combines these with the effect of self-weight or buoyancy and local member accelerations to produce localized member stresses.
9. Ratios of applied stress to specification allowable stress for combined axial and bending loads are computed for selected members and listed. These are called stress ratios or unity checks.
10. A table is given listing those members which have ratios of applied stress to allowable stress greater than unity.

## Conclusion

The use of computer programs provides the means whereby structural and foundational analyses for offshore platforms can be greatly extended and refined.

## References

1. Clough, R.W., "Dynamic Effects of Earthquakes," *Journal of the ASCE Structural Div.*, Apr. 1960, Paper #2437.
2. Clough, R.W., "Effects of Earthquakes on Underwater Structures," *Proceedings of the 2nd World Conf. on Earthquake Engineering, Vol. II,* 1960, pp. 815-831.
3. Clough, R.W., "Earthquake Response of Structures," (Reprint), *Earthquake Engineering,* Prenctice-Hall, Inc., 1969, pp. 1428-1455.
4. Moskowitz, L. and Pierson, W.J., "A Proposed Spectrum Form for a Fully Developed Wind Sea Based on the Similarity Theory of S.A. Kitaigorodskii," *Journal of Geophysical Research,* Vol. 69, No. 24, (1964) pp. 5181-5190.
5. Byrd, R.C.; Rye, H.; Torum, A., "Sharply Peaked Wave Energy Spectra in the North Sea," 1974 OTC Preprints, Vol. 2, Paper #2107.
6. Maddox, R., "Fatigue Analysis for Deep Water Fixed Bottom Platforms," 1974 OTC Preprints, Vol. 2, Paper #2051.
7. Hudson, D.E., "Equivalent Viscous Friction for Hysteretic Systems with Earthquakelike Excitations," *Proceedings of the 3rd World Conf. on Earthquake Engineering,* 1965, Vol. 2, pp. 185-206.
8. Jacobsen, L.S., "Steady Forced Vibrations as Influenced by Damping," *ASME Transactions,* AMP-52-15, 1930.
9. Medearis, K. and Young, D., "Energy Absorption of Structures Under Cyclic Loading," *Journal of the ASCE Structural Div.,* Vol. 90, St-1, Feb. 1964.
10. "Nuclear Reactors and Earthquakes," Lockhead Corporation and Holmes and Narver, Inc., Aug. 1963, TID-7024, Y3. At 7:22/TID-7024.
11. Bathe, K.J.; Ozdemir, H.; Wilson, E.L., "Static and Dynamic Geometric and Material Nonlinear Analysis," Report #UCSESM 74-4, Univ. of Calif., Berkeley, 1974.
12. Paulling, J.R., "Wave Induced Forces and Motions of Tubular Structures," 8th Symp. on Naval Hydrodynamics, 1970, pp. 1083-1110.
13. Kaul, M.K. and Penzien, J., "Stochastic Analysis of Yielding Offshore Towers," *J. of the ASCE Eng. Mech. Div.,* Vol. 100, No. EM-5, Oct. 1974, pp. 1025-1038.
14. Cho, D.M. and Selna, L.G., "Nonlinear Dynamic Response of Offshore Structures," 1971 OTC Preprints, Paper #1402.
15. Biggs, J.M., *Introduction to Structural Dynamics,* McGraw-Hill Book Co., 1964.

16. Malhotra, A.K. and Penzien, J., "Nondeterministic Analysis of Offshore Structures," *J. of the ASCE Eng. Mech. Div.,* Vol. 96, No. EM-6, Dec. 1970, pp. 985-1003.
17. Chakrabarti, S.K., "Discussion to Nondeterministic Analysis of Offshore Structures," *J. of the ASCE Eng. Mech. Div.,* Vol. 97, No. EM-3, Je. 1971, pp. 1028-1029.

# Chapter 10

# *Tubular Joint Design and Fatigue Analysis*

## TUBULAR JOINTS

A *tubular connection* is that portion of a structure where the cross sections of one or more tubes serving as braces are joined by fusion welding to the undisturbed exterior surface of another tube serving as a chord member. A *hollow structural member* may be of either circular (CHS) or square and rectangular (RHS) cross section. Sometimes, hybrid connections are formed by framing tubular braces into wide-flange chords (tube-to-H) or employing wide-flange bracing with tubular chords (H-to-tube).

Technically, a tubular connection is the localized portion of a structure where the joining of the tubular sections occurs, and a *tubular joint* is the interface created by the two intersecting member surfaces. In practice, the terms "connection" and "joint" are used interchangeably.

Tubular sections are preferred over open sections for structural use because of their high-torsional rigidity, symmetry of sectional properties, simplicity of shape, minimum surface for painting, and pleasing appearance. Tubular sections possess great advantages as structural elements, but their use was for many years hampered by the difficulties in joining the members. The best and cheapest method of joining today is to directly weld the contoured end of one tube onto the undisturbed outside of the other tube.

The stresses in tubular joints, as in most structural elements, are combinations of simple stresses such as tension, compression, bending or shear. Theoretical determination of the stress state in a general tubular joint has yet to be demonstrated because of the complexity of the problem. However, stresses can be determined to a satisfactory degree of accuracy by semi-empirical means which will be discussed in this chapter.

Various criteria for failure have been used for tubular joints, namely: reaching the elastic limit of the material, the material yield strength, detection of first cracking in a tension joint, and the ultimate load a joint will sustain in compression before gross deformation occurs.

Structural tubular joints exhibit several types of failure depending on the type of joint, the joint parameters, and the loading conditions. Possible modes of failure are:

1. Failure of the wall of the chord (punching shear).
2. Crack initiation separating the tension brace from the chord.
3. Local buckling of a chord wall in the vicinity of the brace loaded in compression.
4. Shear failure of the overall cross section of the chord.
5. Lamellar tearing of a chord wall in the vicinity of the tension brace.

It is important in the proper design of a tubular joint that the joint possess enough deformation and/or rotation capacity to permit the stresses to redistribute within the joint itself and generally throughout the structure as heavier and heavier loads are applied. Many tubular connections are joined by full-penetration groove welds, especially those in jacket structures of offshore platforms. Sometimes, in other applications, fillet welds are sufficient. When fillet welds are used to join two tubes, the throat thickness of the weld should equal the wall thickness of the thinner tube. Smaller-size welds lead to a large and uncertain reduction in joint strength.

A full-penetration groove weld is achieved by welding from one side only, without backing, and is a complete penetration and fusion of weld metal and base metal throughout the depth of the joint. The weld is actually formed by localized coalescence of the different metals produced by heating to suitable temperatures—with or without the application of pressure and with or without the use of filler metal. If used, the filler metal has a melting point approximately the same as the base metals, generally above 800°F (427°C).

The making of satisfactory welded tubular connections requires the designer to use considerable judgment to avoid elaborate arrangements which are awkward, hard to fabricate, and may be unsatisfactory in long-term service.

## Types of Joints

There are innumerable configurations for tubular joints if one considers three-dimensional geometry. Even restricting consideration to in-plane connections (where the axes of all the tubes lie in the same plane), there are still many configurations.

### Round (CHS) Tubes

In-plane tubular joints are designated as T, double T, Y, K, N, etc., depending on the positions of the braces. Figure 10-1 shows some of the possible geometries of typical joints.

## Tubular Joint Design and Fatigue Analysis 151

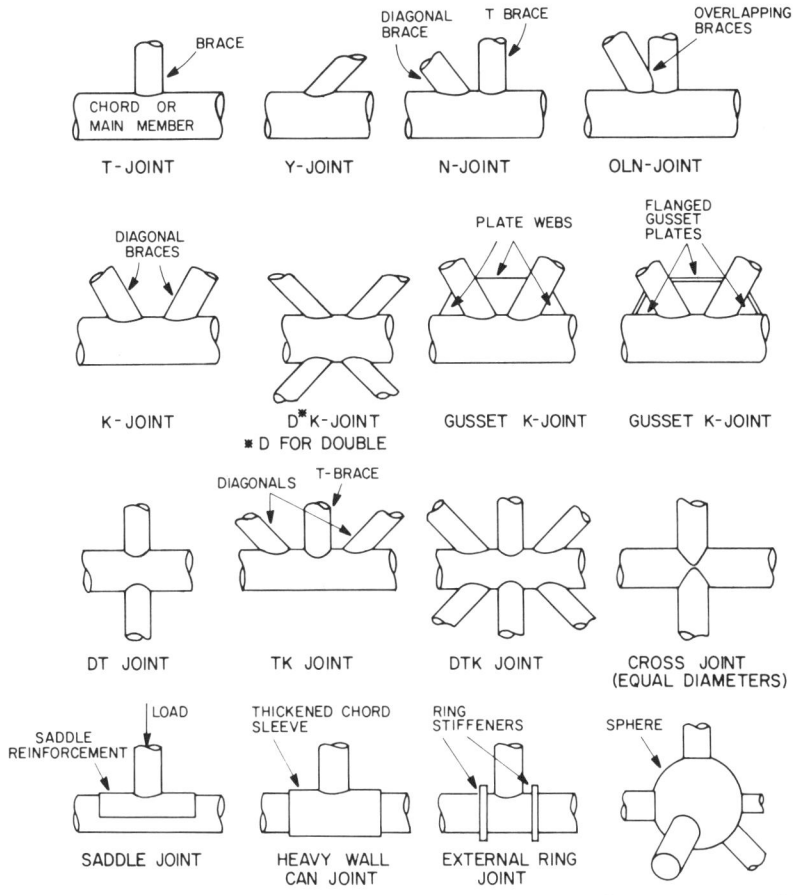

**Figure 10-1.** Typical tubular connections showing some of the many possible geometries.

A tubular joint made of two tubes which form a right angle is called a *T joint*. If the brace meets the chord at an acute angle, the connection is a *Y joint*. If two braces come together on the side of a chord such that the centerline of each brace forms an acute angle with the axis of the chord, this connection is a *K joint*. If one of the braces is perpendicular to the chord and the other meets the chord at an acute angle, this connection is an *N joint*. When two brace tubes frame into the chord from opposite sides with all three tubes in a plane, the connection is an *X joint* or *cross joint*.

A simple connection has no stiffening rings or gussets at the joint. In a tubular connection with more than one brace the joint is simple if the distance on the chord surface between the toes of the brace welds is more than the thickness of the chord. For most tubular joints, the part of the chord surface inside the brace (covered up in the welding process) remains intact. This part is called the *plug*. Sometimes, to permit flooding of the members, a small hole is made in the plug, but it is seldom removed entirely as that would drastically reduce the stiffness of the chord. The saddle joint, see Figure 10-1, is rarely used. It works satisfactorily only as long as the brace load is compressive. The spherical joint has been used with very small braces; it is uneconomical in larger sizes.

### Eccentric Connections

In an in-plane joint where longitudinal axes of two or more braces intersect, the distance perpendicular to the axis of the chord from the point of intersection of the brace axes to the chord axis is defined as eccentricity. (See Figure 10-2.) Eccentricity is negative if measured on the side of the chord axis toward the braces and positive if measured on that side of the chord axis away from the braces. Negative eccentricity tends to cause the braces to overlap; positive eccentricity promotes separation of the braces on the chord. For thin-wall chords with static loads, negative eccentricity connections have increased load-carrying capacity over zero eccentricity connections. However, there may be a decrease in the fatigue life of joints with overlapping braces compared to similar joints without overlapping. The offshore industry today tends to favor the use of heavy-wall chord sections at joint locations with nonoverlapping braces and no gusset plates.

### Offset Connections

When a brace is framed into a chord such that the centerlines do not lie in a single plane, the resulting structural joint is called an *offset connection* (Figure 10-3). Unless an offset joint is symmetrical so that the forces at the ends of the braces cancel each other without introducing any torsion in the chord, this form of connection is likely to be a weaker joint than an in-plane connection. Offset connections are also very difficult to form, especially in circular tubes.

### Rectangular Hollow Section (RHS) Joints

In recent years the use of rectangular tubes in the construction of offshore structures has increased considerably. Such sections are used for deck substructure trusses and lighter above-water framing, e.g., pedestrian bridges between adjacent platforms. Rectangular tubes are available only in relatively small sizes. They are not used more widely because of the nonavailability of appropriate design criteria, although these are being developed.

For above-water framing, rectangular tubes have several advantages over circular tubes and open sections. Their flat surfaces allow rapid fitting and

Tubular Joint Design and Fatigue Analysis 153

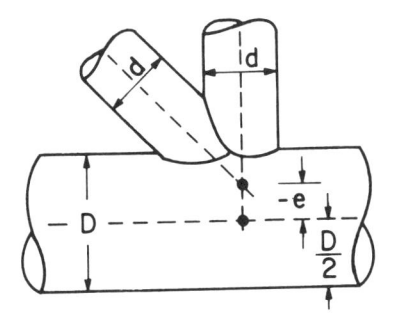

**NEGATIVE ECCENTRICITY JOINT**

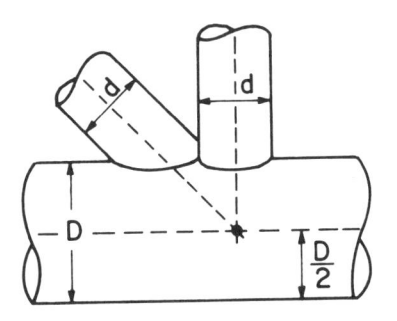

**ZERO ECCENTRICITY JOINT**

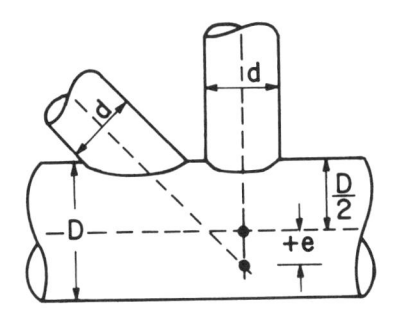

**POSITIVE ECCENTRICITY JOINT**

*Figure 10-2. Joint eccentricity.*

**154** Introduction to Offshore Structures

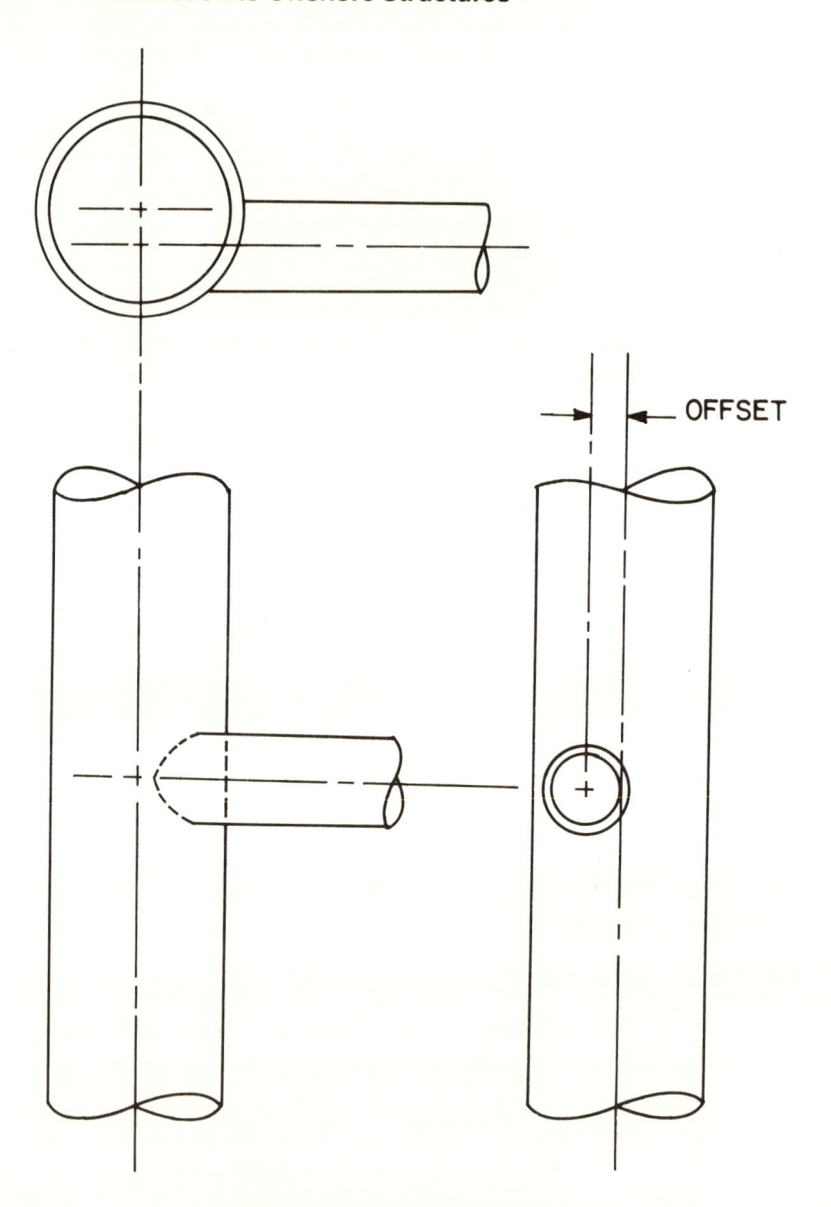

*Figure 10-3.* Offset connection.

## Tubular Joint Design and Fatigue Analysis 155

welding of joints. They are easily painted and more aesthetically pleasing in appearance than the other sections. For approximately the same size and weight per foot, rectangular tubes have larger torsional load capacity than circular tubes.

Figure 10-4 depicts the essential features of an RHS welded T joint. Because of the corner radii of the tubes, it is necessary to speak in terms of effective depth and width of section. The effective depth of the chord is $2a$; the effective width of the chord is $2b$. The ratio $\lambda$ is the effective width of the branch divided by the effective width of the chord. The ratio $\eta$ is the effective depth of the branch divided by the effective width of the chord.

Many geometries are possible for RHS joints just as with CHS joints. RHS joints tend to be planar joints, or complex joints in planes which make right angles on the chord because of the shape of the section.

Regardless of the ratio of inplane brace moment (in the plane of the two member centerlines) to column shear $M/V$, tests on square hollow-tube T joints with $\lambda = 1$ develop approximately the same maximum moment values. The vertical side walls of the brace tube carry almost all of the shear, and this force is transmitted directly into the two faces of the column with which the brace sides are flush.

The strength and stiffness of a rectangular tubular T connection is drastically reduced when the brace frames into a chord face that is wider than the width of the brace. Because of the flexibility of the connected chord face, the in-plane rotation of the brace is markedly increased and the connection may not develop the yield moment. For example, for a brace-to-chord T joint with all welds ground off flat and $\lambda = 0.66$, the maximum moment developed will be on the order of 40% of that developed when $\lambda = 1.0$. For a T connection with $\lambda = 0.66$ and a final weld contour approaching full penetration plus a 45° fillet with throat equal to brace thickness, the maximum moment developed may be as high as 90% of that when $\lambda = 1.0$.

The flexibility ratio (the chord or main member width divided by the wall thickness of the chord) appears to be more important that the $\lambda$ ratio. The flexibility of the corners where the connected chord face turns and becomes the free-side faces is also quite important. The wider the free-side faces, the less the restraint exerted on the edges of the connected chord face. The influence of chord width to chord wall thickness may be stated as follows: If the ratio is doubled, as from 12 to 24, the maximum moment attainable in the connection is approximately halved. The ratio of chord wall thickness to brace wall thickness is a relatively unimportant parameter as long as it is unity or larger.

Rectangular section joints are discussed in several of the references at the end of the chapter.[2,12,13,21] Thus, this type of joint will not be discussed further here.

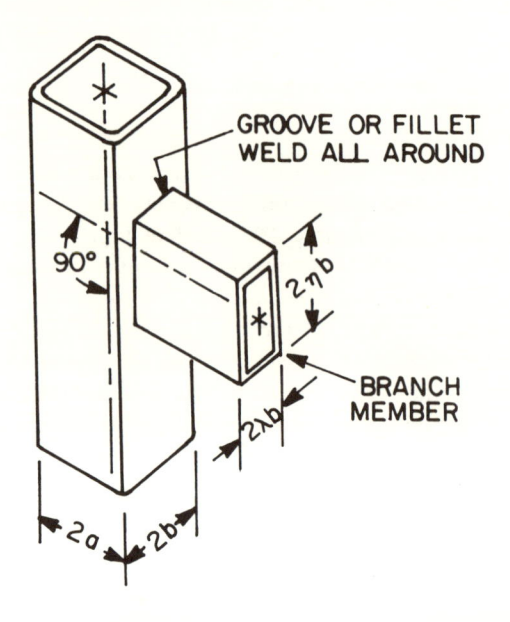

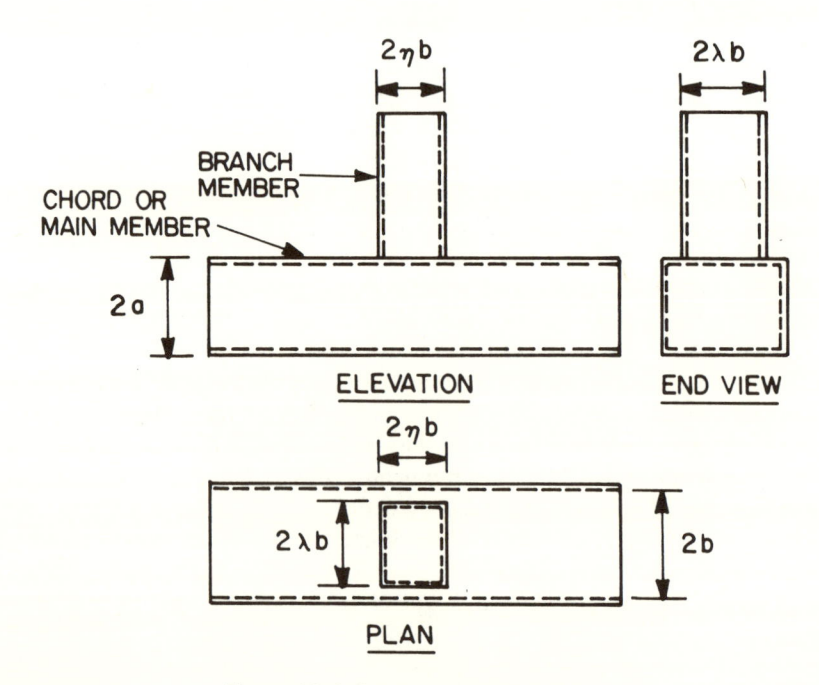

**Figure 10-4.** Rectangular hollow joint.

## Multiplane Connections

When braces from several directions are framed into the chord at a common location such that all of the tube centerlines do not lie in a single plane, the resulting structural joint is called a *multiplane connection*. This connection is sometimes referred to as a *congested connection*, a *complex connection*, or simply a *three-dimensional connection*. Figure 10-5 shows a typical multiplane connection. Use of beta less than 0.7 will relieve some of the undesired overlap conditions between adjacent planes.

Figure 10-6 shows the node for a multiplane connection to be used in a North Sea platform. Because of the stringent environment regulations promulgated for North Sea template platforms, it is often necessary to fabricate nodes of the tubular joints and stress relieve them to dissipate the welding residual stresses. Relieving the residual stresses, which may easily reach the magnitude of the material yield stress, lengthens in-service fatigue life.

Another feature of the node shown in Figure 10-6 is important. Notice that the chord diameter was enlarged for that length which was to be the joint can. With an enlarged can, more surface area is available for attaching braces without the necessity for overlapping. Sometimes, braces carrying lesser loads (so-called secondary braces) are shifted away from the joint somewhat to avoid overlapping. This spreading out of the secondary braces causes no difficulties if the moments produced by the eccentricity of their relocated centerlines are considered in the design analysis.

Figure 10-7 shows another form of node for a North Sea platform. The node pictured rests on a movable platform awaiting its turn to be rolled into the stress-relieving oven.

## Brief History of Tubular Joints

Right after World War II, as the oil industry began to move into offshore waters, there was almost no information on the performance of tubes welded together as structural joints. Manufactured tubes were available only in small sizes, and there were few facilities for rolling and fabricating larger-diameter tubes.

The importance of proper welding was not initially appreciated, and sometimes the sizes and types of welds varied depending on the judgment of the fabrication yard personnel. There was a recurring problem of tubular joint failure among early platforms, but redundancy of braces and on-site repairs coped successfully with the situation.

Hurricanes Hilda (1964) and Betsy (1965) forced a reconsideration of the techniques being used to design structural joints. The American Welding Society, with the encouragement and assistance of the offshore industry,

**158** **Introduction to Offshore Structures**

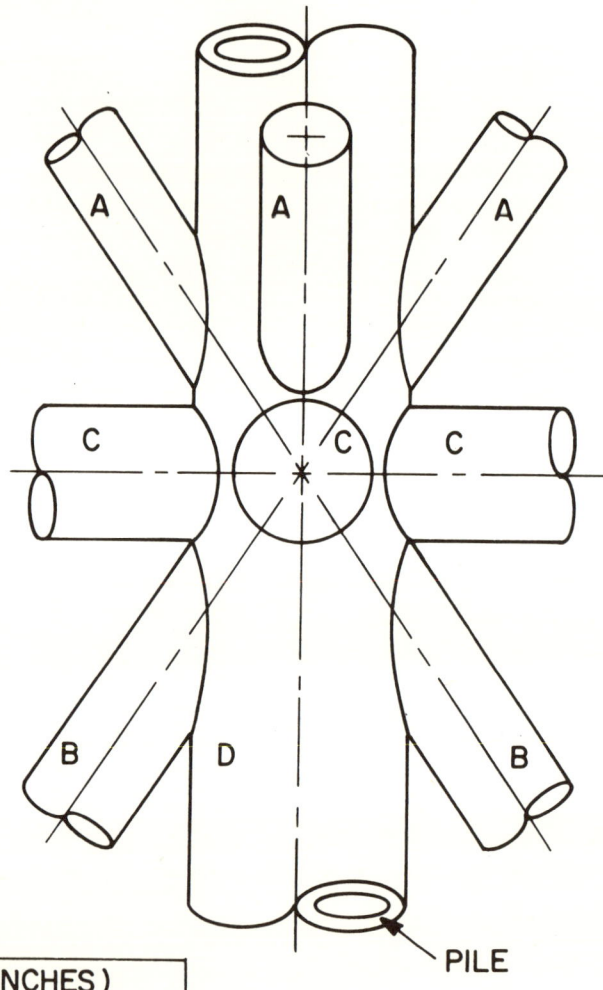

| SIZES (INCHES) | | |
|---|---|---|
| | DIAMETER | WALL |
| A | 16 | 0.500 |
| B | 20 | 0.500 |
| C | 24 | 0.375 |
| D | 40 | 0.750 |
| PILE | 36 | 1.000 |

*Figure 10-5. Multiplane connection.*

Tubular Joint Design and Fatigue Analysis 159

*Figure 10-6. Node for a multiplane connection for a platform for the North Sea Forties field. (Source: D.B.J. Thomas paper, International Symposium on the Integrity of Offshore Structures, Institution of Engineers and Shipbuilders in Scotland, Glasgow, April 1978.)*

**Figure 10-7.** Node for a multiplane connection on a railroad-wheeled platform just outside the stress-relieving oven at the McDermott construction yard, Inverness, Scotland.

formed a committee to develop specifications for proper welding and for the qualification of welders. The concept of a full-penetration groove weld evolved. The oil companies sponsored tubular joint experimental investigations at several universities and indirectly encouraged projects at others. The availability of useful design criteria for circular tubular joints rapidly increased.

The American Petroleum Institute issued the first edition of its *Recommended Practice for Planning, Designing, and Constructing Fixed Offshore Platforms,* API-RP-2A, in October 1969.[1] The American Welding Society incorporated the design of tubular structures into its new *Structural Welding Code,* AWS D1.1, first issued in 1972.[2] In parallel with the AWS committee, The American Society of Civil Engineers formed a national committee for tubular structures with overlapping participation by people already involved in the API and AWS efforts. Technical articles on tubular joints began to appear, and continue to appear, in the *Journal of the Structural Division, ASCE.* With the beginning of the annual Offshore Technology Conference in 1969, yet another forum was provided for sharing technical information on tubular joints and welded tubular structures in marine environments.

## Tubular Joint Design and Fatigue Analysis

Long before the all-out effort of the late 1960s to develop tubular joint design criteria, the owners and fabricators of platforms were making changes in an effort to strengthen the jacket joints. Gusset plates were first tried (Figure 10-1); these were welded in-between the brace ends and the chords during the period of the late 1950s and early 1960s. Sometimes pass-through gussets (Figure 10-8) were used. By about 1965, it was determined that gusset plates produced undesirable stress concentrations which shortened the fatigue life characteristics of the joint, and the industry trend was toward the reinforcing of joints with external ring stiffeners. (See Figure 10-8.) These were placed on the chord to strengthen the chord wall against collapse. Sometimes they were added at intervals along the lengths of the braces to preclude ovalization of the cross section and subsequent buckling of the compression braces.

During the early 1960s it became common practice to overlap braces (Figures 10-1 and 10-2) in order to transmit part of the load directly from one brace to another. However, stress concentrations at the common weld tended to reduce fatigue life performance of the joint. The industry trend in recent years includes no gussets, a minimum of brace overlapping, and strengthening of the chord by increasing its wall thickness. The heavy-wall chord sections, called *cans,* are often made of high-strength, fine-grain steel of better quality than the tubular goods between the joints.

Employment of internal ring stiffeners (Figure 10-8 and Figure 10-9) came into use in the early 1970s as much larger jacket legs began to be utilized for platforms in the North Sea.

### Parameters of an In-Plane Tubular Joint

Figure 10-10 shows the geometry of a typical in-plane tubular joint. Large membrane stresses are developed in the chord of a tubular joint. Regardless of the loading transmitted through the brace, large stress concentrations are produced at certain points along the welds. These stress concentrations cause the stresses at certain critical locations to be several times higher than they would be in the absence of such effects. The locations of the stress concentrations are called *hot spots.* (See Figure 10-10.)

Experience has proven that greater design loads can be safely carried than those determined with elastic theory. Local strains beyond the elastic range at the hot spots in the chord cause local yielding and produce a redistribution of the membrane stresses. (See Figure 10-11).

To facilitate the description of a tubular joint, three geometric ratios are used. The most significant is the diameter ratio $d/D$, also called the $\beta$ (beta) ratio. It is the brace diameter divided by the chord diameter. Whether outside diameters or mean diameters are used makes little difference ordinarily. The $\gamma$

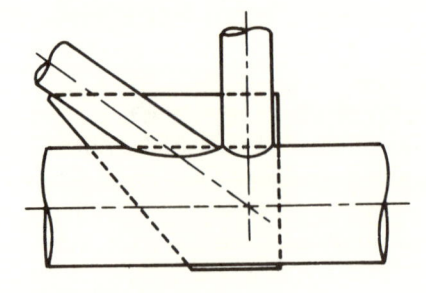

PASS-THROUGH GUSSET PLATE

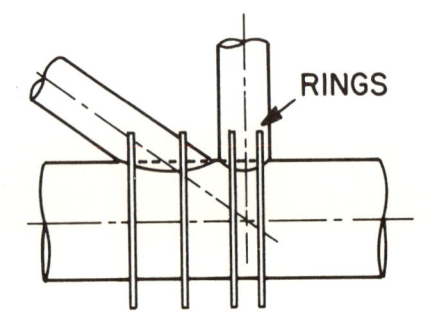

EXTERNAL RING STIFFENERS

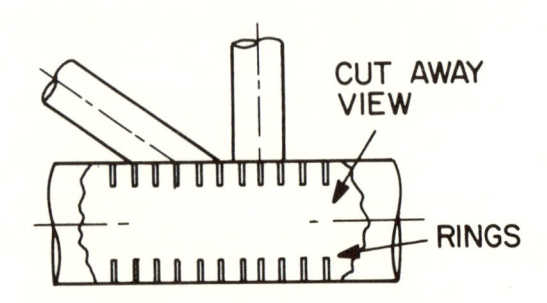

INTERNAL RING STIFFENERS
FOR LARGE DIAMETER CHORDS

*Figure 10-8.* Stiffened joints.

**Figure 10-9.** Internal stiffening rings in a multiplane node. (Source: D.B.J. Thomas, International Symposium on the Integrity of Offshore Structures, Institution of Engineers and Shipbuilders in Scotland, Glasgow, April 1978.)

**164  Introduction to Offshore Structures**

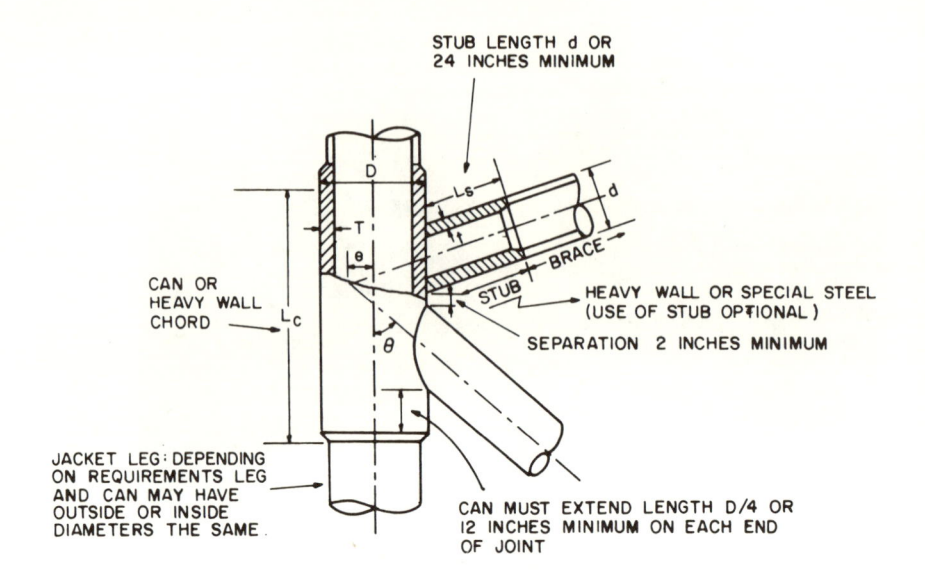

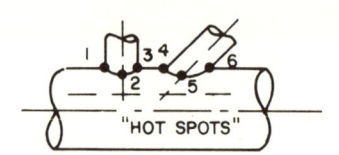

*Figure 10-10. Detail of an in-plane tubular joint.*

(gamma) ratio is the chord radius divided by the chord wall thickness and gives an indication as to the thinness of the chord. In boiler and pressure vessel work a tank is considered a thin cylinder if the ratio of wall thickness to diameter is less than 0.07, that is, a $\gamma$ ratio greater than about 7.0. In classical elastic theory it is assumed that a $\gamma$ of 10 denotes the minimum ratio for a thin shell. $\tau$ (tau) is a ratio of the wall thickness of the brace to that of the chord. It is an index of the likelihood that the chord wall will fail before the brace cross section fractures.

For tubular connections with beta less than 0.3, failure occurs by punching in or pulling out the plug from the side of the chord. This is called a *punching shear failure*. When beta exceeds approximately 0.8, the chord fails by collapse. In the range in-between the designer must estimate the interaction of punching shear and general chord collapse.

Tubular Joint Design and Fatigue Analysis 165

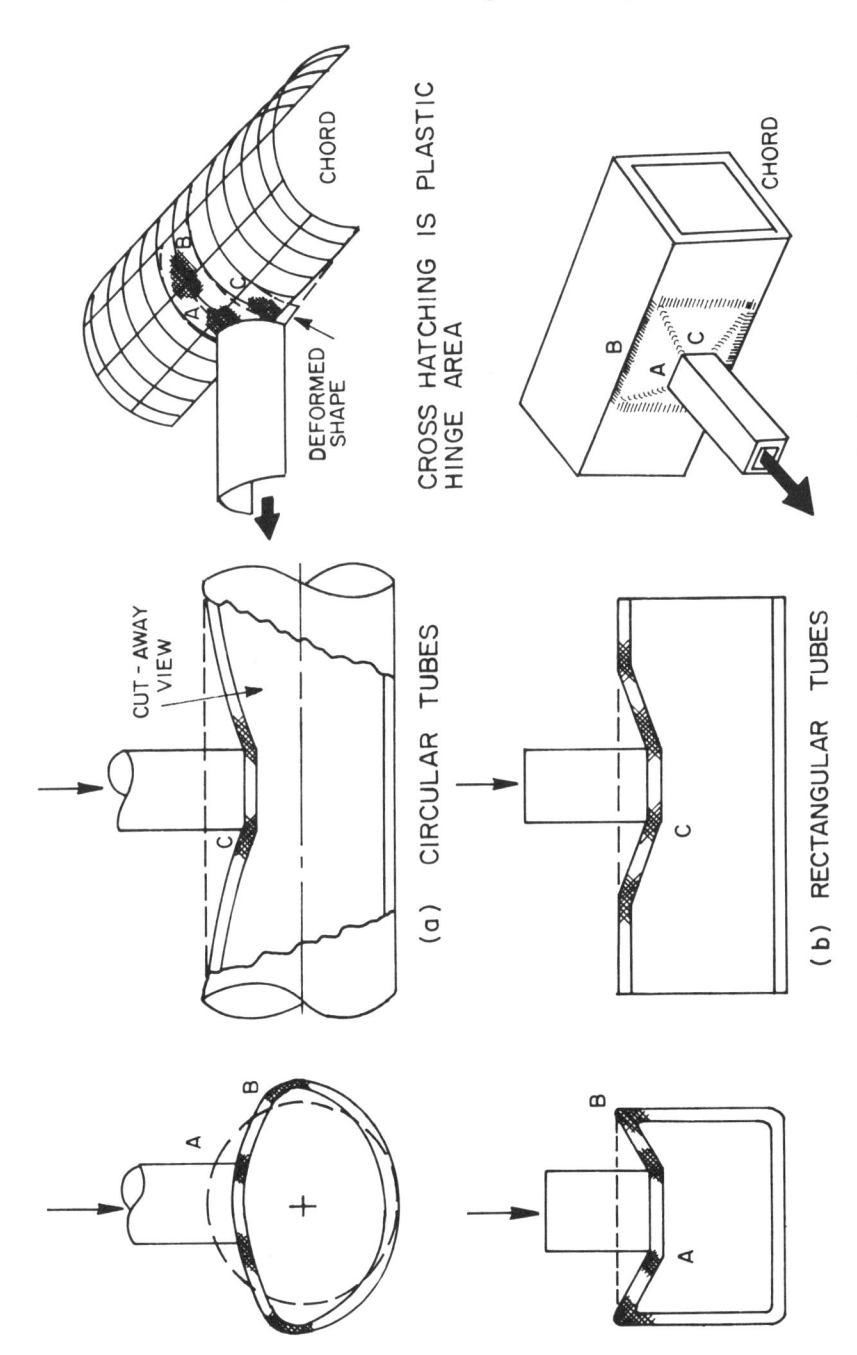

Figure 10-11. Distortion patterns and yield regions in chord.

**166 Introduction to Offshore Structures**

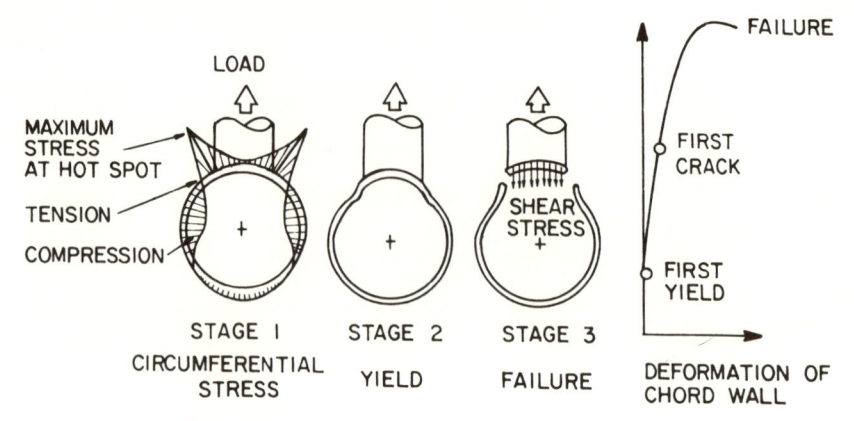

*Figure 10-12. Reserve strength of a tubular joint.*

Many designers proportion tubular connections in a gamma range of 7-15 and a beta range of 0.4-0.7. Experiments have indicated that connections with tau ranging from 0.5 to 0.7 have increased load capacities over connections with larger tau ratios. For tau greater than 0.5, there is little likelihood of the stresses in the brace being critical compared to the stresses in the chord. As a rule of thumb, a ratio of tau over beta that does not exceed unity is considered by some to be a well-proportioned design.

In Figure 10-10, when the load in the brace is tension, compression, or bending circumferentially about the chord axis, the hot-spot stress is at point 2 or 5. For bending of the brace in the plane of the axes of the chord and brace, the hot spot is at one of the points 1, 3, 4, or 6. Tension or compression in the brace combined with bending applied circumferentially about the chord axis produces the most severe stresses in the chord.

For axial force in the brace, there is a large difference between the magnitude of loading that causes initial yielding at the hot spot and that required to cause initial cracking (incipient failure) which occurs at the same point. Thus, tubular joints have tremendous reserve capacity beyond the point of first yield (Figure 10-12).

In Figure 10-12 consider stage 1. The brace load is very small, and the elastic distribution of outside surface circumferential stresses in the chord is shown by radial lines about the circumference. Because the brace is in tension, the peak stress at the hot spot on each side is tension. Depending on the geometrical ratios, stress concentration at the hot spot can be as high as 10 times the nominal stress in the brace. As the load continues to increase beyond yield, the connection deforms as shown in stage 2. Finally, as shown in stage 3, the connection fails at a load 2.5-8.0 times that of first yield. The connection fails by pullout of a plug from the chord for tension or by localized collapse of the chord for compression. The load-deformation curve shown to the right of stage 3 illustrates the relative levels of load for stages 1, 2, and 3 in relation to initial yielding and initial cracking of the chord wall.

## Elastic Stress Distribution

While the designer is ultimately concerned with the fatigue life of welded tubular joints, he is also interested in the distribution of certain elastic stresses in the chord surface related to the first yielding of the material and the initial cracking (incipient failure) adjacent to the weld joint.

The single stress component most useful in the design of T, Y, and K joints is the circumferential stress $\sigma_s$ directed along the $s/a$ axis as shown for each type of joint in Figure 10-13. When two or more braces are involved, the $s/a$ axis is the circumferential axis corresponding to the largest $\sigma_s$ stress distribution about the circumference of the chord.

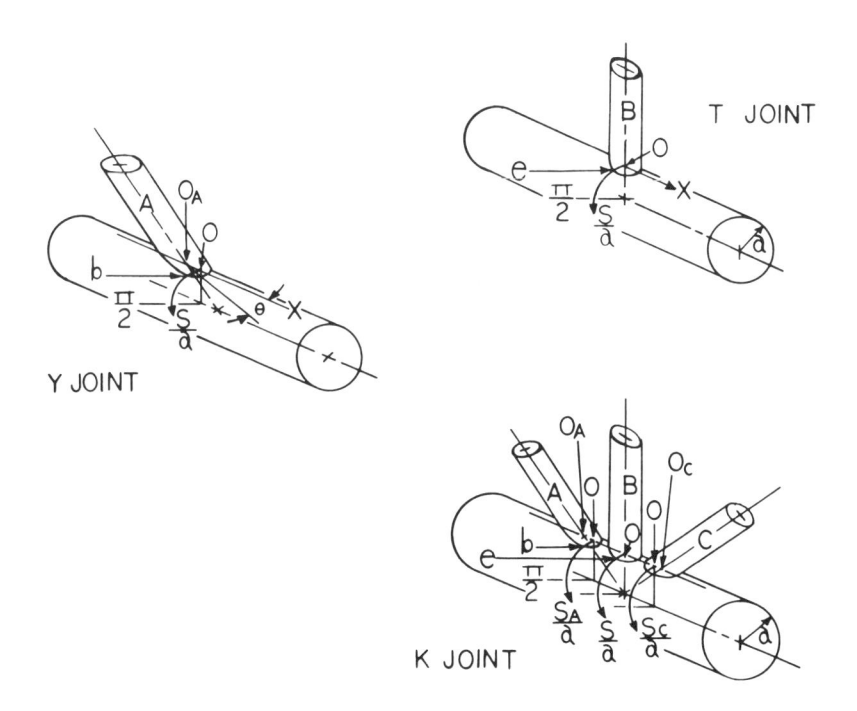

*Figure 10-13.* Nomenclature for tubular joints. (Copyright © 1970 Offshore Technology Conference.)

**168    Introduction to Offshore Structures**

This stress $\sigma_s$ is the maximum principal stress in the joint for some configurations. Even when it is not the maximum principal stress, it approaches in magnitude the maximum principal stress which is at some angle $\phi$ relative to the $x$ axis. The angle $\phi$ ranges from 69° to 90° for T joints,[3,4] from 34° to 135° for Y joints,[3,5] and is in a direction $\phi$ which is approximately normal to the line of the weld toe and within about 45° either side of the $s/a$ axes for K joints.

Regardless of the exact direction of the maximum principal stress, it is the stress $\sigma_s$ about the particular $s/a$ axis in Figure 10-13 that is the critical stress in a given configuration. This is the stress to be considered for design work. This critical stress $\sigma_s$ is a maximum at location $e$ for T joints, location $b$ for Y joints, and at either $b$ or $e$ for K joints, depending on the type of loading. Axially loaded braces cause the greatest stresses. Regardless of the type of loading, $\sigma_x$ at the intersection of the $x$ axis and the toe of the weld reaches the yield condition subsequent to yielding at $e$, but before initial cracking begins at $e$.

Figure 10-14 shows the elastic distribution of chord stress $\sigma_s$ for T joints (either tension or compression-loaded) with a brace load sufficient to cause a stress in the brace of 1 ksi.[3,4,5] Tests were performed using $\alpha$ ratios of 7.7, 11.5,

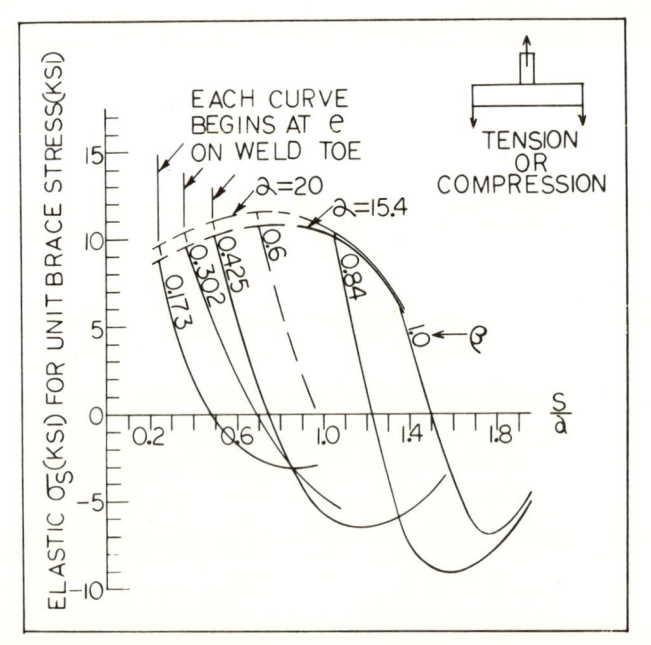

*Figure 10-14. Distribution of circumferential stress in T joint chord due to axial load in brace. (Copyright © 1970 Offshore Technology Conference.)*

## Tubular Joint Design and Fatigue Analysis 169

and 15.4. The $\alpha$ ratio is the length of the chord divided by the mean chord radius. The $\gamma$ ratio for the tests was 25 where $\beta$ and $\tau$ for the specimens had these values, respectively: 0.173, 0.872; 0.302, 0.904; 0.425, 1.0; 0.840, 1.0; and 1.0, 1.0. The test results were extrapolated to $\alpha = 20$ to be more realistic of practical application; hence, Figure 10-14 shows stress distributions for only $\alpha = 15.4$ and 20.

The rate of change of $\sigma_s$ becomes very high as point $e$ is approached. Thus, in some tests the maximum $\sigma_s$ may have entered slightly the plastic stress-strain region. Because the plastic zones are small, the stress distribution becomes elastic very rapidly as $s/a$ increases. As brace load is increased, the principal change in the stress distribution occurs between the crossing point for tension and compression stress on the $s/a$ axis and point $e$, although all of the stress magnitudes increase. The shapes of the stress distributions remain the same as $\alpha$ changes from 15.4 to 20—only the points representing maximum elastic stress at $e$ are increased.

Figure 10-15 shows the elastic $\sigma_s$ distributions for the Y joints tested.[3] For all three specimens, the $\alpha$ and $\gamma$ ratios were 15.4 and 25. The $\beta$ and $\tau$ ratios were, respectively: 0.173, 0.872; 0.302, 0.904; and 0.840, 1.0. Figure 10-16

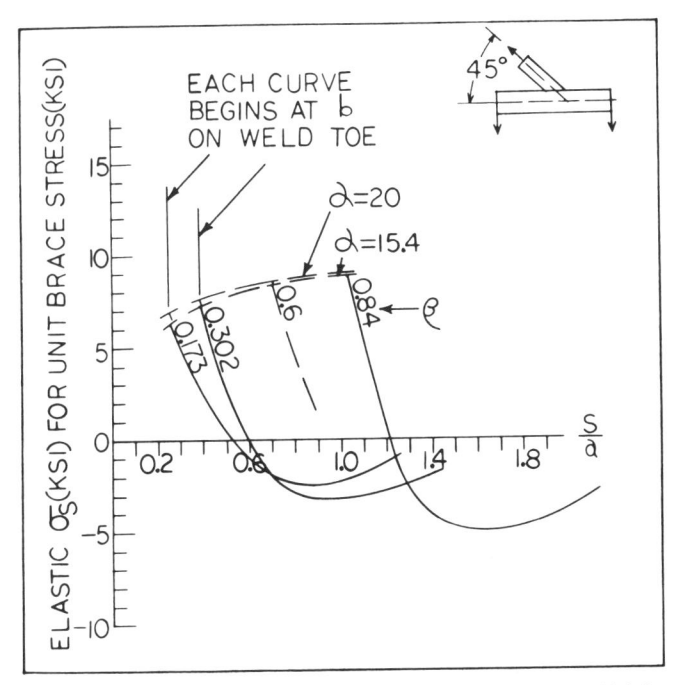

*Figure 10-15.* Distribution of circumferential stress in Y joint chord due to axial load in brace. (Copyright © 1970 Offshore Technology Conference.)

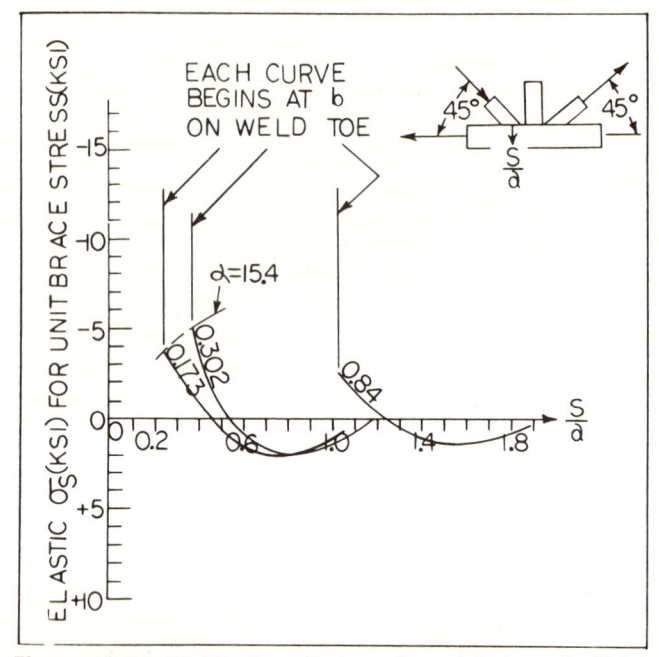

**Figure 10-16.** Distribution of circumferential stress in K joint chord due to axial compression in one diagonal and equal axial tension in other diagonal with longitudinal chord reaction. (Copyright © 1970 Offshore Technology Conference.)

shows the elastic $\sigma_s$ distributions for K joints subject to axial compression in one diagonal and equal axial tension in the other with a resultant longitudinal chord reaction. For all three specimens, the $\alpha$ and $\gamma$ ratios were 15.4 and 25. The $\beta$ and $\tau$ ratios were, respectively: 0.173, 0.872; 0.302, 0.904; and 0.840, 1.0. All of the chords had diameters of 12.75-inches OD and 0.250-inch wall thicknesses.

## Punching Shear Stress[1]

Where a connection is made by simply welding the brace to the chord, local shear on a potential failure surface on the chord may limit the usable strength of the joint. The shear stress at which failure occurs depends not only on the strength of the steel in the chord, but also on the geometry of the joint, particularly the radius-to-thickness (gamma) ratio of the chord. The punching shear stress on the potential failure surface is determined by dividing the component of the brace load perpendicular to the chord by the shear area of the chord. The shear area for computing shear stress in simple connections is the chord wall thickness times the actual length of the intersection measured along the toe of the connecting weld on the chord.

## Tubular Joint Design and Fatigue Analysis

For simple tubular joints without overlap of braces and with no gussets, diaphragms, or ring stiffeners, the allowable punching shear stress in the chord wall is given by

$$V_p = Q_q \times Q_p \times Q_f \left( \frac{F_y}{0.9(\gamma)^{0.7}} \right)$$

where:

$\gamma$ = $R/T$
$F_y$ = specified minimum yield strength (ksi) of the chord at the joint, but not more than two-thirds of the tensile strength
$R$ = radius of the chord, inches
$T$ = thickness of the chord wall, inches

The factor $Q_q$ accounts for the effects of type of loading and geometry. Values of $Q_q$ are given in Table 10-1. The factor $Q_p$ accounts for the plastic reserve strength of the joint and is given by

$$Q_p = \cos\left[ (90°) \left( \frac{f_a}{f_a + f_b} \right) \right] + \left( \frac{f_a}{f_a + f_b} \right)$$

Where $f_a$ and $f_b$ are the axial and bending stresses in the brace, respectively. The factor $Q_f$ accounts for the axial load in the chord

$Q_f$ = 1.0 for $A \leqslant 0.44$
$Q_f$ = 1.22 - 0.5$A$ > 0.44

where

$$A = \frac{|f_a'| + |f_b'|}{0.6 F_y}$$

and $f_a'$ and $f_b'$ are the axial and bending stresses in the chord, respectively.

The method of calculating the applied punching shear stress is given in Figure 10-17.[1] The allowable stress value calculated by

$$V_p = Q_q \times Q_p \times Q_f \left( \frac{F_y}{0.9(\gamma)^{0.7}} \right)$$

should be larger than that calculated using Figure 10-17.

### Table 10-1
### Values for $Q_q$ Factor

| Line | Type of joint | Geometry* | Axial tension | Axial compression | In-Plane bending | Circumferential bending |
|---|---|---|---|---|---|---|
| 1 | K | $0 < \zeta < 0.15$ | $1.3 - 2\zeta$ | $1.3 - 2\zeta$ | 2.25 | 1.0 |
| 2 | | $\zeta > 0.15$ | 1.0 | 1.0 | 2.25 | 1.0 |
| 3 | | $\beta > 0.6$ | ** | ** | 2.25 | ** |
| 4 | | $\eta > 1.0$ | $\frac{1 \text{ times}}{\eta}$ line 3 | $\frac{1 \text{ times}}{\eta}$ line 3 | $\frac{\text{line 3}}{\eta} > 1.5$ min. | $\frac{\text{line 3}}{\eta} > 1.0$ min. |
| 5 | T and Y | $\beta \leqslant 0.6$ | 1.4 | 1.0 | 2.0 | 1.0 |
| 6 | | $\beta > 0.6$ | *** | $Q_\beta$ | 2.0 | $Q_\beta$ |
| 7 | | $\eta > 1.0$ | $\frac{\text{line 6}}{\eta}$ | $\frac{\text{line 6}}{\eta}$ | $\frac{\text{line 6}}{\eta} > 1.5$ min. | $\frac{\text{line 6}}{\eta} > 1.0$ min. |
| 8 | Cross joints | $\beta \leqslant 0.6$ | 1.0 | 1.0 | 1.4 | 0.7 |
| 9 | | $\beta > 0.6$ | $Q_\beta$ | $0.7 Q_\beta$ | $1.4 Q_\beta$ | $0.7 Q_\beta$ |
| 10 | | $\eta > 1.0$ | $\frac{\text{line 9}}{\eta}$ | $\frac{\text{line 9}}{\eta}$ | $\frac{\text{line 9}}{\eta} > 1.0$ min. | $\frac{\text{line 9}}{\eta} > 0.7$ min. |

* Consult Figure 10-10 for $\beta$, $\gamma$, $\tau$, $\eta$, and $\zeta$.

** Use line 2 or beta factor $Q_\beta = \left( \frac{0.3}{\beta(1 - 0.833\beta)} \right)$ if larger.

*** Use 1.4 or beta factor $Q_\beta$ if larger.

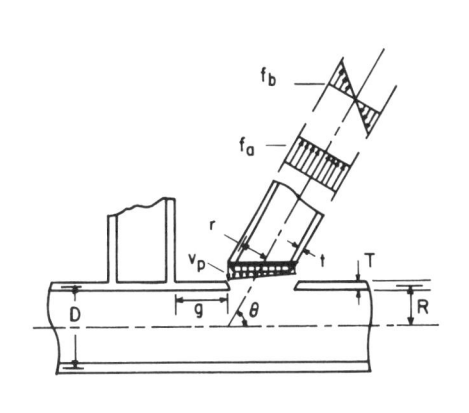

# Tubular Joint Design and Fatigue Analysis

$f_a$ = AXIAL STRESS, ksi
$f_b$ = BENDING STRESS, ksi
$t$ = BRACE THICKNESS, IN.
$T$ = CHORD THICKNESS, IN.
$r$ = BRACE RADIUS, IN.
$R$ = CHORD RADIUS, IN. (1/2 D)
$g$ = GAP, IN.

$$\tau = \frac{t}{T}$$

$$\beta = \frac{r}{R}$$

$$\gamma = \frac{R}{T}$$

$$\eta = \beta / \sin\theta$$

$$\zeta = \frac{g}{2R}$$

JOINT GEOMETRY IS DEFINED BY $\tau, \beta, \gamma$ AND $\theta$ AS WELL AS BRACE STRESSES $f_a$ AND $f_b$.
PUNCHING SHEAR STRESS MAY BE CALCULATED FROM:

$$v_p = \tau \left( \sin\theta \; \frac{f_a}{K_a} + \frac{f_b}{K_b} \right) \text{ k.s.i.}$$

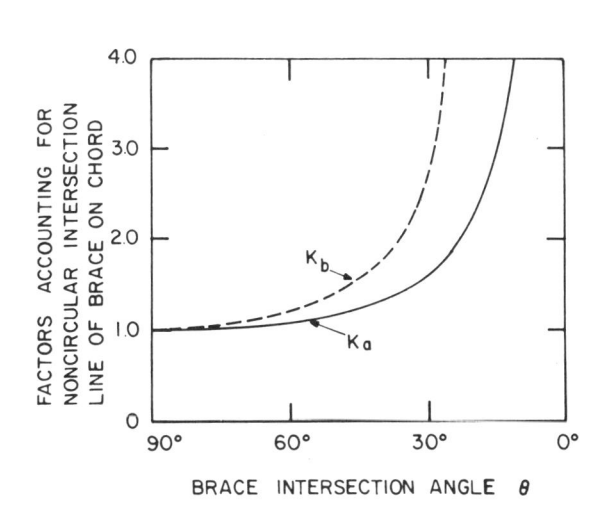

**Figure 10-17.** *Punching shear stress.*

## Overlapping Braces

In overlapping joints the braces intersect each other as well as the chord, and part of the load is transferred directly from one brace to another through their common weld. One advantage of such joints is that since the chord no longer transfers the entire load, its thickness can be reduced. The amount of overlap can be controlled by adjusting the eccentricity of the brace centerlines. Negative eccentricity can be used to increase the amount of overlap and the static load capacity of the connection; however, fatigue behavior of the overlapping braces may not be good.

The allowable load component transverse or perpendicular to the chord is designated $P_T$ (kips). It is expressed as

$$P_T = (v_p\, TL_1) + 2\,(v_w\, t_w\, L_2)$$

where:

- $v_p$ = calculated punching shear stress, ksi
- $T$ = chord wall thickness, inches
- $L_1$ = actual length of that portion of the brace which contacts the chord, inches
- $v_w$ = allowable shear stress for the common weld between the braces, ksi (Unless the common weld is specifically designated as a full-penetration groove weld, the value of $v_w$ is taken as 0.3 times the specified minimum tensile strength of the weld metal or electrode.)
- $t_w$ = lesser of the effective throat dimension of the comon weld or the thickness of the thinner brace.
- $L_2$ = projected distance perpendicular to the chord axis from the lowest point of the common weld around the side of the chord to the highest point of the weld between the braces. (See Figure 10-18.)

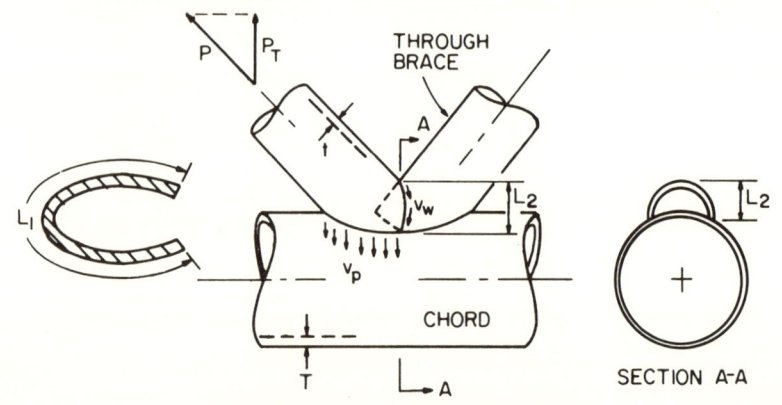

**Figure 10-18.** *Overlapping joint detail.*

# Tubular Joint Design and Fatigue Analysis 175

Overlapping joints are preferably proportioned so that the overlap itself will take at least 50% of the acting $P_T$. For good design, brace wall thicknesses should not exceed the chord wall thickness. However, in an overlapping joint one brace wall thickness may be greater than the other, since one brace may carry a substantially greater load than the other. Where different brace wall thicknesses occur, the member with the heavier wall should be the through-member which welds throughout its entire periphery onto the chord surface. The thinner brace is coped to fit over the thicker brace and onto the contour of the chord surface.

## Stress Concentration

Since the early 1960s, the stress analysis of tubular joints for offshore structures has received considerable attention. Both experimentally and analytically the problem has been studied by many investigators. Because of its complexity, the problem is still far from solved.

It is well known that the applied loads on tubular joints cause stresses at certain points along the intersection weld to be many times the nominal stress acting in the members. This multiplier applied to the nominal stress to reach the peak or maximum stress at the critical location (hot spot) is called the *stress concentration factor* (SCF).

The stress concentration factor is different from one joint geometry to another and is a measure of the joint strength, particularly its fatigue strength. Thus, since it is imperative to design offshore tubular structures for long fatigue lifes, it becomes important to determine accurately the proper stress concentration factor for each joint geometry.

One study has described the design of tubular joints for offshore structures as an iterative procedure.[6] The process begins with the sizing of the jacket piles according to the requirements of the specific soil/foundation needs of the platform. Sizing determines the diameter of the jacket legs and allows clearance for the piles to go through them. Once the trusswork geometry is selected, the column buckling characteristics determine the diameters of the various jacket braces. The initial wall thicknesses of chords and braces are determined by structural analysis. The next cycle of calculation involves increasing the chord wall thicknesses with heavy joint cans to ensure sufficient static strength to meet code or specification requirements. The next iteration involves calculating the fatigue strength of the joint to determine if it is compatible with the service life requirements of the platform. Depending on the method of fatigue analysis used, allowable stress concentration factors must be either specified for each joint or built-in to the method of analysis.

There are three classes of tubular joints. The *simple joint* was defined in the section on joint configurations as a nonoverlapping joint without any stiffening rings, gussets, or other reinforcement. Stress transmission through

## 176  Introduction to Offshore Structures

the joint depends largely on the bending characteristics of the chord. The *locally stiffened joint* employs reinforcing rings and sometimes longitudinal stiffeners to strengthen the weak areas in the chord. The bending action of the chord wall serves to extend the area over which the stiffening is effective. One ring stiffener is effective over an axial distance along the chord of approximately one-half times the square root of the product of the chord wall thickness and the chord radius. In the *fully stiffened joint* internal longitudinal and circumferential diaphragms are used to achieve stress transmission primarily through membrane action. The effect of bending is of secondary importance. The stress concentration factors are different in each case.

In summary, the various methods of providing reinforcement to a tubular joint are:

1. Negative eccentricity (increasing the overlap of the braces coming together at a joint)
2. Use of a heavy-wall can
3. Filler gussets between tubes
4. Gussets that pass through either the brace, the chord, or both
5. Outside ring stiffeners, internal ring stiffeners, internal diaphrams
6. Internal longitudinal stiffeners
7. Penetrating braces with internal structure in the chord
8. Flared ends on braces

Considerable information is available on stress concentration factors for unstiffened tubular joints.[6-10] Papers are available that discuss stress concentration factors for stiffened joints, although these papers are few in number and hard to find.[11-13]

The subject of stress concentration factors is too detailed to go into any deeper here. References 6, 9 and 11 are suggested for further reading.

### Chord Collapse and Ring Stiffener Spacing

**General Chord Collapse**

Most tubular joints are designed on the basis of punching shear as outlined in the section on punching shear stress. Only in special cases where there is reason to suspect the possibility of stability failure of the chord is collapse considered. The possibility of chord collapse is made more remote by the use of local reinforcement such as a heavy-wall can or the employment of ring stiffeners on the chord section at the joint.

General collapse failure of a tubular connection is the gross flattening or distortion of a large part of the cylindrical shell which forms the chord. A conceptual model of this mode of failure is shown in Figure 10-19. Plastic

**GENERAL CHORD COLLAPSE**

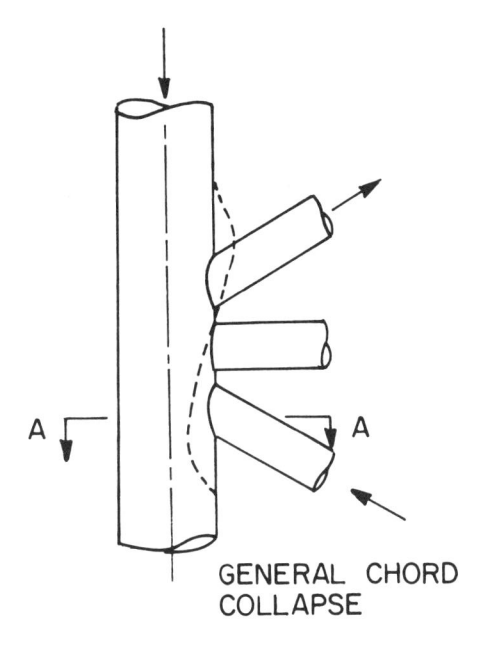

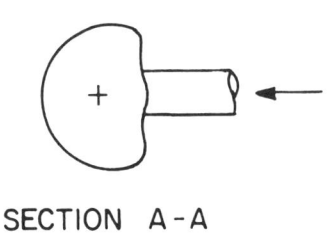

**SECTION A-A**

*Figure 10-19.* Conceptual model of chord collapse.

behavior, triaxial stresses, strain hardening, load redistribution, and large deformation behavior place extraordinary demands on the ductility of the chord material. The actual mode of failure involves interaction between punching shear and general bending of the chord wall.

Because it is so difficult to estimate the strength of tubular joints under realistic loading conditions, some researchers have found it convenient to use the interaction method.[14] The principles of the method as applied to tubular joints are:

## 178  Introduction to Offshore Structures

1. Determine the strength (loading at failure) under each simple loading condition (all punching shear or all moment). This is done through tests or by calculation.
2. The real loading, which is a combination of punching shear and moment, is represented by a load ratio of the form

$$R = \frac{\text{maximum applied load of type } i}{\text{ultimate load of type } i}$$

The term "ultimate" means the load of type i which will cause the chord to collapse.

3. The two load ratios, one for moment and one for punching shear, are plotted as ordinate and abscissa of an interaction diagram. The end of the curve on the ordinate marks the condition of all moment

$$R = \frac{M}{M_{ult}} = 1$$

The end of the curve on the abscissa marks the condition of all punching shear

$$R = \frac{P'}{P_{ult}} = 1$$

For an in-plane tubular joint, the interaction expression takes the same form as that for combined axial loading and bending, considering ideal plasticity. This form is a parabolic curve. (See Figure 10-20.) It is expressed as

$$\left(\frac{M}{M_{ult}}\right) + \left(\frac{P'}{P_{ult}}\right)^2 = 1$$

where:

$P'$ = total applied load component perpendicular to the chord centerline for symmetrical loading, kips; i.e., $P' = 2 P_T$ (See Figure 10-21.)

$M$ = gross in-plane moment at the joint for antisymmetric loading, kips; i.e., $M = P_T h$ (See Figure 10-21.)

$P_{ult}$ = limiting line load $W$ in kips/inch that will cause flattening of the chord times the applicable chord length ($P_{ult} = WL$ kips).

$M_{ult}$ = ultimate moment of the applicable chord length about its midlength; i.e., $M_{ult} = WL^2/4$, kip inches.

Tubular Joint Design and Fatigue Analysis

$$\frac{M}{M_{ULT.}} + \left(\frac{P'}{P_{ULT.}}\right)^2 = 1$$

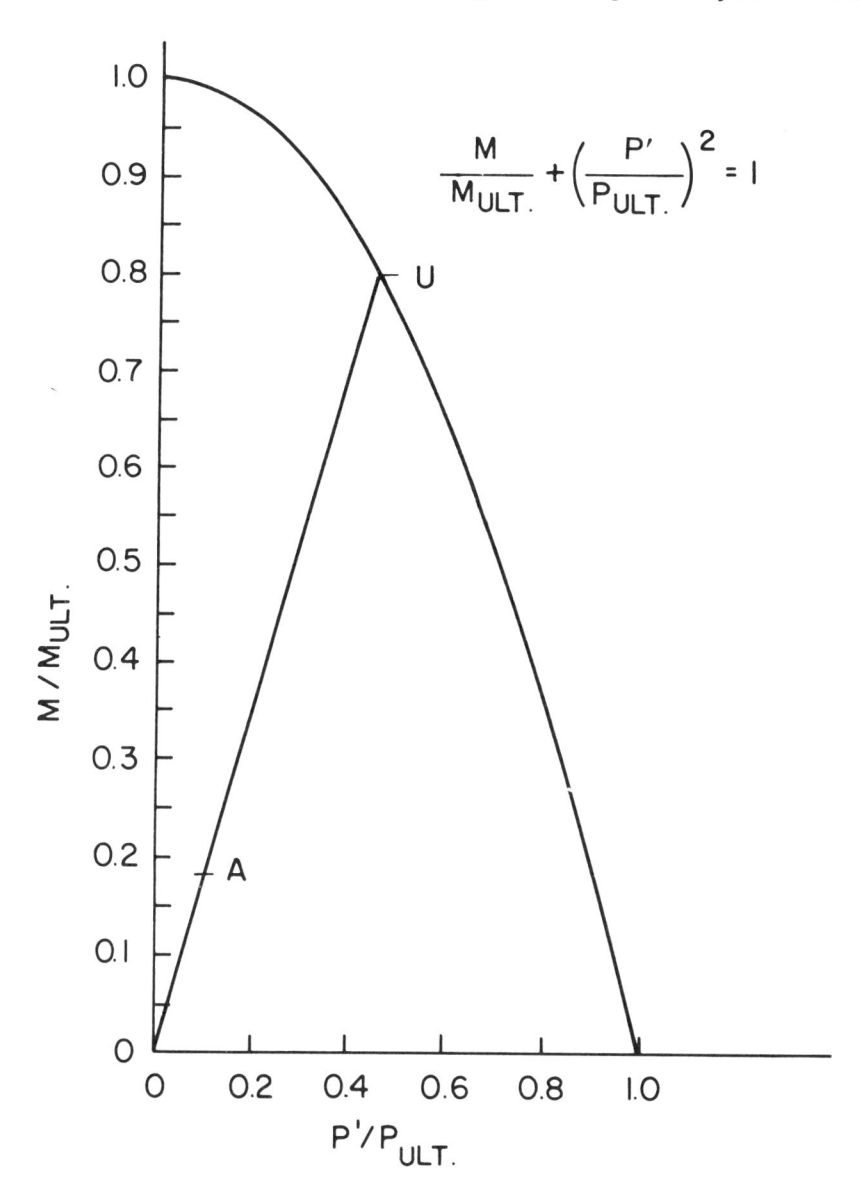

**Figure 10-20.** Interaction diagram.

**SYMMETRICAL**  $P' = 2P_T$

**ANTISYMMETRICAL**  $M = P_T \cdot h$

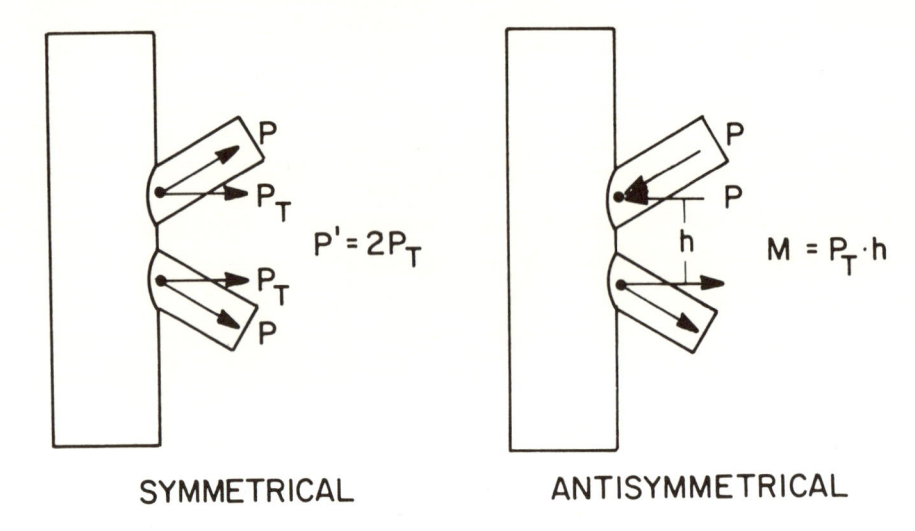

*Figure 10-21. Types of brace axial loading.*

4. The limiting line load mentioned previously comes from the table of formulas for circular rings and arches.[15] For this analysis, the load $P'$ is divided into two parallel line loads in the direction of the chord longitudinal axis. These are considered to be separated circumferentially by a distance equal to the brace diameter. Knowing the yield stress of the material, the full plastic moment of the chord wall can be calculated. From this calculation, using case number 25 formula, the maximum value of $P'$ for the joint can be found.[15] The length $L$ is often taken to be the length of the joint can.

From the interaction diagram, one can estimate the safety factor for general shell stability or collapse graphically by measuring the distance $OU$ and $OA$. (See Figure 10-20.) The safety factor for collapse is taken to be the ratio ($OU/OA$). Point $A$ is located by plotting the two load ratios $M/M_{ult}$ and $P'/P_{ult}$ for the particular joint being analyzed.

The condition of combined shear and bending is characterized by a circular interaction formula.[14] Using this formula as an analog of the tubular joint, the overall safety factor for punching shear and general collapse may be estimated from

$$\left(\frac{1}{\text{F.S. Overall}}\right)^2 = \left(\frac{1}{\text{F.S. Collapse}}\right)^2 + \left(\frac{1}{\text{F.S. Punching Shear}}\right)^2$$

Tubular Joint Design and Fatigue Analysis    181

This method has given reasonable answers when applied to several connections where the loads were known and the connections were judged to have been adequately designed. To develop judgment, the method should be checked against specific connections where known loads have been sustained before it is used to analyze new connections.

### Ring Stiffener Spacing

The influence of ring stiffener spacing on the punching shear capacity of cylindrical shells subjected to line loads is given in Figure 10-22. These curves

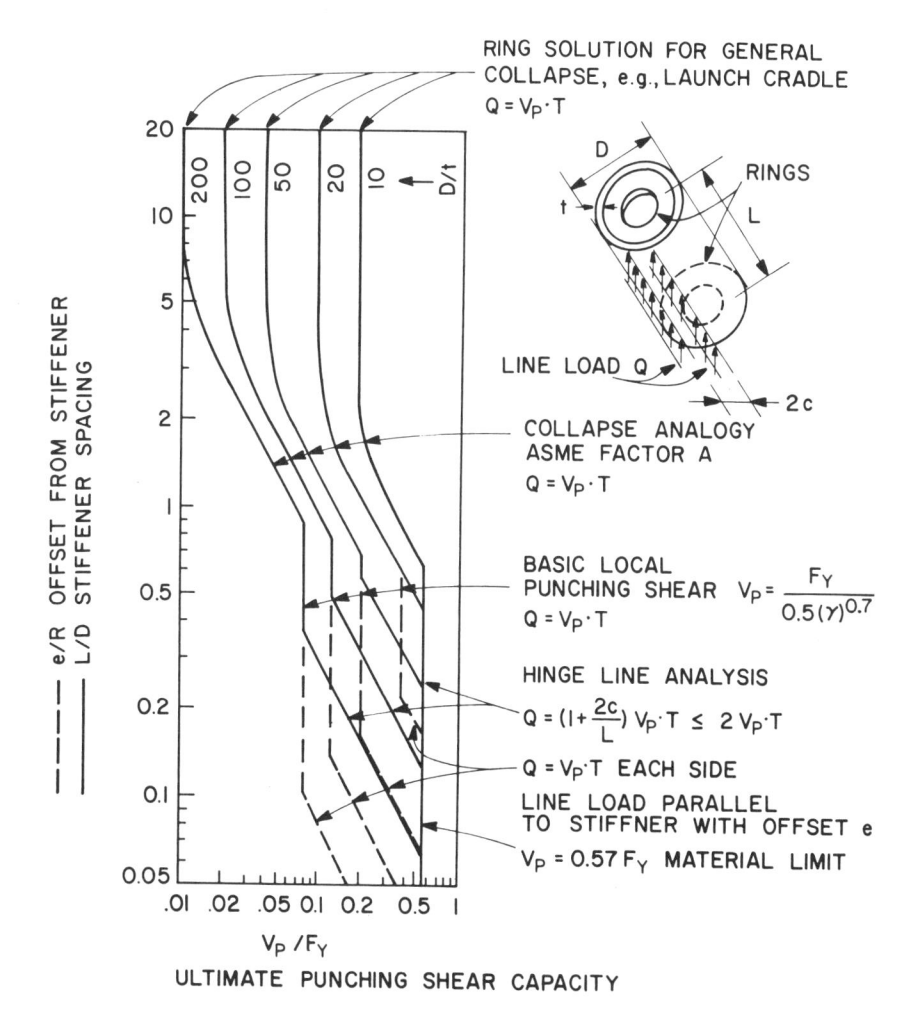

*Figure 10-22. Punching shear capacity for stiffened cylinders. (Reference 12, P.W. Marshall.)*

are a composite of several available solutions and should be considered approximate.

With a chosen ring stiffener spacing, enter Figure 10-22 and determine the ratio of ultimate punching shear stress $v_p$, divided by material yield stress $F_y$. Determine $v_p$ knowing $F_y$. The line load $Q$ in kips/inch of length is given by the appropriate equation in the figure. For very closely spaced rings, a flat-plate yield line analysis gives the following result

$$\text{ultimate } v_p = \frac{4 K_r F_y}{\left(\frac{D}{T}\right)\left(\frac{L}{D}\right)}$$

where $K_r$ is a factor of 1.8 to account for strain hardening and load redistribution. If the loaded line is parallel to the stiffeners rather than crossing the stiffeners, a lower capacity results as indicated by the dashed lines at the bottom of the figure.

## Stiffened Tubes

To improve their resistance to buckling, large-diameter tubes are frequently stiffened with longitudinal and circumferential stiffeners. (See Figure 10-23.) For circumferential stiffening, these stiffeners take the form of circular rings which are continuously welded to the outside or to the inside of the original tube. For longitudinal stiffening, the added strips (stringers) are continuously welded to the outside or to the inside of the original tube. Stringers are positioned perpendicular to the main tube surface with their longitudinal narrow edge against the main tube and their length parallel to the axis of the main tube. Stiffeners may be welded with either fillet welds or full-penetration groove welds.

Unstiffened thin-walled cylinders are prone to sudden and disastrous failures at loads well below the theoretical buckling loads predicted by classical small-deflection theory. The behavior of unstiffened cylinders is especially unpredictable when subjected to axial compression and bending loads. Stiffened cylinders are not characterized by sudden drastic failures as compared to unstiffened cylinders.

### Axial Compression

Initial imperfections caused by fabrication tolerances have a severe effect on the behavior of unstiffened cylinders. Even for stiffened cylinders, the calculated theoretical strength must be reduced by multiplying with an empirically known coefficient to obtain a value which will correlate with experimental results.

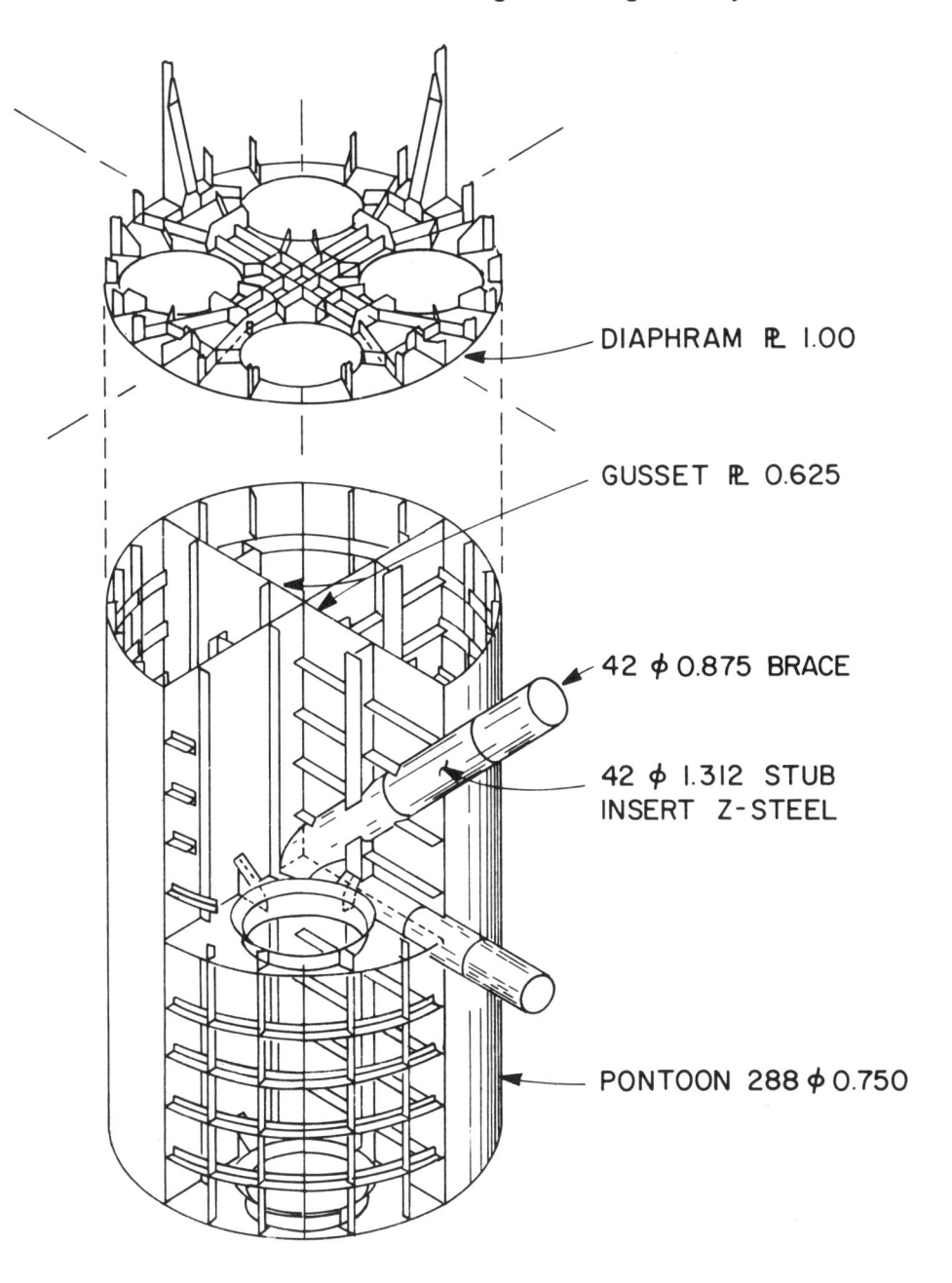

**Figure 10-23.** Cutaway view of Brent A stiffened joint. (Reference 12, P.W. Marshall.)

## 184 Introduction to Offshore Structures

Under axial compression, the possible modes of instability include panel buckling (skin instability), column instability of the effective stringers, torsional instability of the stringers, Euler column buckling, and general instability. Buckling of the skin between any two adjacent stiffeners is called *skin instability;* it is not preceded or accompanied by the failure of the stringers or rings. Buckling of a stringer between two rings is called *column instability.* The effective stringer includes that portion of the skin that acts with the stringer. Torsional instability is rarely a problem, since closed sections are very stiff in torsion. Sometimes the overall cylinder acts as a column; in this case the Euler column buckling criteria must be examined. A general instability failure is one in which the wave form of the buckle is multilobed and has, in general, a wavelength less than the length of the cylinder but greater than the ring spacing. The nature of the general instability failure requires that the rings and stringers fail simultaneously under the critical load.

If the stringers and rings are sufficiently rigid, the shell wall in axial compression can be treated as a series of curved plates supported along four edges. For a simply supported curved panel of width $b$, the critical elastic stress is

$$\sigma_{cr} = k_c \left( \frac{\pi^2 E}{12(1 - \mu_e^2)} \right) \times \left( \frac{t}{b} \right)^2$$

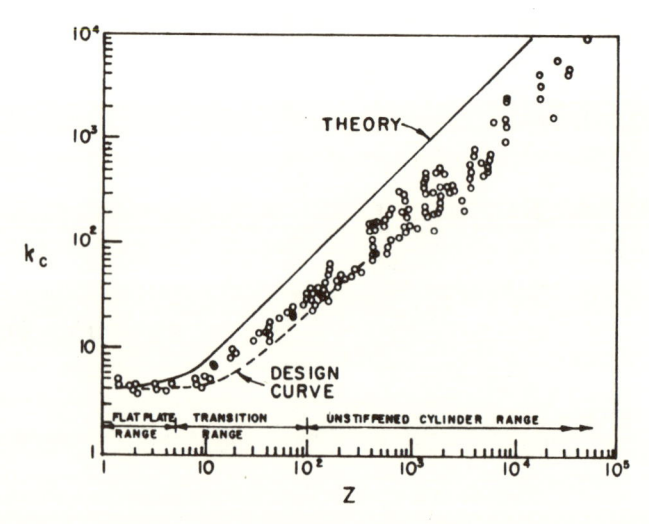

**Figure 10-24.** Comparison of linear buckling theory with test data for circular cylinders under axial compression. (Copyright ©1976 Offshore Technology Conference.)

where:

$E$ = Young's modules
$\mu_e$ = Poisson's ratio in the elastic range
$t$ = the cylinder wall thickness
$b$ = the stringer spacing
$k_c$ = compressive buckling coefficent given by the lower curve in Figure 10-24. This recommended design curve takes into account the effect of geometric imperfections.
$Z$ = curvature parameter.

For axial compression, $Z$ is

$$Z = \frac{b^2}{rt}\sqrt{1-\mu_e^2}$$

where $r$ is the cylinder radius.

The recommended design curve for $k_c$ is conservative over the range of interest for circular cylinders used in offshore platform applications, $50 < r/t < 300$. The critical buckling stress for a cylinder loaded in bending is somewhat larger than that for pure axial compression caused by the stress gradient present in bending. However, the increase is commonly neglected, and the bending buckling stress is assumed to be equal to the axial buckling stress.[16]

### External Hydrostatic Pressure

Under external hydrostatic pressure, the possible modes of instability include panel buckling between rings, torsional ring instability, and general instability.

The buckling of the cylinder skin between rings is called *panel* or *shell instability*. This form of instability occurs when the shell is stiffened by relatively heavy rings; the skin buckles between the rings while the rings remain essentially circular. One form of panel instability not found in offshore platform legs is called *axisymmetric collapse*. This instability occurs when the ring spacing is very small and/or the cylinder wall is very thick. *Asymmetric*, or *lobar buckling*, is the form of panel instability encountered in offshore platform legs. Lobes or waves are formed around the circumference; the minimum number of circumferential buckle waves is two, corresponding to ovalling of the cylinder as it buckles. For closely spaced rings, the number of circumferential waves is usually much greater than two.[17]

Torsional ring instability occurs when the rings twist or buckle out of their plane. Experience indicates that if the rings are proportioned to satisfy the compact section and lateral support requirements of the *AISC Specification for the Design, Fabrication, and Erection of Structural Steel for Buildings*, torsional buckling will not occur.[18]

General instability occurs when the rings and cylinder skin buckle simultaneously at the critical load. This buckle wave form is different from the buckle wave form for general instability under axial compression, in that the longitudinal half-wave length under external pressure is normally equal to the length of the cylinder. General instability is strongly influenced by the slenderness ratio $L/r$ of the cylinder and by the properties of the stiffening rings.

To investigate panel or shell buckling between rings, each cylindrical section between the rings is treated as a short cylinder simply supported between the rings. Neglecting bending of the cylinder wall, the critical hoop stress is

$$\sigma_{cr} = k_p \left( \frac{\pi^2 E}{12(1 - \mu_e^2)} \left( \frac{t}{d} \right)^2 \right)$$

where $k_p$ is the empirical buckling coefficient, and $d$ is the ring spacing. The geometric curvature parameter $Z$ and the parameter $\beta$ are defined as

$$Z = \frac{d^2}{rt} \sqrt{1 - \mu_e^2}$$

$$\beta = \frac{nd}{\pi r} = \frac{\text{ring spacing}}{\text{half-wave buckle length}}$$

where $n$ is the number of circumferential buckle waves. The critical elastic buckling pressure is

$$P_{cr} = \sigma_{cr} \left( \frac{t}{r} \right)$$

The buckling coefficient $k_p$ is shown in terms of the curvature parameter $Z$ in Figure 10-25.

The actual buckling stress will be less than the theoretically calculated buckling stress due to the effects of imperfections and inelasticity. Donnell and Wan have shown that the effect of imperfections on the buckling of thin cylinders under external pressure is not as severe as in the case of axial compression.[19] Accounting for imperfections in manufacture, the curve labeled "recommended design curve" in Figure 10-25 is given as applicable for offshore platform applications.

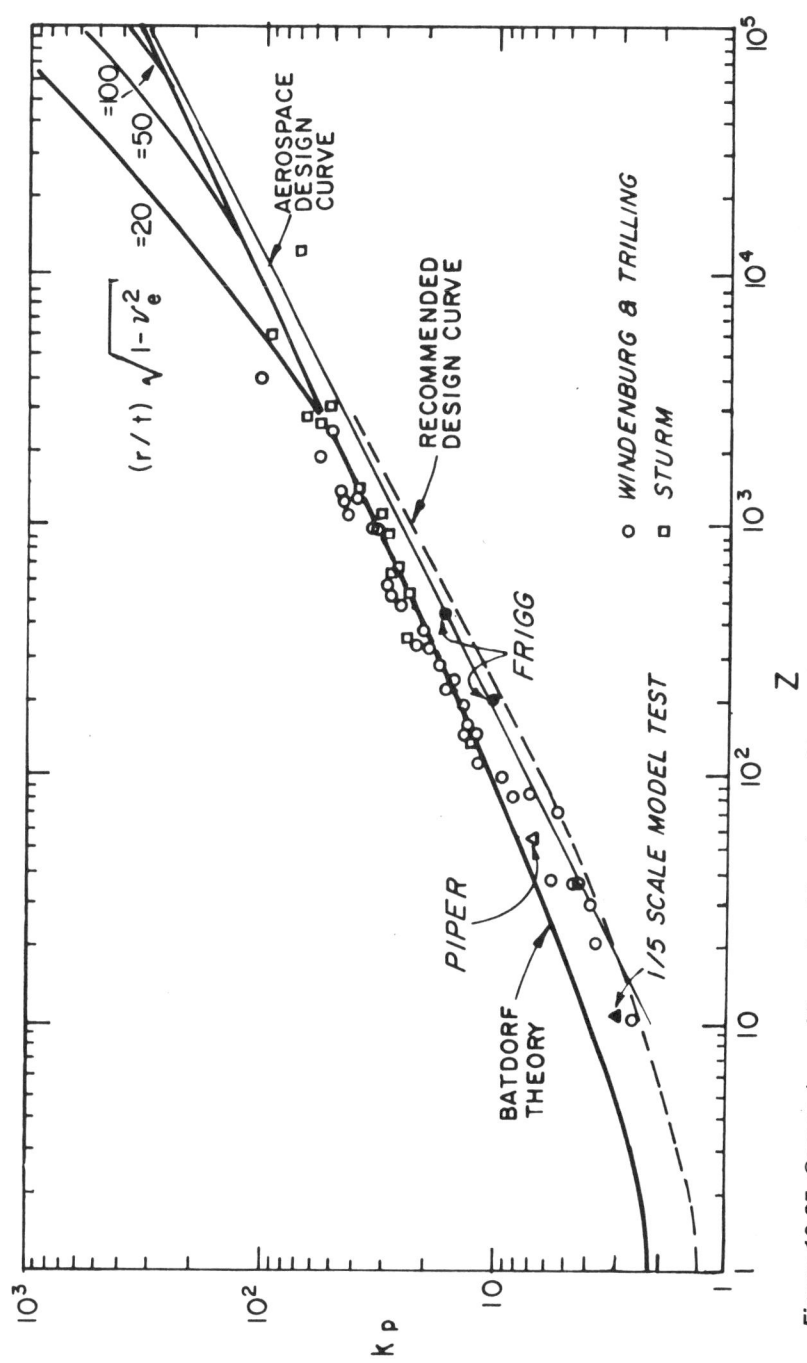

**Figure 10-25.** Comparison of linear buckling theory with test data for shell buckling under external hydrostatic pressure. (Copyright ©1976 Offshore Technology Conference.)

### Buckling From Transverse Shear

Local instability from transverse shear is manifested by the buckling failure of panels in the region of maximum shear stress. A theoretical expression for the critical shear stress $\tau_{cr}$ based on small deflection theory has been developed for curved rectangular panels with simply supported edges. This expression may be applied to stiffened cylindrical shells, assuming that the shell skin is simply supported between the rings and stringers.[20]

The critical shear stress is

$$\tau_{cr} = k_s \left( \frac{\pi^2 E}{12(1-\mu_e^2)} \right) \left( \frac{t}{s} \right)^2$$

where $k_s$ is the buckling stress coefficient, and $s$ is the axial or circumferential dimension of the panel, whichever is smaller. Design curves for $k_s$ are given in Figure 10-26.

## FATIGUE OF TUBULAR JOINTS

### Fatigue Behavior

Experience over the last 60 years and many laboratory tests have proven that a metal may fracture at a relatively low stress if that stress is applied a great number of times. It is known that sometimes a crack will form and grow under the repeated action of applied stresses that are lower than those required for yielding the same material under unidirectional static loading. Such fractures are referred to as fatigue failures. The initially small crack formed at the point of high localized stress grows or spreads until the remaining solid cross section of the load-carrying member is not sufficient to transmit the load and the member fractures.

Before World War II, fatigue investigations dealt primarily with engine or airplane development. Consequently, the maximum applied tensile stress and the mean stress were thought to be the most important parameters of the fatigue problem. Consider an engine with several main bearings that has one bearing slightly misaligned (at a lower elevation). As the crankshaft rotates, a point on its surface is alternately stressed to the same level of tension and compression. The ratio of minimum (compressive) stress divided by maximum (tensile) stress is called the *stress ratio R*. In this case $R$ is -1. The mean stress is zero. For an airplane in level flight, the mean stress on the wing comes from supporting the weight of the craft ("1-g" loading); effects of wind gusts and airplane maneuvers cause stress fluctuations (random). Thus, the mean stress is not zero. If the bending tensile stress fluctuated only (or regularly) between zero stress and some constant maximum stress, the stress ratio would be zero. The term "stress ratio," however, is not used with random loading.

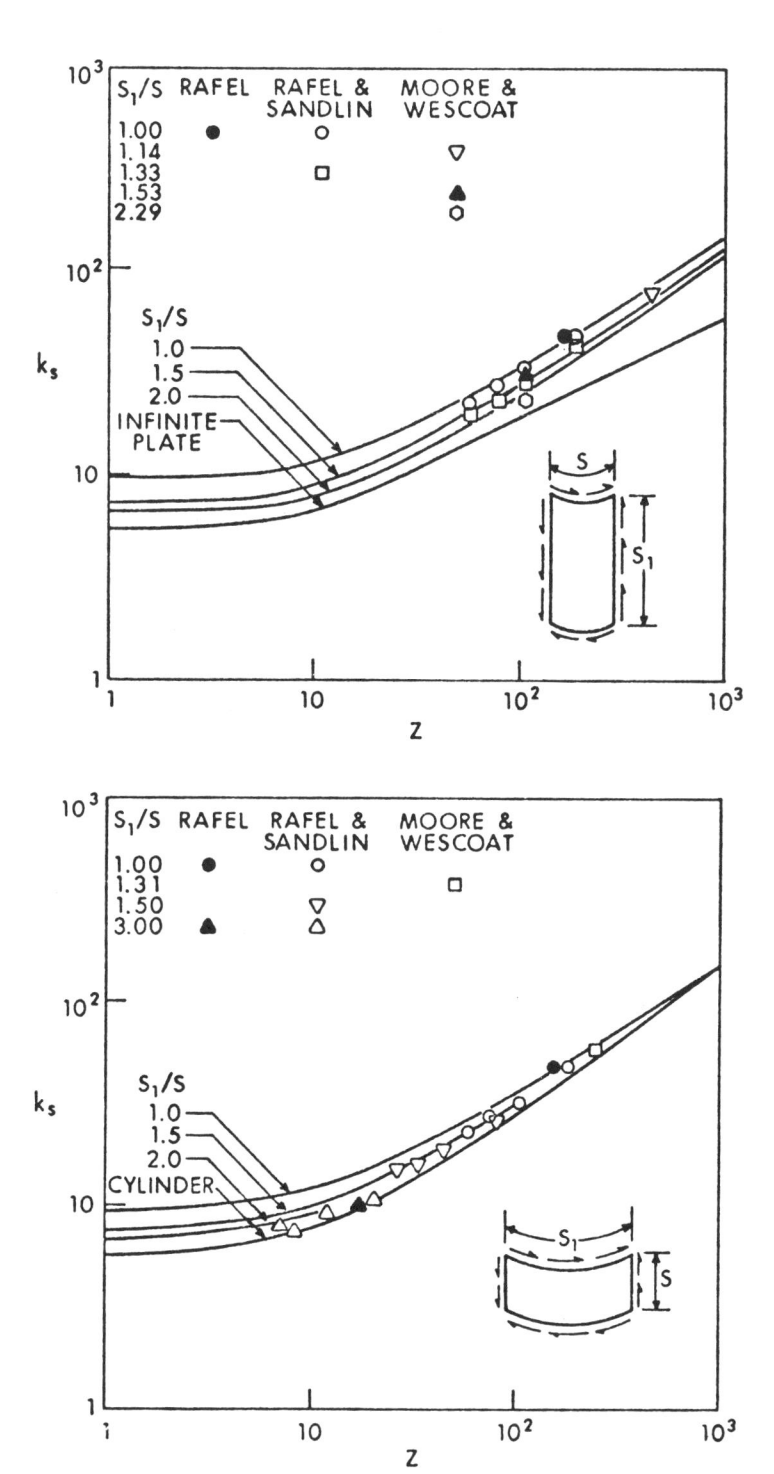

**Figure 10-26.** Theoretical curves and test data for shear buckling coefficient of simply supported plates. (Reference 20.)

Until recent years, the situations involving fatigue behavior were cases where the loading was repeated at fairly high rates (frequencies), e.g., thousands of cycles per minute for an engine crankshaft. As the matter of random loading on offshore platforms by ocean waves came into consideration, two important differences were noted from the usual fatigue situation. First, the rates of loading were vastly slower. Ocean waves pass the offshore structure at the rate of about 10 cycles per minute or 600 per hour. Second, offshore structures are immersed in a corrosive medium. Fatigue cracks grow because of tensile stresses; corrosion of a metal is accelerated if the metal is subjected to tensile stress. Thus, the effects of corrosion and fatigue are combined in the case of an offshore platform.

## S-N Curves

Metals that are fatigue-tested in air exhibit randomness in the number of stress cycles required to cause fracture at a given stress ratio. Thus, many identical specimens (5-10) must be tested at the same stress ratio to find the normal distribution of the number of cycles to failure at that stress ratio. After this procedure has been done for many stress ratios, a curve can be drawn plotting the maximum applied stress against the mean number of cycles to failure; this figure is called an *S-N curve*. These S-N curves can be found in the technical literature for many metals.

Because the number of cycles to failure in most cases is in thousands or hundreds of thousands of cycles (often millions of cycles), it is convenient to plot the N axis of the S-N curve on a logarithmic scale. The ordinate, or stress axis, may be plotted either as a linear or a logarithmic scale. When stress is plotted on a linear scale, as it customarily was until recent years, the S-N curve exhibits a flattening out such that at sufficiently high numbers of cycles the slope of the curve is zero. The stress level at which the curve becomes flat (the number of cycles to failure becomes infinite) is called the *fatigue limit*. Presumably, a metal part stressed to no higher stress than the fatigue limit will have an infinite lifetime under cyclic loading.

Again, with offshore structures the behavior seems to be different. Because of the rates of loading, the randomness of waves, and the corrosiveness of sea water, metals in the ocean subjected to repeated loading do not appear to have fatigue limits. The downward slope of S-N data seems to continue for as long as the dynamic loading is applied.

A great deal of welding is required to fabricate a tubular joint. The residual stresses left in the joint as a result of welding often reach or exceed the yield stress, and local yielding of the material may occur to redistribute the stresses when the joint is put into service. In studying fatigue data for complex welded specimens it has been found that plotting stress range rather than maximum stress better describes the metal behavior. Thus, for tubular joints the term "S-N curve" means a plot of stress range versus cycles to failure. The stress range

## Tubular Joint Design and Fatigue Analysis

is the difference between the maximum tensile stress and the minimum (quite possibly compressive) stress that may occur in a particular joint due to a variety of applied loads.

**Stress Concentration and Fatigue**

The subject of stress concentration was discussed in an earlier section. It is mentioned here again for a particular reason. Ordinarily, neither stress concentration nor fatigue loading alone constitutes the controlling influence in design, but in combination they are definitely major factors worth consideration. Stress concentration describes the condition in which high localized stresses are produced as a result of the geometry of the structural element. Fatigue failures occur at nominal stress levels lower than the yield stress of a material in areas of high local stress and propagate perpendicular to the direction of maximum applied tensile stress into areas of low local stress. These areas of high local stress are present because of the stress concentrations. The two effects reinforce each other.

The location in a tubular joint where the maximum applied tensile stress occurs is called the *hot spot*. To do a fatigue analysis of certain selected tubular joints in an offshore structure, the stress history of the hot spots in those joints must first be determined. Three basic stress types contribute to the development of hot spots:

1. Primary (type A) stresses are caused by axial forces and moments resulting from the combined truss and frame action of the jacket. In Figure 10-10 stress at hot-spot locations 1, 3, 4, and 6 are most affected by axial forces and in-plane bending moments in the braces. The regions around hot spots 2 and 5 are most affected by axial forces and circumferential moments in the braces.

2. Secondary (type B) stresses are caused by the structural details of the connection such as poor joint geometry, poor fit-up, local variations in stiffness within the joint due to too rigid reinforcement, restraint of braces caused by circumferential welds, etc. These stresses tend to amplify the primary (type A) stresses.

3. Secondary (type C) stresses are caused by metallurgical factors that result from faulty welding practice, i.e., insufficient weld penetration, undercutting, heavy beading, weld porosity, varying cooling rates, etc. Type C stresses tend to predominate at hot-spot location 6, although their effect is significant at locations 1, 3, and 4 also. Because metallurgical factors have such a pronounced effect on hot-spot stresses, elaborate qualification programs for welders have been instituted to ensure consistent high-quality welding.

## Tubular Joint Fatigue Behavior Literature

The literature concerning fatigue in general is vast. References 21 through 34 constitute a list of selected articles directly related to the fatigue behavior of tubular joints. However, even for such a specific topic as tubular joints another equally long, or longer, list of selected references can be compiled.

The list of articles referenced here is representative, since there are several ways of making a fatigue analysis. Fatigue is a random process and, as such, is not well understood. The phenomenon must be studied from a semi-empirical standpoint utilizing rules deduced from experimental data. Among the several ways are a deterministic approach and a probabilistic approach. One approach requires calculating wave power spectral densities. Still another approach uses the principles of linear elastic fracture mechanics. As in the discussion of stress concentration factors, the subject of fatigue is too detailed to treat here. References 2, 21, 22, and 28 are suggested for further reading.

## AWS Fatigue Curves

Once punching shear failure of the chord is prevented by using a heavy-wall can or other suitable reinforcement, the emphasis in joint design shifts toward determining the fatigue life of the connection.[1,2,21] To analyze for fatigue life, one must have a family of S-N curves based on experimental data to represent the way the particular structural element can be expected to behave.

S-N curves for tubular connections were first published in 1972 by the American Welding Society (AWS). The rationale for the AWS S-N curves was published initially in Reference 21. A summary of the AWS S-N criteria is presented in Figure 10-27 and Table 10-2. Curves *A* and *C* are consistent with

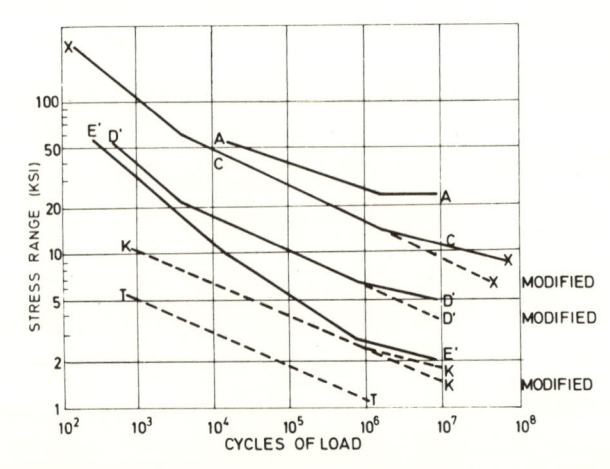

**Figure 10-27.** *American Welding Society S-N curves. (Reference 31.)*

### Table 10-2
### AWS Fatigue Categories

| Stress category | Situation | Kinds of stress[1] |
|---|---|---|
| A | Plain unwelded tube | TCBR |
| A | Butt splices, no change in section, full-penetration groove welds, ground flush, and inspected by X-ray or ultrasonics. | TCBR |
| C | Butt splices, full-penetration groove welds as welded | TCBR |
| D'[2] | Simple T, Y, or K connections with full penetration tubular groove welds | TCBR in branch member (main member must be checked separately per Category K or T) |
| E'[2] | Simple T, Y, and K type tubular connections with partial penetration groove welds or fillet welds; also complex tubular connections in which load transfer is accomplished by overlap (negative eccentricity), gusset plates, ring stiffeners, etc. | TCBR in branch member (main member in simple T, Y, or K connections must be checked separately per Category K or T; must also be checked for shear in weld regardless of direction of loading. |
| X | Main member at simple T, Y, and K connection | Hot spot, stress or strain on the outside surface of the main member at the toe of weld joining branch member—measured in model of prototype connection, or calculated with best available theory |
| X | Unreinforced cone-cylinder intersection | Hot-spot stress at angle change |
| X | Connections whose adequacy is determined by testing an accurately scaled steel model | Worst measured hot spot strain, after shake down |
| K[3] | Simple K type tubular connections in which gamma ratio R/T of main member does not exceed 24 | Punching shear on shear area of main member |
| T[3] | Simple T and Y tubular connections in which gamma ratio R/T of main member does not exceed 24 | Punching shear on shear area of main member |

(1) T = tension, C = compression, B = bending, R = reversal.
(2, 3) Empirical curves based on 'typical' connection geometries; if actual stress concentration factors or hot spot strains are known, use of curve X is to be preferred.
(Source: Reference 31.)

## 194  Introduction to Offshore Structures

the fatigue criteria of The American Institute of Steel Construction and are based on fatigue research conducted during the 1960s.[18] For the simple welds covered by curves $A$ and $C$, the sum of the calculated nominal axial and bending stresses in the member satisfactorily represents the actual stress.

In Figure 10-27 curve $X$ represents the basis for current design practices for offshore structures. The $X$ curve falls on the safe side of 97% of the available test data that refers to complete fracture of the tubular member. The hot-spot stress measured perpendicular to and adjacent to the weld is the relevant stress to use in studying fatigue behavior of tubular connections. As mentioned in the discussion on stress concentration, this stress is many times higher than the nominal member stress and is obtainable only from a detailed experimental or theoretical analysis of the connection. The number of loading cycles of a tubular connection normally breaks into three regions. The low-cycle, high-stress region extends from a very low number of cycles up to about $10^4$ cycles. In this range it is more realistic to analyze in terms of hot-spot strain rather than stress because of possible inelastic stress situations. The intermediate region on the number of cycles axis ranges from $10^4$ to $2 \times 10^6$ cycles. Most of the available test data are in this range. Limited experimental data are available in the low-stress, high-cycle region identified as over $2 \times 10^6$ cycles. Platforms designed to last 20 years are often stipulated to receive $10^8$ cycles of wave loading in that time. A major portion of the load cycles are in the low-stress, high-cycle range. As indicated in the section on S-N curves, because of notches and imperfections inherent in tubular joints and the presence of a corrosive environment offshore, there is no fatigue limit stress; the S-N curve continues to drop off as curve "$X$-modified" shows in Figure 10-27.

Use of the $X$ curve requires knowledge of stress concentration factors and hot-spot stresses for the particular tubular joint. This information is often expensive and difficult to obtain. Thus, other curves are provided in Figure 10-27 for a simplified approach to estimating the fatigue worthiness of a tubular connection. Curves $K$ and $T$ are empirical design curves for use with calculated cyclic punching shear stress. Curves $D'$ and $E'$ are empirical curves for use with calculated nominal stresses in brace members.

### Palmgren-Miner Cumulative Damage Rule

The fatigue life of a welded connection depends on many factors: number and occurrence of waves causing local yielding, metallurgical behavior of the steel, surface appearance of weld (flaws and initial microcracks), geometrical shape of the weld (stress concentration factor), etc. Over the span of its design life the offshore structure is subject to a spectrum of cyclic waves which produce a very large number of load cycles. Thus, cumulative fatigue damage must be considered.

## Tubular Joint Design and Fatigue Analysis

It is the low-stress, high-cycle load situation that is the significant contributor to cumulative fatigue damage. The measure of cumulative damage that is used is the Palmgren-Miner rule

$$D = \sum_{i=1}^{K} \frac{\eta_i}{N_i} \leq 1$$

where:

$\eta_i$ = number of cycles within stress range interval $i$ of the long-term stress range distribution
$N_i$ = number of cycles to failure at the same stress range, derived from the S-N curve
$K$ = total number of stress range intervals
$D$ = cumulative damage ratio. The expression for $D$ is frequently required to be less than unity, perhaps 0.5 or 0.33, to further ensure that the calculated life will be twice or three times the planned life.

The Palmgren-Miner cumulative damage rule assumes that failure will occur when the damage ratio reaches unity, or whatever fraction represents conservatism. The wave spectrum is a cumulative history of frequencies and variable amplitudes. The Palmgren-Miner rule is criticized because it does not account for the order in which the various parts of the wave spectrum occur. Still, the rule is used in all of the approaches to studying fatigue in tubular connections. As Maddox and Wildenstein state, the rule ". . . appears to be adequate in view of the uncertainties involved in the environment, in the loading of the structure, and in the joints themselves."[26]

### Simplified Procedure

A simplified fatigue analysis can calculate punching shear stress, use empirical design curves $T$ and $K$ to judge cyclic punching shear, and employ $D'$ and $E'$ for nominal stresses in a brace member.

The main task in a fatigue calculation is to determine the long-term stress distribution. Each tubular connection and point along each intersection weld has its own stress distribution. In large structures with hundreds of tubular joints it is imperative to have a method of fatigue calculation that will give adequate answers with reasonable expenditures of manpower and computer time. The steps in a simplified, approximate, and yet conservative fatigue analysis are:

1. Determine the operational lifetime in cycles for the structure. Usually, the lifetime in years is a known requirement from the function of the

**196     Introduction to Offshore Structures**

structure. For example, if waves strike the structure at the rate of 10 per minute, and if the structure is to function for 20 years, the lifetime in cycles is $1 \times 10^8$.

2. Determine the long-term wave height distribution. This distribution is derived from the oceanographic data for the site. Normally, the following relationship can be used

$$H = H_L \left(1 - \frac{\log n}{\log n_L}\right)$$

where:

$H$ = wave height
$H_L$ = largest wave height expected in 20 years
$n$ = number of times $H$ is exceeded in 20 years
$n_L$ = total number of waves in 20 years ($1 \times 10^8$).

For both the North Sea and the Gulf of Mexico, the long-term wave distributions for a 20-year lifetime are represented by Figure 10-28. $H_L$ for the North Sea is 95 ft; for the Gulf of Mexico, 60 ft.

3. Select the highly stressed tubular joint for which the fatigue calculation is to be made. Assume all the waves approaching this joint have the same direction.
4. Take a selected sea state (the normal daily operating wave height), and pass that wave through the structure in several selected directions to determine the largest maximum stress and the lowest minimum stress in the joint being considered. Let this difference between maximum and minimum stress, neglecting directional differences, be the stress range.
5. The stress range is then increased by multiplying it by the dynamic amplification factor, especially for deepwater structures.
6. As the cumulative number of waves of normal daily operating wave height is known, plot this point on a set of log-log coordinates of stress range versus cumulative cycles.
7. Assume the long-term stress range distribution curve diminishes in the same way as the long-term wave height distribution curve.
8. Prepare graphs of S-N curves $D'$ or $E'$ for the brace and $T$ or $K$ for the punching shear stress in the chord. Read from the long-term stress range distribution curve stress range values at different cycle decades, i.e., $10^4$, $10^5$, etc., and plot these on the graph which contains the appropriate S-N curve. If the curve thus formed for the joint lies below the appropriate $D'$, $E'$, $T$, or $K$ curve, the fatigue behavior of the joint is satisfactory.

9. The cumulative damage ratio $D$ can be calculated by dividing the stress range distribution into a large number of blocks, assuming a constant stress range within each block and applying the Palmgren-Miner rule.

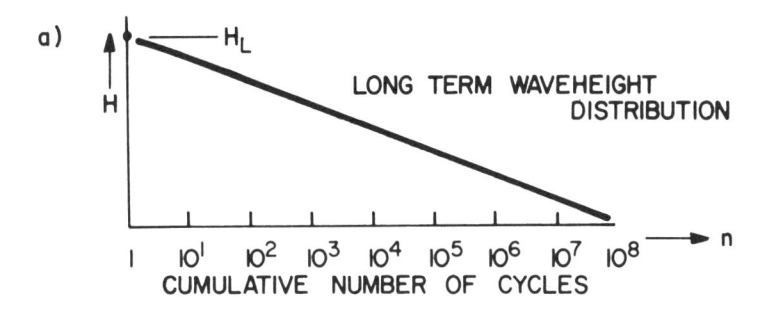

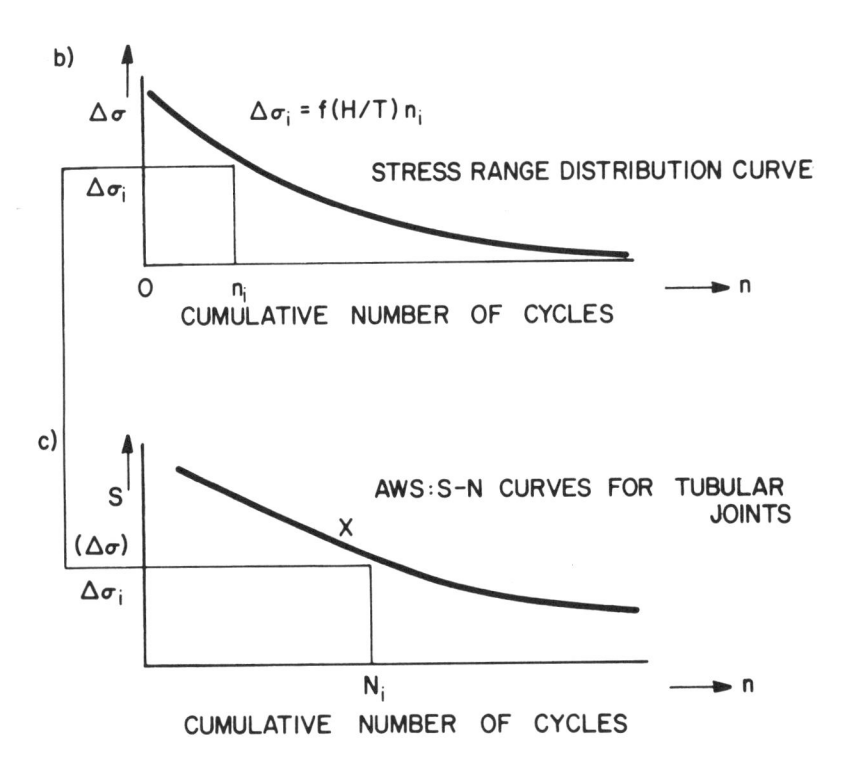

*Figure 10-28. Long-term distribution of wave heights. (Reference 30.)*

## References

1. *Recommended Practice for Planning, Designing, and Constructing Fixed Offshore Platforms,* American Petroleum Institute, Dallas, Texas, API-RP-2A (revised annually).
2. *Structural Welding Code,* American Welding Society, Miami, Florida, AWS-D1.1 (first edition in 1972 with revisions periodically since).
3. Graff, W.J., "Design Correlation of Elastic Behavior and Static Strength of Zero Eccentricity T, Y, and K Tubular Joints," 1970 Offshore Technology Conference Preprints, Paper 1310.
4. Noel, J.S., Beale, L.A., and Toprac, A.A., "An Investigation of Stresses in Welded T Joints," Department of Civil Engineering, University of Texas at Austin, Report S.F.R.L. P-550-3, March 1965.
5. Beale, L.A., and Toprac, A.A., "Analysis of Inplane T, Y and K Welded Tubular Connections," Department of Civil Engineering, University of Texas at Austin, Report S.F.R.L. P-550-9, April 1967.
6. Kuang, J.G., Potvin, A.B., and Leick, R.D., "Stress Concentration in Tubular Joints," 1975 Offshore Technology Conference Proceedings, Paper 2205. (See also: "Stress Concentration in Tubular Joints," *Society of Petroleum Engineers Journal,* August 1977.)
7. Dundrova, V., "Stresses at Intersection of Tubes: Cross and T Joints," Department of Civil Engineering, Structures Fatigue Research Laboratory, University of Texas at Austin, S.F.R.L. Technical Report P-550-5, July 1965.
8. Holliday, G.H., and Graff, W.J., "Three-Dimensional Photoelastic Analysis of Welded T-Connections," 1971 Offshore Technology Conference Preprints, Paper 1441.
9. Kwan, C.T., and Graff, W.J., "Analysis of Tubular T-Connections by the Finite Element Method: Comparison with Experiments," 1972 Offshore Technology Conference Preprints, Paper 1669.
10. Hans, D., Visser, W., and Zunderdorp, H.J., "The Stress Analysis of Tubular Joints for Offshore Structures," *Proceedings of Second Annual Meeting of the Society of Petroleum Engineers of AIME,* London, April 1973, Paper SPE-4342.
11. Visser, W., "On The Structural Design of Tubular Joints," 1974 Offshore Technology Conference Preprints, Paper 2117.
12. Marshall, P.W., and Graff, W.J., "Limit State Design For Tubular Connections," *Proceedings of the First International Conference on the Behavior of Offshore Structures* (BOSS-76), Trondheim, Norway, August 1976.
13. Marshall, P.W., "A Review of Stress Concentration Factors in Tubular Connections," Head Office Production, Civil Engineering Department, Shell Oil Company, Houston, Texas, Report CE-32, April 1978.
14. Shanley, F.R., *Strength of Materials,* (McGraw-Hill Book Co., New York, 1957) pp. 633-643.
15. Roark, R.J., *Formulas for Stress and Strain,* 4th edition, (McGraw-Hill Book Co., New York, 1965) pp. 172-180.

## Tubular Joint Design and Fatigue Analysis 199

16. Kinra, R.K., "Structural Considerations in Large Diameter Tubes," Shell Oil Company Report CE-6, September 1974, p. 5.
17. Kinra, R.K., "Hydrostatic and Axial Collapse Tests of Stiffened Cylinders," *1976 Offshore Technology Conference Proceedings,* Paper 2685.
18. *Manual of Steel Construction, 7th edition,* American Institute of Steel Construction, New York, N.Y., 1970.
19. Donnell, L., and Wan, C., "Effect of Imperfections on Buckling of Thin Cylinders Under External Pressure," *Journal of Applied Mechanics, ASME,* December 1956, pp. 569.
20. Kinra, R.K., "Design of Stiffened Cylinders," Shell Development Company, Technical Progress Report, BRC-EP 22-73-F, November 1973.
21. Marshall, P.W., "Basic Considerations for Tubular Joint Design in Offshore Construction," *Welding Research Bulletin,* No. 193, April 1974. (*See* also: Marshall, P.W., and Toprac, A.A., "Basis for Tubular Joint Design," *Welding Research Supplement,* Vol. 39, no. 5, May 1974, pp. 192-S through 201-S.)
22. Becker, J.M., Gerberich, W.W., and Bouwkamp, J.G., "Fatigue Failure of Welded Tubular Joints," 1970 Offshore Technology Conference Preprints, Vol., II, Paper 1228. (*See* also: "Fatigue Failure of Welded Tubular Joints," *Journal of the Structural Division, ASCE,* Vol. 98. No. ST-1, January 1971.)
23. Maddox, N.R., "A Deterministic Fatigue Analysis for Offshore Platforms," *Journal of Petroleum Technology,* July 1975, pp. 901-912. (*See* also: Maddox, N.R., "Fatigue Analysis for Deepwater Fixed-Bottom Platforms," 1974 Offshore Technology Conference Preprints, Paper 2051.)
24. Mukhopadhyay, A., Itoh, Y., and Bouwkamp, J.G., "Fatigue Behavior of Tubular Joints in Offshore Structures," *1975 Offshore Technology Conference Proceedings, Vol. I,* Paper 2207.
25. Maison, J.R., and Holliday, G.C., "Comparison Between Predicted and Experimentally Determined Low Cycle Fatigue Life of Welded Tubular Connections," *1975 Offshore Technology Conference Proceedings, Vol. I,* Paper 2208.
26. Maddox, N.R., and Wildenstein, A.W., "A Spectral Fatigue Analysis for Offshore Structures," *1975 Offshore Technology Conference Proceedings,* Paper 2261.
27. Vughts, J.H., and Kinra, R.K., "Probabilistic Fatigue Analysis of Fixed Offshore Structures," *1976 Offshore Technology Conference Proceedings,* Paper 2608.
28. Pan, R.B., Maddox, N.R., and Plummer, F.B., *Fatigue Analysis of Offshore Structures, Offshore Drilling and Producing Technology,* (Petroleum Engineer Publishing Co., Dallas, Tex. 1976) pp. 55-61.
29. Pan, R.B., and Plummer, F.B., "A Fracture Mechanics Approach to Non-Overlapping Tubular K Joint Fatigue Life Prediction," *1976 Offshore Technology Conference Proceedings, Vol. III,* Paper 2645.
30. Roren, E.M.Q., and Furnes, O., "Behaviour of Structures and Structural Design: Concepts, Principles and Fatigue," *First International Conference on the Behaviour of Offshore Structures (BOSS-76) Proceedings, Vol. 1,* August 1976, Trondheim, Norway, pp. 70-111.

31. Walter, J.C., Olbjorn, E., Alfstad, O., and Eide, G., "Safety Against Corrosion Fatigue Offshore," *Det Norske Veritas Publication No. 94,* April 1976.
32. Marshall, P.W., "Preliminary Dynamic and Fatigue Analysis Using Directional Spectra," *Journal of Petroleum Technology,* June 1977, pp. 715-721.
33. Wylde, J.G., and McDonald, A., "The Influence of Joint Dimensions on the Fatigue Strength of Welded Tubular Joints," *Second International Conference on the Behaviour of Offshore Structures (BOSS '79) Proceedings,* August 1979, London, England, Paper 42.
34. Minas, A.N., and Graff, W.J., "A Survey of Investigations of Tubular Joints," Annual ASCE Convention and Exposition, Portland, Oregon, April 14-18, 1980, Preprint 80-034.

# Chapter 11

# *Fabrication and Installation*

An offshore platform is a specialized structure uniquely developed to serve a special purpose. The concept and design of an offshore platform are based almost entirely on the method of installation of the structure at the offshore site. Architectural or aesthetic matters are scarcely considered.

The entire platform is designed to withstand the environmental and operational loads described earlier. Wind and wave loads act as lateral or horizontal loads on the topside facilities. Waves act as lateral loads on the jacket. The weights of the topside facilities plus the loads resulting from the operations to be performed (drilling/production) act as vertical loads transferred from the various deck modules through the deck substructure and jacket into the pile foundation. Sufficient vertical and lateral soil resistance must be developed in the pile foundation to equal these applied loads.

In addition, all platform components and modules must be designed to withstand the forces imposed on them during fabrication, load-out, transportation to the offshore location, and launching or lifting from the barge into final position. This chapter will describe the processes of fabrication and installation so that the nature of these additional forces may be better understood.

### Jacket Fabrication

Fabrication begins with the ordering of long lead-time items like steel plate, tubes, and wide-flange sections from the steel mill or supplier. Steel plates for rolling or forming heavy-walled tubulars are ordered in specific dimensions to produce the required tube sizes without waste. Tubular goods (of the proper steel specification) in diameters less than about one meter are usually ordered in standard lengths because the contouring of the brace ends requires different

## 202  Introduction to Offshore Structures

increments of length dependent on location of the brace in the jacket framework. If possible, all wide-flange sections for deck beams, all deck plates, and grating are usually ordered cut-to-size.

While waiting for the delivery of the steel, the jacket lofting is completed. Lofting is the process of drawing patterns of exactly how each end of each brace is to be contoured to fit within the structure. The shape of the cut, the extent of beveling (for welding), and the correct length of each brace must be determined graphically or by computation. If the jacket is to be fabricated by first constructing all the welded joint nodes (a node is a short section of jacket leg, called a *can*, to which stubs of braces are welded), much of the lofting work goes into the making of the node patterns. Even so, the correct brace lengths must be determined between opposite ends of brace stubs.

Prior to beginning fabrication of the jacket two parallel skid runners must first be constructed perpendicular to the quayside and extending far enough away from the quay edge to accommodate the height of the jacket. The quay is a 300-400 ft (90-120 m) long sheet-pile bulkhead along the water edge. The skid runners must be placed on very firm foundations capable of supporting the final jacket weight without imparting detrimental differential deflections on the surface as it slides on its way to the launch barge during load-out.

The jacket legs are among the first items of the structure to be fabricated. Careful attention must be given to roundness and straightness. Usually, crawler-mounted tractors with side booms transport the jacket leg sections to the place of final assembly. The sections are joined on leveled ground or on a level timber rack so that the legs can be rolled to facilitate welding and enable the internal shims (for proper alignment of the piles later on) to be installed (welded) in the horizontal position.

The jacket legs are next placed on leveled pedestals or blocks in proper position for installing the braces. As many of the jacket components as possible are fabricated lying horizontally on the ground. Each major planar truss or bent across the narrow dimension of the jacket is fabricated flat on the ground. The first two built are the two that have the parallel launch trusses (Chapter 8) attached to the sides of the jacket legs. After completion, these two bents are rotated upright and placed on the skid runners. This position causes the jacket, when completed, to lie on its side with the legs that contain the parallel launch trusses mated with the skid runners embedded in the ground of the fabrication yard. The skid runners are faced with heavy timbers before the first two bents are put in position. The first bent is held in position with guy wires; the second is likewise guyed before installation of the braces between the trusses begins. The braces between the bents are held in place with cranes, and the fitters and welders work either from scaffolding or out of baskets supported by cranes. If the full length of the bent is too heavy to be rotated into the vertical position all at once, it is constructed in sections, and the sections are welded end-to-end after rotation into the vertical plane.

## Fabrication and Installation 203

Fabrication of the jacket braces is started as soon as the tubulars and necessary patterns are available. This work is usually done on a special rack. The pipe is placed on the rack, the specified distance between the end patterns is measured off, the patterns are wrapped around the pipe in the proper orientation, and lines are drawn representing the end cuts. The line marked on the pipe represents the intersection of the inside surface of the brace and the outer surface of the chord it frames against. The cut is made with a manually controlled or programmable cutting torch. Care must be taken to provide the appropriate amount of bevel to the edge for welding; this amount varies continuously around the cut. As the fabrication of the braces is completed, they are marked and moved to a storage area adjacent to the jacket fabrication site.

As the other bents are finished on the ground, they are rotated upright, located in position, and temporarily supported while connecting braces are welded between them and the monolithic structure on the skid runners. The braces are welded to the jacket legs and other chord members from the outside only using full-penetration groove welds.

While the jacket is being fabricated, many smaller ancillary structures are also being fabricated such as conductor tube guides and guide framing, pile guides, boat landings, barge bumpers, walkways, buoyancy tanks, piping components for the decks, pipeline riser guides, handrails, lifting eyes, stiffener rings for the jacket legs, etc. As many as possible of these items should be incorporated into the bents before they are rotated into the vertical position. Some of these items weigh several tons, and proper positioning and welding of them high in the air, often under adverse weather conditions, can be very difficult.

With jacket fabrication complete, auxiliary buoyancy tanks are attached, or inserted in the pile guides if there are to be skirt piles. These are securely welded to the jacket at a point as high as possible on the framework (the highest elevation when the jacket is in its normal vertical position). This position permits easy cutting loose at removal time after launching and upending of the jacket.

The jacket is constructed on the skid runners with its bottom, or large end, facing the quay edge. The jacket is loaded onto the barge so that its bottom is at the opposite end of the barge from the rocker arms.

Figures 11-1 through 11-9 show the sequence of fabrication of a deepwater jacket. The two concrete skid runners are shown in Figure 11-1 with half of the first interior bent already rotated into the vertical plane and held in position by cranes. Figure 11-2 shows the second half of the second interior bent being rotated into the vertical plane. The extra trusswork needed for launching is visible. This bent is shown in final position in Figure 11-3.

Two bottle sections have been welded into position in Figure 11-4 at the large end of the jacket thus marking where the next two bents are to be

**204   Introduction to Offshore Structures**

**Figure 11-1.** Fabrication yard skid runners and the first half of the first interior bent. (Courtesy J. McDermott Incorporated.)

**Fabrication and Installation** 205

*Figure 11-2.* Raising the second half of the second interior bent. (Courtesy J. McDermott Incorporated.)

positioned. Also, one section of one outside bent complete with conductor guides and skirt-pile guides is shown being rotated into position beside the structure on the skid runners. By comparing the height of the men to the bottle in Figure 11-5, one may judge the diameters of the horizontal brace being positioned for welding and the skirt-pile conductor holes. The canvas enclosure shown at the left provides wind protection for the welders.

In Figure 11-6 one may see in the foreground a bottle and a portion of a jacket corner leg ready for lifting into the opening provided in the top of the trusswork to complete the outside leg of the jacket. Because of the weight of the bottle, it was actually cut in two perpendicular to the axis of the jacket leg and raised into position in two lifts. Figure 11-7 shows the second half of the bottle leg being lifted into position. Also visible in Figure 11-6 are the anodes of the cathodic protection system attached at several locations along each tubular member.

Figures 11-8 and 11-9 are two views of the completed jacket lying on skid runners ready for load-out onto the launch barge. Note the auxiliary buoyancy tubes positioned in the skirt-pile guides along the upper half of the jacket corner legs.

**Figure 11-3.** *The two interior bents with launch bracing rest on the skid runners. (Courtesy J. McDermott Incorporated.)*

Fabrication and Installation 207

*Figure 11-4. Beginning fabrication of the outside bents. (Courtesy J. McDermott Incorporated.)*

**208 Introduction to Offshore Structures**

*Figure 11-5. Close-up view of a bottle section. (Courtesy J. McDermott Incorporated.)*

**Fabrication and Installation** 209

*Figure 11-6. Bottle section ready to be lifted into position. (Courtesy J. McDermott Incorporated.)*

210     Introduction to Offshore Structures

*Figure 11-7.* Bottom half of bottle section being lifted into position. (Courtesy J. McDermott Incorporated)

Fabrication and Installation 211

Figure 11-8. Side view of completed jacket. (Courtesy J. McDermott Incorporated.)

## 212 Introduction to Offshore Structures

*Figure 11-9. Jacket ready for transferring onto the transportation barge. (Courtesy J. McDermott Incorporated.)*

A four-leg jacket for shallow water is shown in Figure 11-10. Note the mudmats at the large end, and the piles stored on deck between the tie-down points protruding from under the small end of the jacket.

### Deck Substructure

Construction of the deck substructure begins with the fabrication at ground level of decks, trusses, box girders, walkways, stairways, etc. Frequently, these components are built inside a fabrication building to avoid delays from bad weather.

The deck substructure may be fabricated in many ways. One common method is to construct on the ground the four planar trusses (each including two deck substructure columns) that go across the narrow dimension of the jacket. Numbering from one end of the jacket, these are called Row 1, 2, 3, and 4 trusses. The two deck substructure trusses running along the long dimension of the jacket, each containing four column legs, are called Row A and B trusses. Figure 11-11 shows a portion of row A or B truss. The interior braces between Row 1 and 2 trusses are fabricated onto Row 2 truss while it is still horizontal. Cranes then rotate Row 2 truss and the attached bracing into the vertical position and place the column ends on skid runners. Other cranes

**Figure 11-10.** *Four-leg jacket with mudmats. (Courtesy J. McDermott Incorporated.)*

**214  Introduction to Offshore Structures**

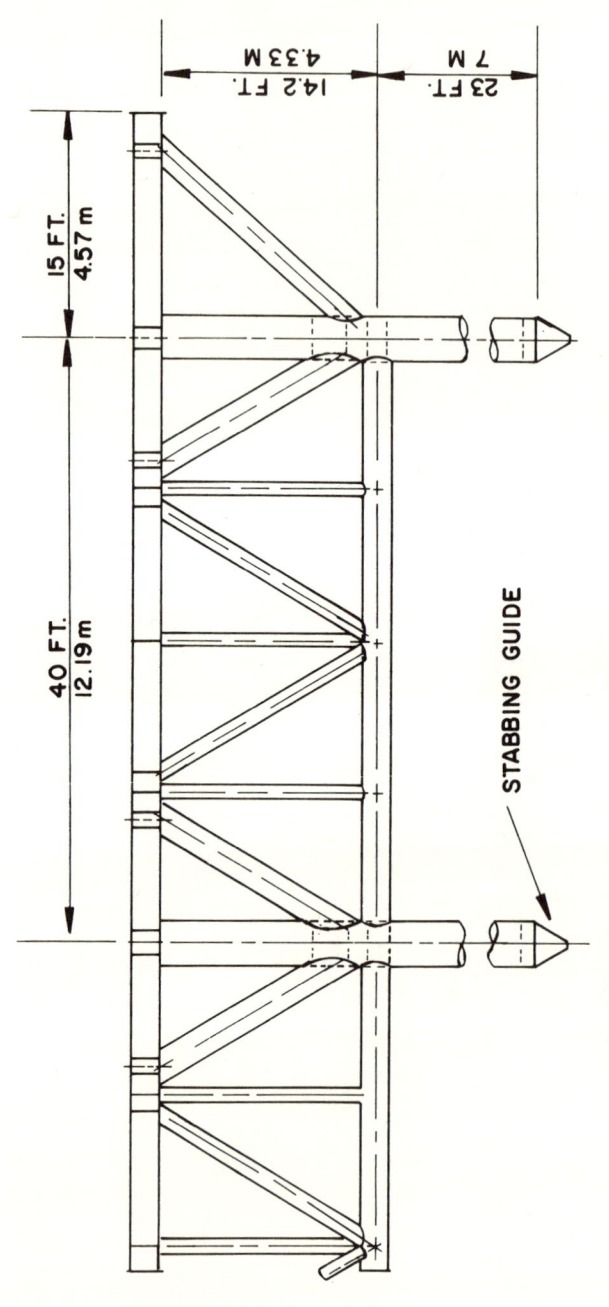

*Figure 11-11. Portion of row A or B truss.*

COLUMN LEG 3 FT. DIA. BY 1.0 IN. WALL THICKNESS (0.915 mm × 25.4 mm)
LOWER CHORD OF TRUSS 20 IN. DIA. BY 0.75 IN. WALL THICKNESS (0.5m × 19mm)
UPPER CHORD OF TRUSS 28 IN. WIDE FLANGE BEAM (0.71m)
TUBULAR BRACES 24 IN. × 0.5 IN. W.T.; 18 IN. × 0.5 IN. W.T.
12.75 IN. × 0.5 IN. W.T.; 10.75 IN. × 0.365 IN. W.T.

raise Row 1 truss, position it at the ends of the bracing, and place the column ends on skid runners. After welding, the four-legged space frame is self-standing on its columns. The remainder of the section of Row A and B trusses are next added to the space frame. The space frame of Row 3 and 4 trusses is constructed in the same way. That part of the deck substructure between Row 2 and 3 trusses is fabricated as a separate framework. Seats are provided on the facing sides of Row 2 and 3 trusses so that at installation offshore the center framework can be placed as a third lift between each four-legged deck substructure space frame.

If the deck substructure is not too heavy, it may be possible to attach the in-between framework to one of the four-legged space frames onshore and have only two lifts at the final offshore installation. (See Figure 11-12.) It is desirable to install into the deck substructure as much of the final equipment and piping as possible while these space frames are still standing on the skid beams in the fabrication yard. This procedure reduces offshore fabrication hook-up time, thereby reducing cost. The lifting capacities of the available derrick barges determine the number of lifts and how much equipment can be mounted within the substructure framing while still on shore.

Figures 11-13 through 11-15 show three stages in the fabrication and installation of a 750-ton cellar deck for a two-deck Gulf of Mexico production platform. In Figure 11-13 the finished cellar deck is shown resting on dollies mounted on railroad tracks leading to the quay. Figure 11-14 shows the cellar deck securely welded to the deck and bulkhead framing of the transportation barge. In Figure 11-15 two derrick barges with a combined lifting capacity of 900 tons are shown lifting the cellar deck into position over the ends of the piles protruding from the tops of the jacket legs. Note that this production platform is being constructed adjacent to an existing drilling platform. Figure 11-16 shows a diagram of the lifting arrangement.

**Deck Plating and Modules**

Design of the framing of the deck beams and flooring is given in Chapter 8. Those deck areas not covered by modules are surfaced with either steel plating or grating. The thickness of the plating is a function of the deck loads; however, the plates are from ½ to ¾ inches (13 to 19 mm) thick. Figure 11-17 shows a typical arrangement of deck beams and plating.

The deck equipment packages are fabricated in modules of approximate size 10-15 ft x 50-60 ft x 16-20 ft (3-4.6 m x 15-18 m x 4.9-6.1 m) high. Sometimes the size reaches 35 ft x 75 ft x 21 ft (11 m x 23 m x 6.5 m) high. The lifting capacity of the available derrick barge determines the size and/or weight of the module. References 1 through 4 discuss selection and/or fabrication of components for decks. In particular, Reference 2 presents guidelines for module sizes and weights. Figure 11-18 summarizes the

**216 Introduction to Offshore Structures**

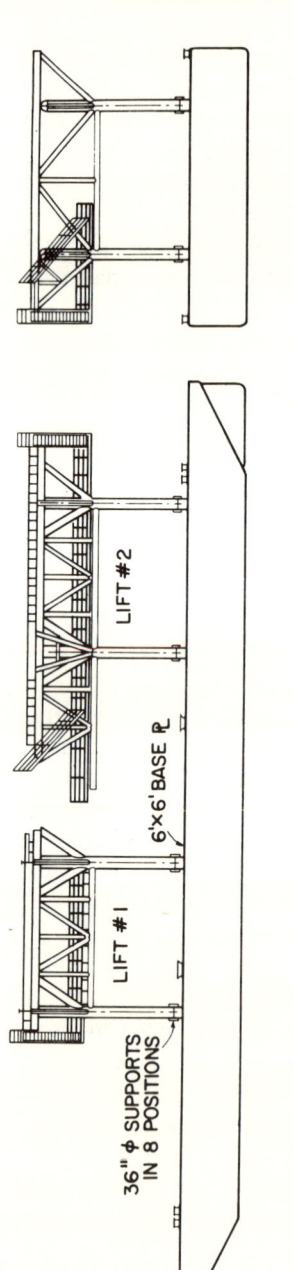

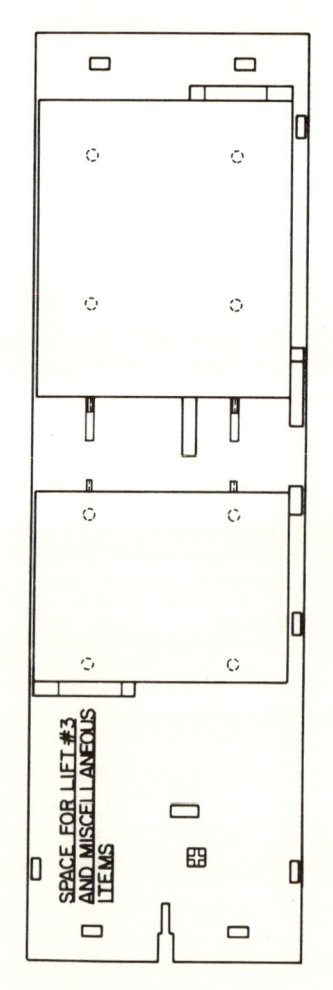

*Figure 11-12. Load-out of deck substructure for two lift installation.*

Fabrication and Installation 217

*Figure 11-13.* 750-ton cellar deck for a two deck production platform. (Courtesy J. McDermott Incorporated.)

*Figure 11-14.* A 750-ton cellar deck is towed to the production site in the Gulf of Mexico. (Courtesy J. McDermott Incorporated.)

**Figure 11-15.** Two derrick barges with a combined lifting capacity of 900 tons hoist a 750-ton cellar deck into position. (Courtesy J. McDermott Incorporated.)

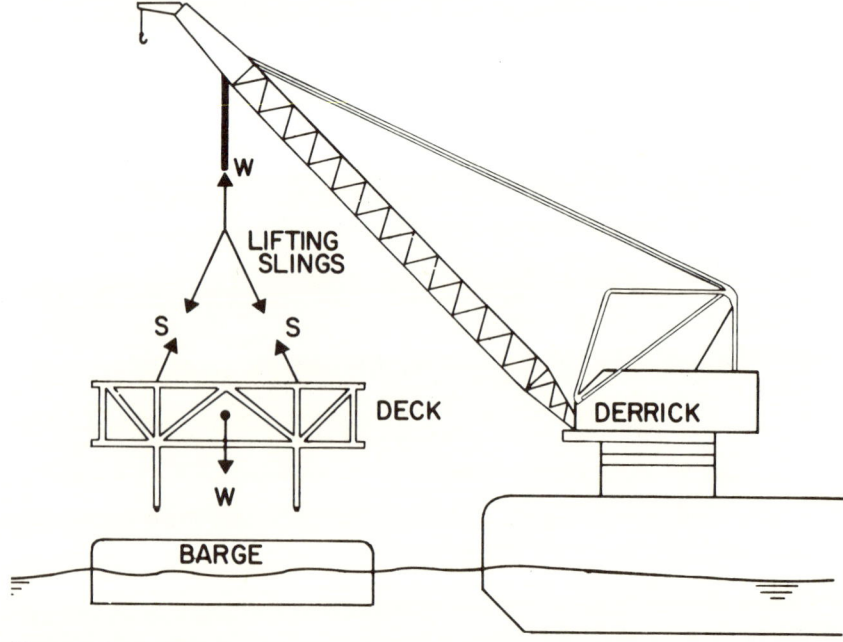

**Figure 11-16.** Lifting of a deck structure from a transportation barge. (Courtesy Brown & Root, Inc.)

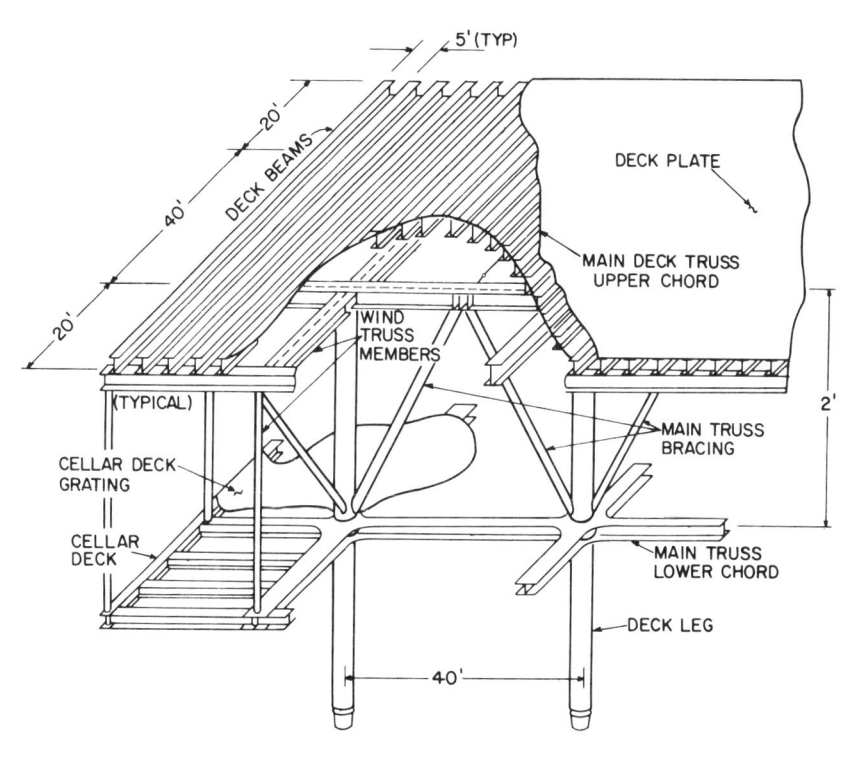

**Figure 11-17.** Cutaway view of deck structure. (Courtesy Brown & Root, Inc.)

## engine generator package

### plan

major equipment uses to establish dimensions:

(2) 2,200-hp dc electric power units

(2) 500-kW diesel power ac electric generator units

Estimated lift weight of package = 300 tons

### side elevation

### end elevation

## pump package

major equipment used to establish dimensions:

(2) 1,000-hp triplex mud pumps powered by electric motors

(3) 60-hp mud mixing and charge pumps

estimated lift weight of package = 350 tons (includes crane weights)

### side elevation

### end elevation

## drilling, cooling, potable water

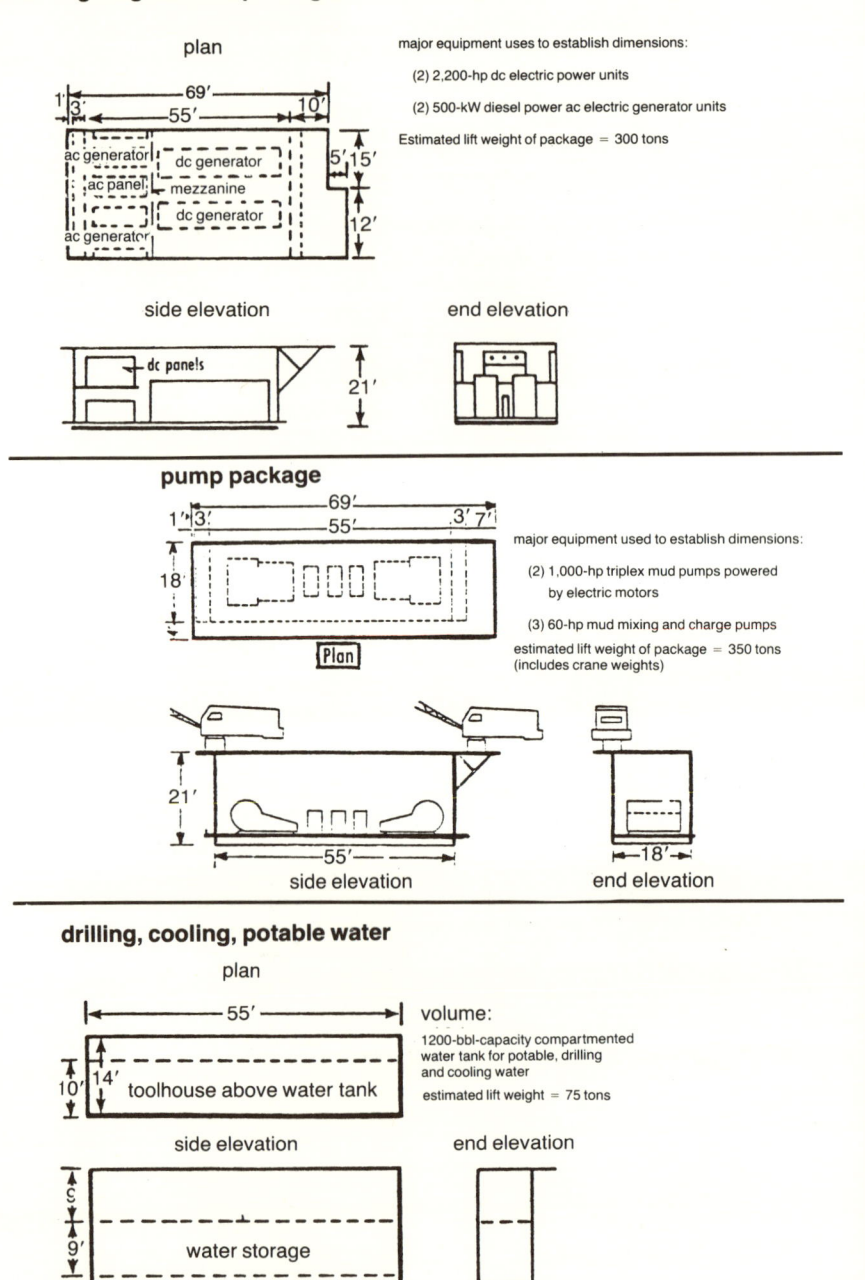

### plan

volume:

1200-bbl-capacity compartmented water tank for potable, drilling and cooling water

estimated lift weight = 75 tons

### side elevation

### end elevation

**Figure 11-18.** *Modular dimensions for deck equipment packages. (Courtesy Penn Well Publishing Co.)*

## dry storage package

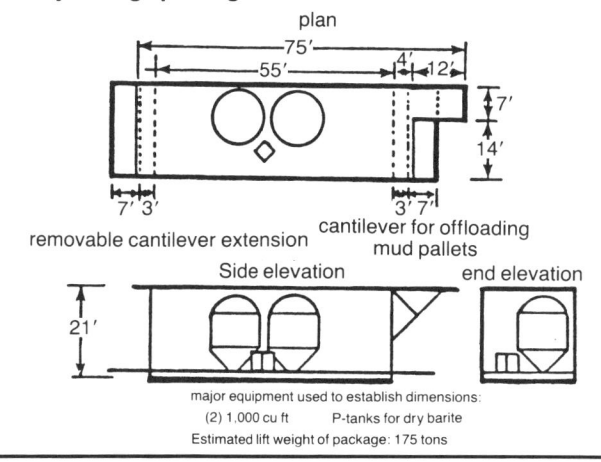

### plan

*removable cantilever extension*
*cantilever for offloading mud pallets*

### Side elevation

### end elevation

major equipment used to establish dimensions:
(2) 1,000 cu ft    P-tanks for dry barite
Estimated lift weight of package: 175 tons

## active mud system package

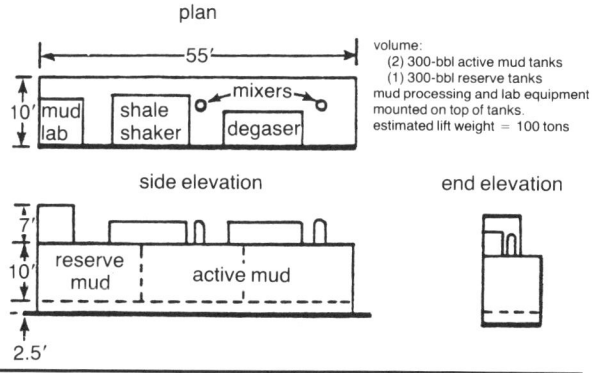

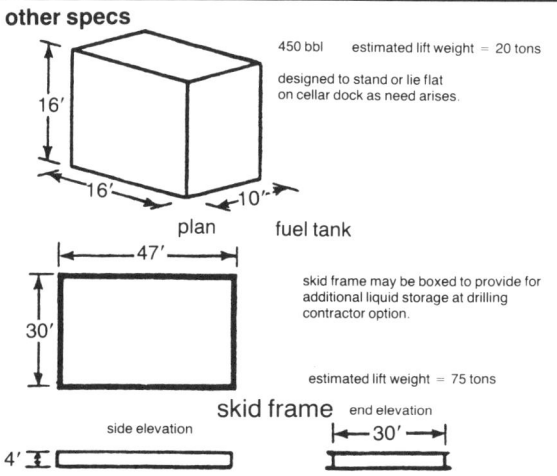

### plan

volume:
(2) 300-bbl active mud tanks
(1) 300-bbl reserve tanks
mud processing and lab equipment mounted on top of tanks.
estimated lift weight = 100 tons

### side elevation

### end elevation

## other specs

450 bbl    estimated lift weight = 20 tons

designed to stand or lie flat on cellar dock as need arises.

### plan    fuel tank

skid frame may be boxed to provide for additional liquid storage at drilling contractor option.

estimated lift weight = 75 tons

### skid frame    end elevation

side elevation

*Figure 11-18 (continued)*

## 222 Introduction to Offshore Structures

modular dimensions. Modules are built on pads of girders and beams called skids. They are designed to be self-supporting and to span between major trusses in the deck substructure. Other skids of all sizes accommodate smaller and lighter pieces of equipment intended to rest on the platform floor system.

A module must be designed to support its maximum combined fixed and variable weights. According to Reference 2, the maximum module lifting weight should not exceed 350 tons (317 tonnes); however, module lifting weights of 500 tons (454 tonnes) are not uncommon.

For minimum self-contained drilling platforms, the module sizes must be compatible with the arrangement and size of the drilling rig package. Several schemes are available for packaging drilling rig components. Figures 11-19 through 11-21 show a rig package designed and constructed following the Reference 2 guidelines.

Deck modules may be fabricated at any remote construction site adjacent to the water and brought to the offshore location by barge. Important considerations include planning the proper width for the skid beams for load-

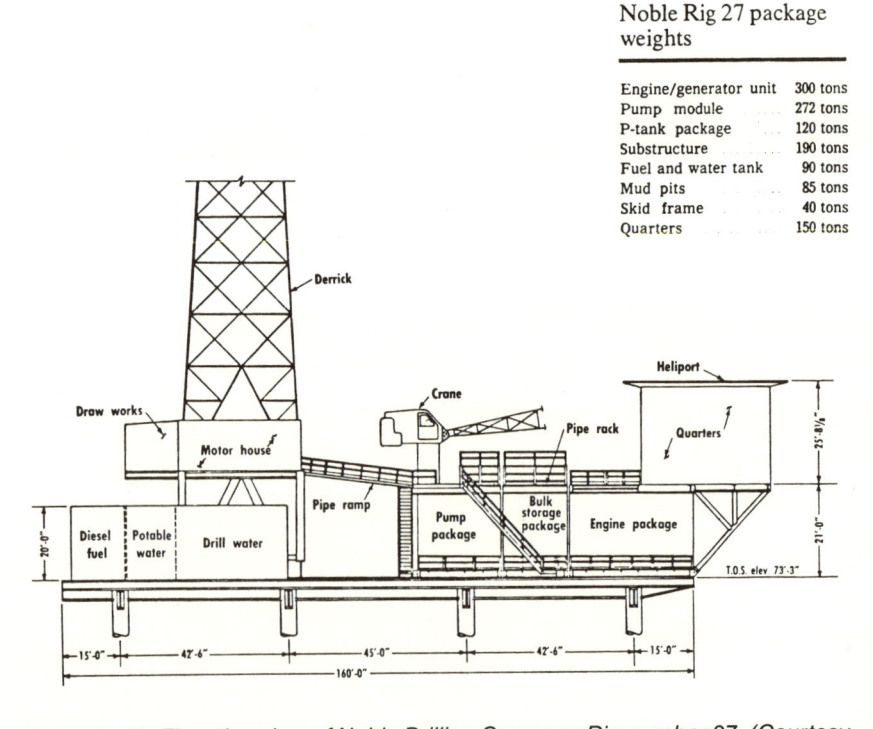

Noble Rig 27 package weights

| | |
|---|---|
| Engine/generator unit | 300 tons |
| Pump module | 272 tons |
| P-tank package | 120 tons |
| Substructure | 190 tons |
| Fuel and water tank | 90 tons |
| Mud pits | 85 tons |
| Skid frame | 40 tons |
| Quarters | 150 tons |

*Figure 11-19.* Elevation view of Noble Drilling Company Rig number 27. (Courtesy Penn Well Publishing Co.)

Fabrication and Installation 223

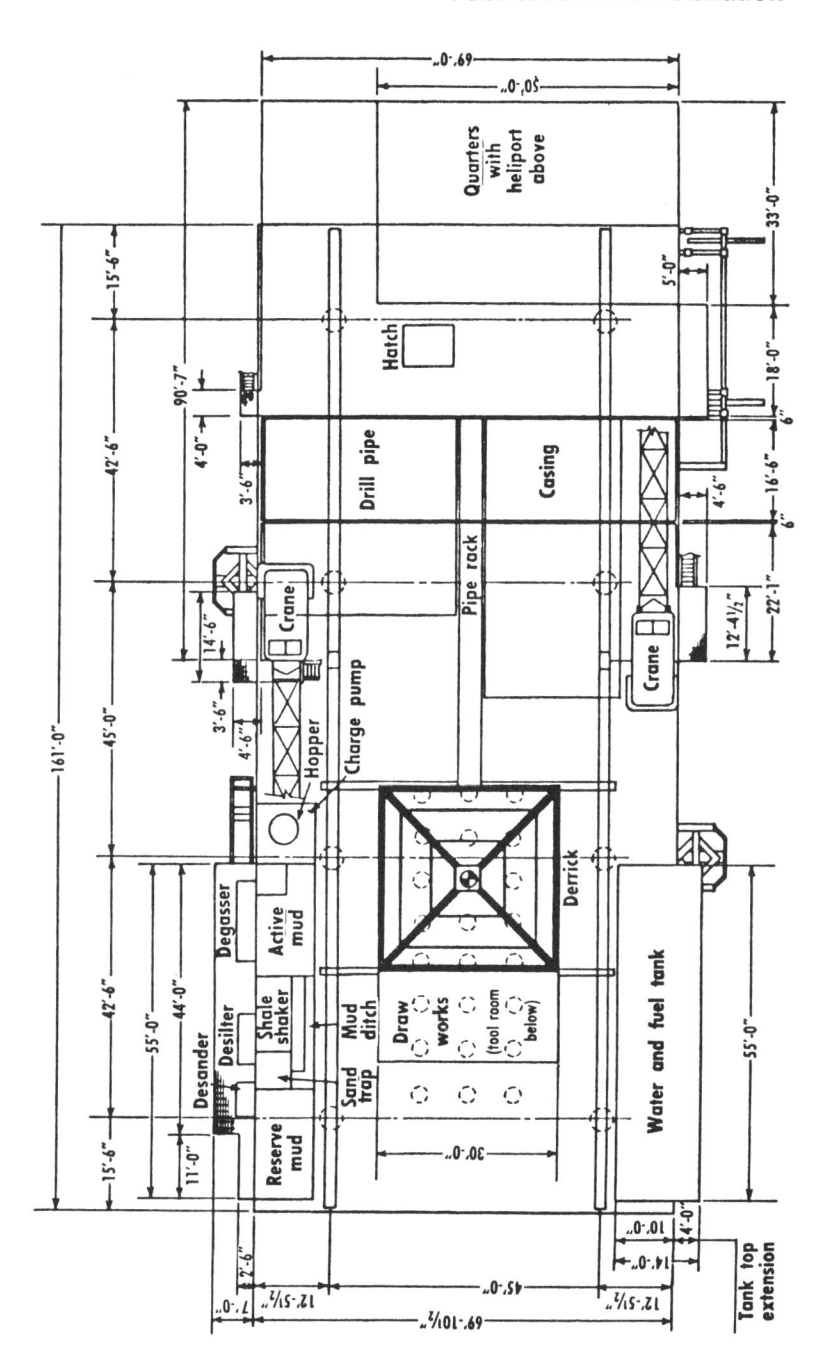

Figure 11-20. Plan view of top deck of Noble Drilling Company Rig number 27. (Courtesy Penn Well Publishing Co.)

**224  Introduction to Offshore Structures**

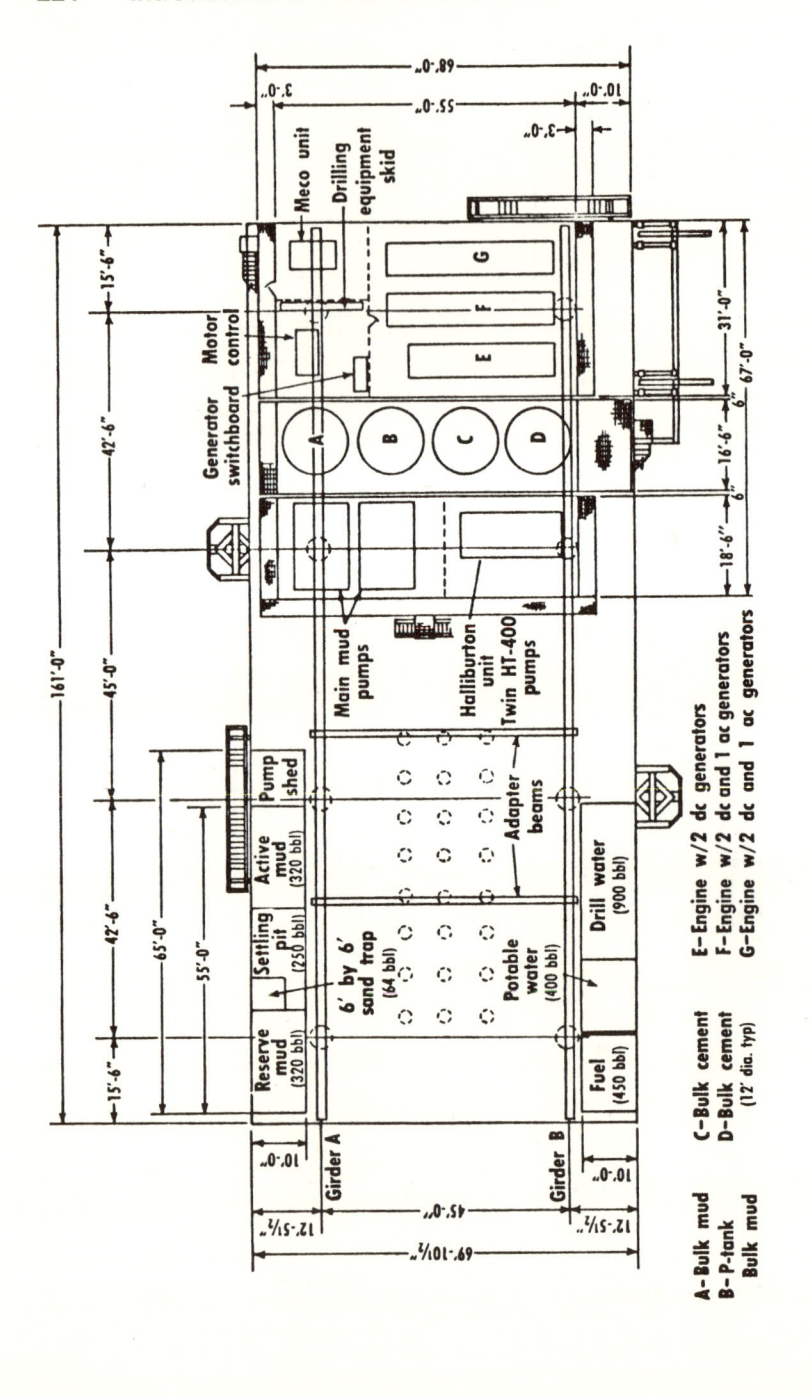

*Figure 11-21. Plan view of lower deck of Noble Drilling Company Rig number 27. (Courtesy Penn Well Publishing Co.)*

A–Bulk mud     C–Bulk cement     E–Engine w/2 dc generators
B–P-tank       D–Bulk cement     F–Engine w/2 dc and 1 ac generators
Bulk mud       (12' dia. typ)    G–Engine w/2 dc and 1 ac generators

out onto the barge, design and location of lifting eyes, and adequacy of tie-down braces during ocean transportation on the barge.

### Piles

For the typical platform, the piling alone amounts to many tons of weight. Piles are made of high-strength steel. Large plates up to 2.5 inches thick are rolled or formed into tubular shapes and welded longitudinally with full-penetration groove welds. These tubulars are then welded end-to-end, with care not to align any two longitudinal welds, to form a section of the pile. Full-penetration groove welds are used to join the short tubulars which are about 10 ft (3 m) long. Enough sections are fabricated so that when welded end-to-end the sum of the sections will constitute the pile. The pile may be 400-800 ft (122-244 m) long. The length of a pile section depends on the lifting capacity and working height of the derrick barge that will lift the pieces into place. As each section is added to the pile being driven, it is field-welded with full-penetration groove welds. Great care is taken in the fabrication of piles with regard to roundness and straightness.

Pile sections are fabricated on a horizontal piling rack. The rack contains rollers so that tubular pieces can be rotated relative to one another. Each pile section is provided with a stabbing guide for mating during offshore operation. The pile sections are marked and put in storage until time for load-out to a transportation barge.

Figure 11-22 shows how the pile wall thickness may vary in relation to the location of the mudline.

### Jacket Load-out and Installation

Major structural components such as deck modules and major items of operational equipment are fabricated or assembled onshore and are installed on the platform by lifting onto the deck substructure after the jacket has been piled to the ocean floor. Jackets for deep water are launched from barges, floated to the offshore site employing their own buoyancy, or floated to the offshore site attached to a floatation pontoon. For water depths less than 150 ft (45 m), the jacket may be launched from a launch barge or lifted from an ordinary barge by derricks on one or two derrick barges and lowered into the water. The deck substructure, often in pieces, is lifted into place above the pile ends protruding from the tops of the jacket legs. Consequently, all the major structural components must be designed for forces encountered during installation. Figures 11-23 through 11-26 show different methods of installing jackets.

## 226 Introduction to Offshore Structures

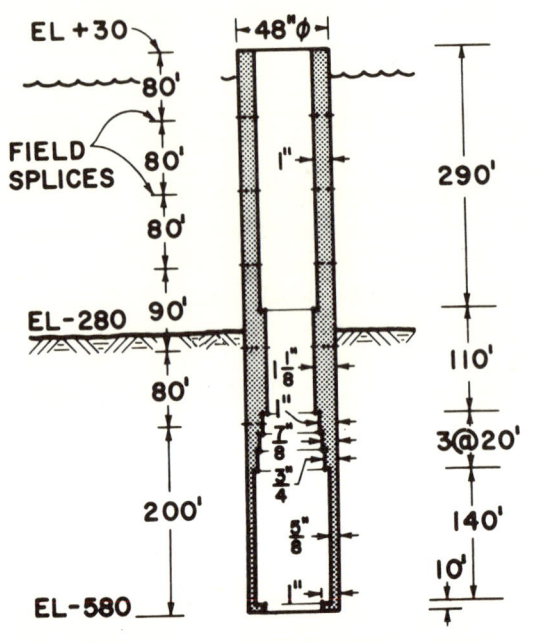

**Figure 11-22.** Wall thickness and assembly design for 48-inch diameter pile. (Source: McClelland, B., Focht, J.A., and Emrich, W.J., Proceedings of the First Civil Engineering in the Oceans Conference, 1968, p. 603.)

### Launching from Barge

Figure 11-23 shows the various steps in this method. The jacket, including stairs, pile guides, conductor tube guides, sacrificial corrosion anodes, etc., must be analyzed for loads induced during transfer from the skid runners in the fabrication yard to the skid beams of the launch barge which will transport it to the offshore location. This analysis is performed on the basis of barge movement, either up or down, which induces deflections into the framework. The jacket must be strong enough to resist the stresses due to the maximum deflection expected as the changes in elevation occur in moving from the shore to the barge. Launch barge ballasting must be carefully controlled.

The launch barge must be equipped with pumps for ballasting the barge, skid beams, rocker beams, two winches with a single-line pull of appproximately 75 tons (68 tonnes), and a pulley block arrangement capable of exerting a pull of 800-1000 tons (726-908 tonnes) on the jacket. The jacket is pulled onto the barge with the hoist and block arrangement on the launch barge. The surfaces of the timbers on the skid runners are sealed with paraffin, and grease is applied to the timbers to reduce friction.

**Fabrication and Installation** 227

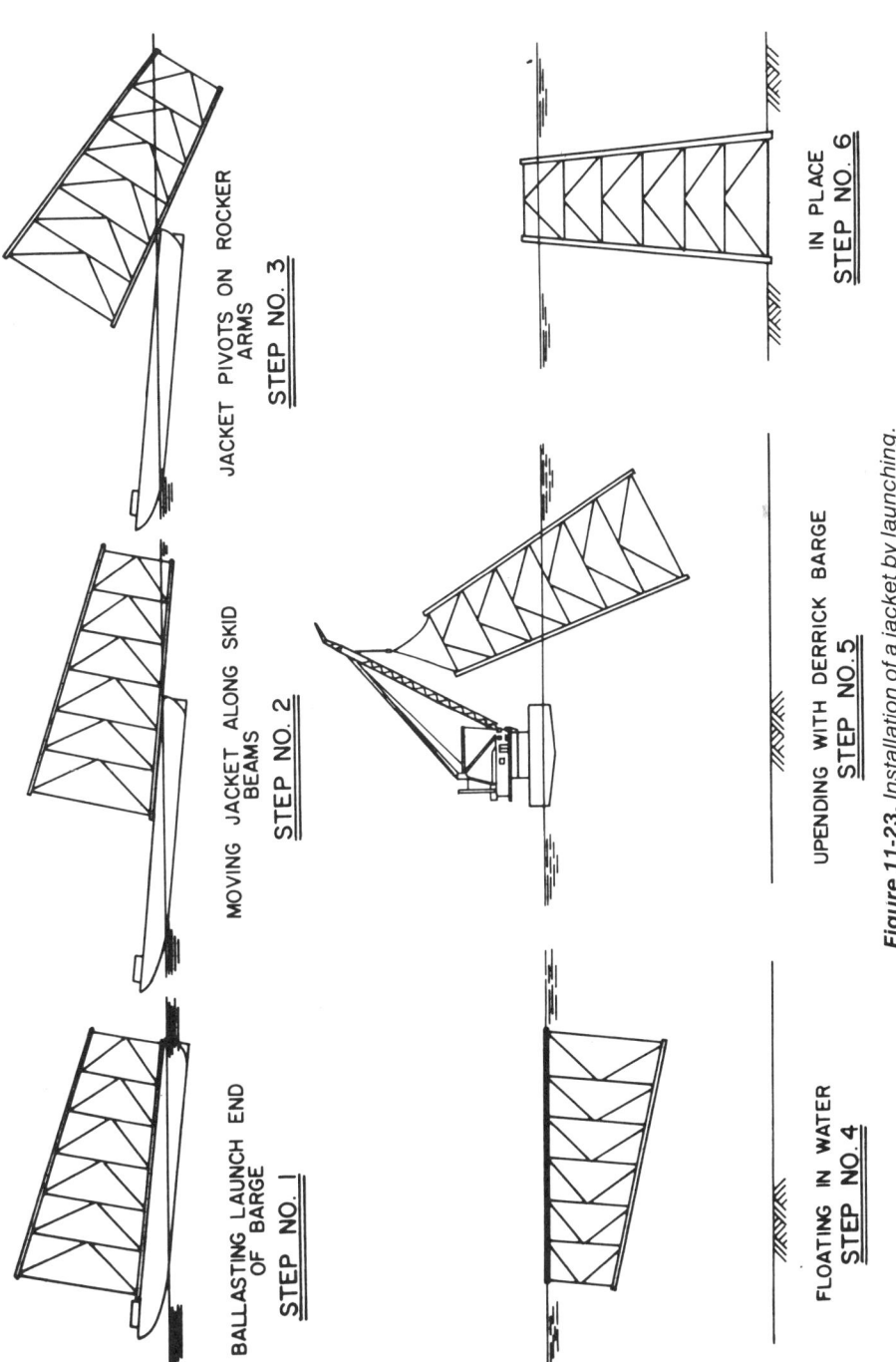

*Figure 11-23. Installation of a jacket by launching.*

**228** Introduction to Offshore Structures

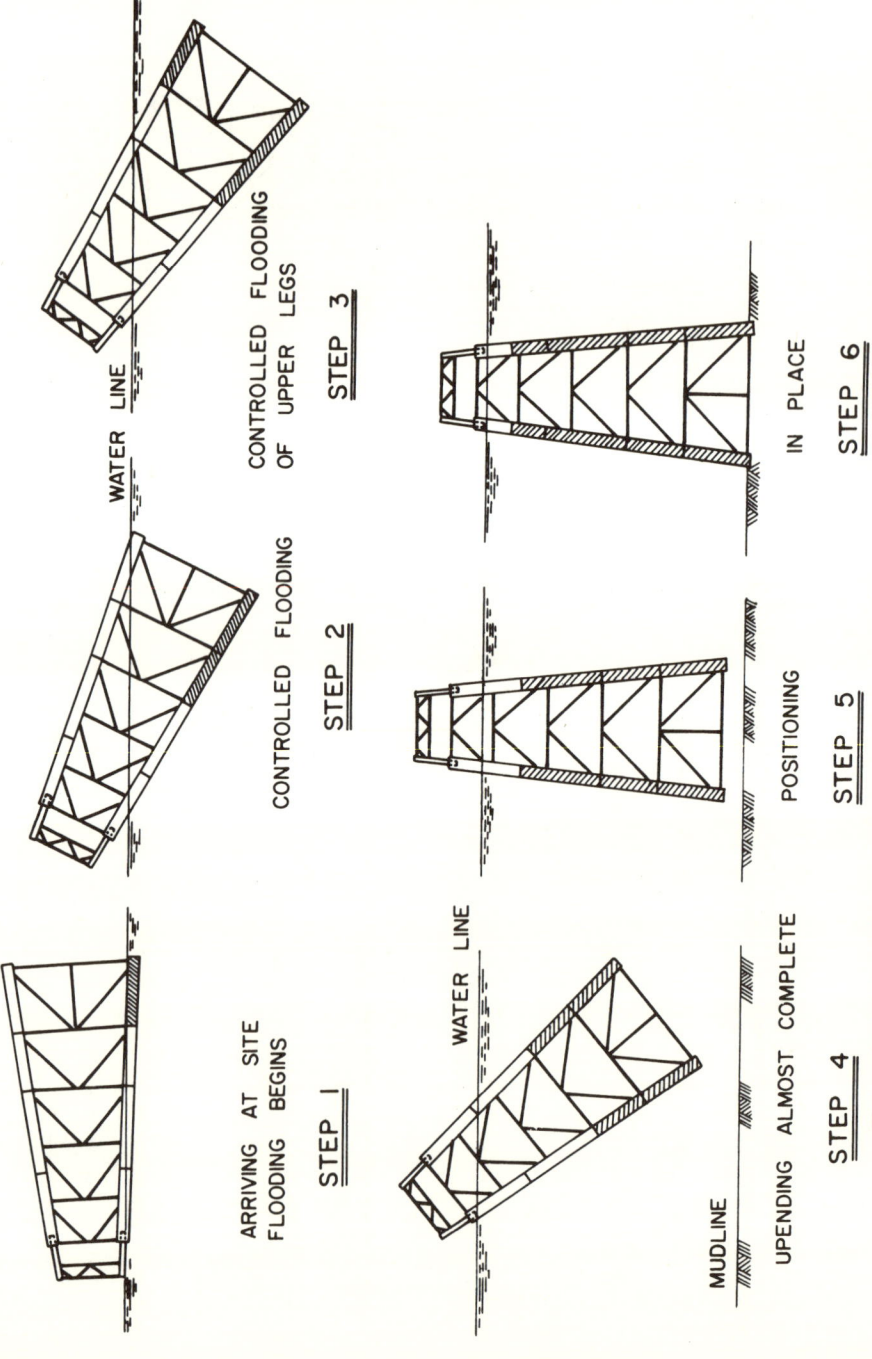

**Figure 11-24.** Installation of a self-floating jacket. (Courtesy J. McDermott Incorporated.)

**Fabrication and Installation** 229

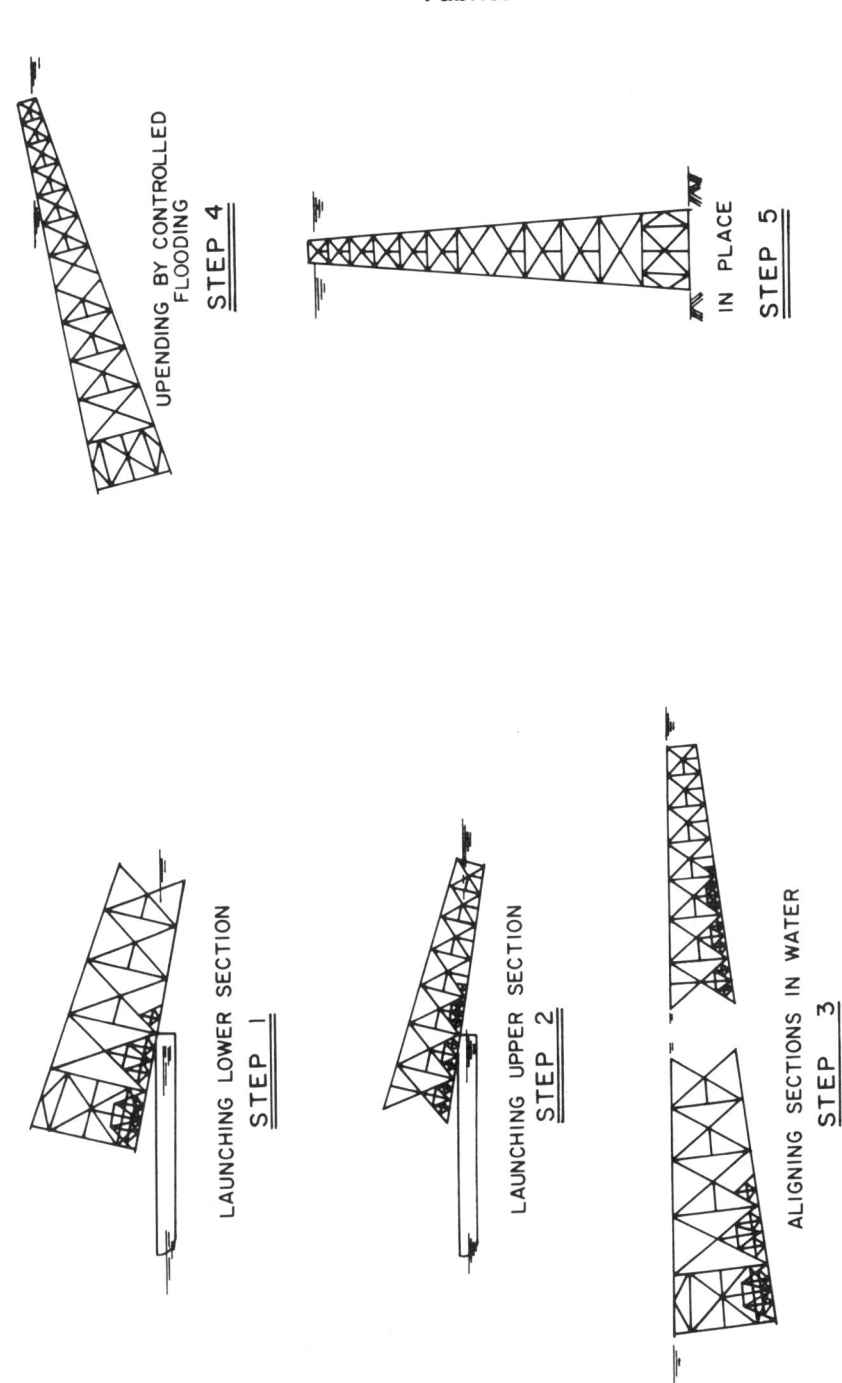

*Figure 11-25. Installation of a horizontally connected sectionalized jacket. (Courtesy J. McDermott Incorporated.)*

**230** Introduction to Offshore Structures

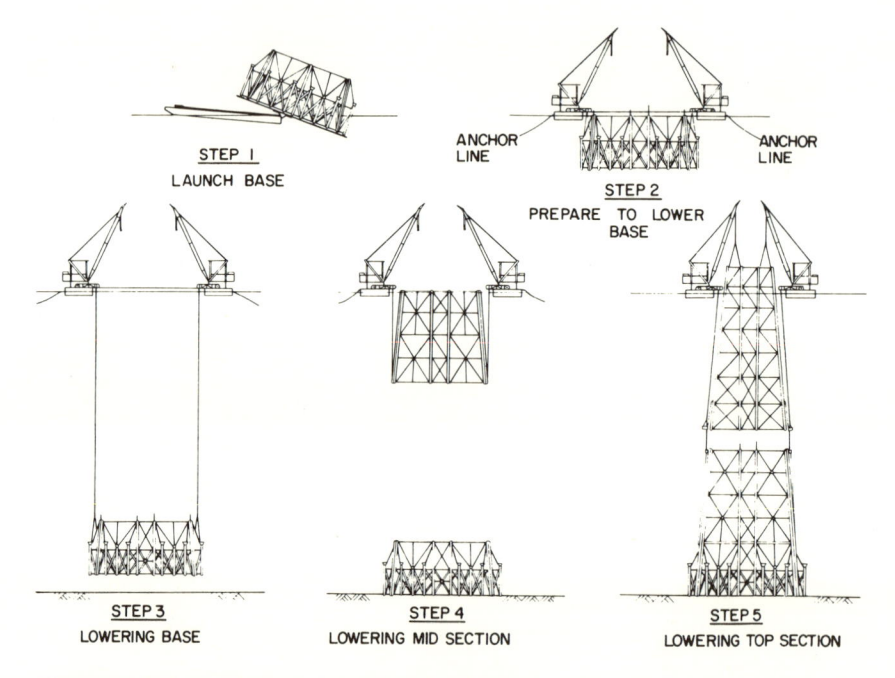

*Figure 11-26.* Installation of a vertically connected sectionalized jacket. (Courtesy J. McDermott Incorporated.)

## Fabrication and Installation 231

For load-out, the launch barge must be ballasted to a deep draft so that the skid runners on the barge align with those on the quayside. During transfer of the jacket to the barge, barge tanks must be selectively deballasted to maintain the skids at the desired elevations. At the beginning of the design, one must make sure the jacket is fabricated at an elevation that will allow convenient load-out and sufficient water depth for the barge to be ballasted to the required draft. The typical plan view size of a launch barge for transporting a 200 ft (61 m) tall jacket is about 100 x 340 ft (30 x 104 m). When the jacket is positioned correctly on the barge, it is then securely fastened to withstand the barge motion during transportation by welding tie-down braces between selected points on the barge and the jacket. Because of barge movements such as pitching or rolling, and because of waves impacting the fastened components (in storm situations), stresses are induced in the tie-downs. The tie-downs must be designed to withstand the environmental forces expected during transportation. The jacket structure in the vicinity of each tie-down must be analyzed so that it is strong enough to resist the localized reactions, and the barge deck must be analyzed to ensure that the attachment points of tie-downs are over bulkheads, or other locations stiff enough to take the applied force without excessive deflection.

The jacket must also be analyzed for stresses induced during launching from the launch barge. At one end of the launch barge, a long section of each skid beam pivots in a vertical plane about its mid-length; these are called rocker arms. After the launch barge and jacket have arrived at the offshore location, the tie-down braces are removed, the barge is ballasted by flooding compartments to the proper draft and trim angle, and the jacket is made to slide along the launch beams with either hydraulic jacks or cable winches, or both, until at some point it moves under its own momentum. As the jacket moves toward the end of the barge, the barge's trim angle increases until finally the jacket's center of gravity passes the pivots on the rocker arms. The jacket then rotates and slides rapidly into the water.

During launch the entire weight of the jacket is supported by the two jacket legs that contain the launch trusses. At the time of tilting over the pivots in the rocker arms, the weight becomes concentrated along a short segment of each leg. This concentration of reactions to the gravity forces is the reason why the launch trusses are added to the bents that lie on the skid runners. The jacket legs containing the launch trusses must be reinforced externally at the joints with ring stiffeners to prevent collapse. There is added weight on the jacket at this time because of the auxiliary buoyancy tanks on the upper end of the jacket as it rests on the launch barge.

The auxiliary buoyancy tanks must be designed to withstand the hydrostatic forces imposed during upending of the jacket in the water. The jacket is ordinarily stable while floating on the water surface; it must remain stable while being rotated into a vertical position. Ballasting reduces the buoyancy of the jacket. Thus, the auxiliary buoyancy tanks provide not only the added buoyancy for stability during upending, but also the buoyancy necessary to maintain proper mudline clearance after the jacket reaches its vertical position.

Before load-out, a temporary work platform is installed on the jacket in-between and fastened to the barge bumpers, which are at approximately the waterline elevation when the jacket is in its vertical position. These barge bumpers are on the two parallel legs on the side of the jacket that does not contain the launch trusses. The control valves for ballasting the various members of the jacket and the auxiliary buoyancy tanks are all located near this work platform.

Accurate weight and buoyancy calculations are required early in the design to ensure that the jacket will float after launching and to determine where the auxiliary buoyancy tanks will be located—if they are needed. With the auxiliary buoyancy tanks positioned correctly, the jacket will come to rest in the water at an attitude somewhat like that which it had on the barge except that the four-legged side of bracing without the launch runners will be parallel to and essentially in the plane of the water surface. At this time, a carefully trained crew of workmen boards the jacket, and, by ballasting (flooding) selected jacket members and auxiliary tanks from valves on piping all brought to the location of the temporary work platform, upends the jacket to its vertical position. A program must be devised early in the design delineating the sequence of ballasting. The derrick barge usually lifts on the upper end of the jacket to assist in the upending.

By controlling buoyancy, the jacket is kept slightly off the ocean bottom and is towed and/or rotated into its correct position. Finally, it is ballasted until it rests on the ocean bottom. The weight on the ocean bottom must be carefully regulated by ballasting so that the bearing capacity of the soil beneath the structure is not exceeded. Sometimes, when the soil conditions at the mudline are semi-solid or soft, the lowest level of horizontal bracing within the jacket will be provided with heavy timber planking, called mudmats, to temporarily support the jacket until piles can be driven.

At the offshore location, the derrick barge is positioned with the location for the platform just off the stern of the barge. The launch barge is positioned about 800 ft (244 m) behind the derrick barge. The jacket is launched upper end first in the span of water between the derrick barge and the launch barge. Thus, with a cable that is brought from the derrick barge to the jacket and attached to the lifting sling after launch, the derrick barge pulls the jacket to the installation site and assists in the upending. A tugboat moves in right after launch and secures cables to the lower ends of the two outside jacket legs. As the derrick barge pulls the jacket toward the platform location, the tug exerts a backward pull on the jacket for better control of movement.

**Self-floating Jackets.** Deepwater jackets are made self-floating when they become too heavy for the available launching equipment. The legs of a self-floating jacket, at least those that will lie on or just below the water surface, are made larger in diameter than otherwise necessary to provide the buoyancy for floating. These legs may have diameters of 20-30 ft (6-9 m) or more.

Because of the larger legs, self-floating jackets require more steel than a conventional launched jacket. Greater wave forces are generated on the larger diameter legs. This effect requires extra piles and more structural bracing. In addition, the large legs require both internal ring stiffening and internal longitudinal stiffening to provide structural stability. To protect against hydrostatic collapse, the large leg members are flooded after installation.

Figure 11-24 shows the steps in installing a self-floating jacket. During tow-out the jacket normally floats at a draft of 10-20 ft (3-6 m).

Another alternative to a launched jacket (a variation of the self-floating type) is the jacket that is fabricated on a specially constructed floatation pontoon. The pontoon must be built before the jacket. The construction must occur in an area called a *graving dock* which can be flooded for moving the combined structure out to sea. After the jacket and pontoon are rotated into the vertical and placed on the ocean bottom, the pontoon is ballasted to a neutral buoyancy and released from the jacket by actuating hydraulically operated connecting pins.

The weight of the pontoon is approximately the same as that of the jacket. Construction of this special device increases the project cost. The pontoon is reusable only for jackets of similar configuration in about the same water depth. For these reasons, the pontoon concept is not used very often for transporting and installing deepwater jackets.

**Horizontally connected sectionalized jacket.** Conventional launched jackets and self-floatation jackets have been installed in approximately 500-ft (152 m) water depths. At water depths somewhat deeper than this it becomes necessary to separate the jacket into sections. Two methods have been developed for sectional jackets.

Faced with earthquake criteria in the Santa Barbara Channel, California, the Exxon Company decided in 1975 to use the conventional launched jacket concept for a platform to be placed in 850 ft (259 m) of water but to transport the jacket in two sections. It was concluded that the self-floatation type jacket was not suitable for earthquake-prone areas because of the large virtual mass of the system, i.e., the great steel mass of the large-diameter legs, the mass of entrapped water in these legs upon flooding after installation, and the large amount of added mass from the submerged big legs.

With proven fabrication yard techniques, the jacket was built as one long structure. At the location for separation, the jacket legs were fitted with external stabbing cones for alignment and remotely operated flange type connections for watertight sealing. The jacket was then cut apart and each section was transported on its own barge to the offshore site and launched.

The two sections were pulled together while floating horizontally, aligned, and joined with the flange connections. Locating watertight bulkheads inside each leg a short distance back in each direction from the flanged connections and providing access tubes to these locations made it possible to create dry habitats after dewatering. Welders working in these habitats rejoined the jacket legs permanently with full-penetration groove welds. After welding, the installation proceeded just as with any other launched jacket. Figure 11-25 shows the steps involved in this method of installation.

**Vertically connected sectionalized jacket.** This concept was developed for the *Cognac* platform discussed in Chapter 5. To stack and join the jacket sections in vertical position, the base section first had to be lowered to the mudline, leveled, and fully piled to the ocean bottom. This procedure included grouting the piles in their skirt sleeves, since all of the piles were skirt piles. Pile driving was done with underwater hammers. Figure 11-26 shows the steps in this method of installation; greater detail is given in Chapter 5.

### Piles and Well Conductors

The vast majority of steel template offshore platforms are of the launched jacket type. Thus this discussion of pile installation emphasizes methods used with launched jackets.

The sections of piling and conductor tubes are loaded-out with cranes onto transportation barges. The plan view size of such a barge is commonly about 72 x 240 ft (22 x 73 m). In loading-out the piles consideration must be given to the order in which sections will be picked up for use, since the wall thicknesses of the sections are not all the same. After the pile driving begins, the time spent in repositioning pile sections on the barge is very costly.

As soon as the jacket has been set on the bottom, a pile is inserted in each of the four center jacket legs. The jacket is deballasted selectively until it assumes the proper level in the water, and the initial piles are driven to the desired penetration. For jackets with skirt piles, the main piles are driven first, then the skirt piles, and finally the conductor tubes.

Skirt piles are connected to the jacket by grouting the annular space between the piles and the sleeves which are integral parts of the jacket structure. After driving the skirt pile, the part of the pile contained in the pile guides from the top of the skirt-pile sleeve to the water surface, called the *follower,* is cut free from the pile just above the sleeve by divers. At the lower end of the skirt-pile sleeve, a ring type device inside the sleeve, called a *packer,* can be actuated hydraulically to expand a hard rubber O-ring to close the annular space between the pile and the sleeve. Cement grout is then pumped through preinstalled grouting lines into the cylindrical space above the packer. The grout is pumped until all the water in the space is displaced, and grout of a customer-specified consistency emerges from the top of the sleeve.

Conductor tubes are normally about two ft (0.6 meters) in diameter. They are driven from 100 ft to 200 ft (30-60 m) into the ocean floor through the conductor tube guides fabricated into the interior of the jacket framework.

Pile hammers are either steam or diesel-powered. Commonly, single-acting steam hammers with rated energies of 60,000; 120,000; and 180,000 foot-pounds (81,300; 162,600; 244,000 Joules) are used for driving piles with diameters of 42-60 inches (1-1.52 m). The hammer size is increased as the driving resistance increases. Larger-size hammers are available. Ordinarily, at least one hammer is kept in ready reserve. This hammer is usually of the 120,000 or 180,000 foot-pound class. Hammers with rated energies of 48,000-60,000 foot-pounds (65,000-81,300 Joules) are often used for driving the conductor tubes. Figure 11-27 shows a steam hammer driving a pile through one of the legs of the jacket in the foreground. Deck substructure is shown in the background ready for installation on completion of pile driving.

After all of the piles have been driven to the desired penetration, the annular spaces between the piles and the insides of the guides or legs are filled with cement grout. The grout produces a permanent bond between the piles and the jacket which creates a single, rigid structure. If the design calls for skirt piles, these are usually grouted after the deck substructure and deck modules are in place.

The pile ends extending out the tops of the jacket legs are cut off in such a manner that the deck substructure will be level after installation—even if the jacket top is not. The tops of the piles are fitted with internal cones to match the pointed ends on the deck substructure columns. The rims of the pile tops are beveled so that full-penetration groove welds can be made with the columns after placement of the deck substructure.

## Deck Substructure Installation

The deck substructure is constructed and transported in its normal upright position. It may be necessary to install temporary erection bracing to avoid overstressing the framework while it is being transferred from fabrication yard skid runners to transportation barge skid runners or lifted into final position offshore. The lower ends of the deck substructure columns are made cone-shaped so that they may be more easily stabbed into the pile ends protruding from the tops of the jacket legs. (See Figure 11-11.)

The deck substructure houses a considerable amount of equipment and piping. This equipment and piping may be built as part of the framing of the

**236** Introduction to Offshore Structures

**Figure 11-27.** Derrick barge holding a steam hammer for pile driving. (Courtesy J. McDermott Incorporated.)

deck substructure. However, care must be taken to keep the weight of the particular subassembly comprising a unit within the lifting capacity of the available derrick barge. The location of the center of gravity must be controlled also. Often, the magnitude of equipment and piping is so large that modularized units are fabricated on skids, and special runners are provided within the deck substructure so that these units may be either lowered into the framework at location or placed on the runners and then winched or jacked horizontally into position. When planning to winch units along runners into position, attention must be given to the order of lifting the items from the transportation barge.

In the fabrication yard steel brackets are made for the cone-shaped lower ends of the substructure columns to provide flat bearing surfaces on the skid runners. The deck substructure is skidded along the skid runners onto the transportation barge in a manner similar to that used for moving the jacket. The deck substructure, because of its high center of gravity, must be securely fastened to the transportation barge with the tie-down braces. Before leaving the fabrication yard, the deck substructure must be fitted with a lifting sling designed not to overstress the structure when it is lifted into final position at the offshore location.

## Deck Modules

Deck modules or pieces of equipment weighing more than several hundred tons are generally fabricated on long, parallel skid runners positioned to lead to the quay edge where the barge is tied. To load-out the deck module, the module skid is moved from the shore onto the barge along greased skid runners by winches or jacks. Modules are fastened to the barge with welded tie-down braces.

Within the equipment modules the piping and piping supports must be fitted with temporary braces or supports to withstand the forces from barge movements during transportation. The number of deck modules placed on a particular barge depends on the barge capacity and the ability of the derrick barge to reach the modules for lifting. The skid beneath each module is provided with stabbing guides to ensure proper placement on the deck substructure. If a module has to be moved into final position on the deck substructure by sliding it on runners with hydraulic jacks, the stabbing guides may become positioning holes for drift pins driven in place to hold the module while it is being welded into position.

At the offshore location, the modules are lifted into place with a derrick barge. Sometimes two derrick barges are necessary to pick up large, heavy modules. Design of the lifting eyes on deck modules requires considerable care. The modules must be designed to avoid deleterious deformations during the skidding and lifting operations. It is desirable to distribute the masses on a particular module so that the center of gravity is as close as possible to the geometric centroid of the module skid. Throughout the design of a module, it is necessary to repeatedly check to see that the module weight does not exceed the capacity of the available derrick barge.

## References

1. *Recommended Practice for Planning, Designing, and Constructing Fixed Offshore Platforms,* API-RP-2A, Am. Pet. Inst., 10th Edition, 1979.
2. *Specifications for Drilling Rig Packaging for Minimum Self-Contained Platforms,* API-Spec-2E, Am. Pet. Inst., 1974.
3. *Recommended Practice for Production Facilities on Offshore Structures,* API-RP-2G, Am. Pet. Inst., 1974.
4. *Recommended Practice for Planning, Designing, and Constructing Heliports for Fixed Offshore Platforms,* API-RP-2L, Am. Pet. Inst., Dec. 1978.

# Chapter 12

# *Corrosion*

Corrosion is the deterioration of a solid body through interaction with its environment, that is, destruction through unintentional chemical or electrochemical reaction beginning at its surface. Both metals and nonmetals are covered by this definition. Usually, the term "corrosion" describes the destruction of metals; however, the deterioration of paint and rubber by sunlight or chemicals is also corrosion. Deterioration of the ceramic lining in a furnace or the hardening and cracking of plastics in heat or sunlight may be considered forms of corrosion.

Nonmetallic materials are usually resistant to the attack of mildly corrosive media such as water and the atmosphere. However, they are also attacked by certain chemical agents under certain specific conditions. Almost all metals suffer corrosion to some extent due to the action of water and the atmosphere. Metals often corrode from electrochemical action because of their crystallographic nature (electrons are free to move through their lattice structures). Electrochemical corrosion involves the presence of an electrolyte in contact with the metal. An electrolyte can be any conducting substance, but it usually is an aqueous solution of a salt, acid, or alkali. Because the electrolyte is moist, electrochemical corrosion is also called *wet corrosion.*

As nonmetallic materials do not have the characteristic of free movement of electrons through their structures, they are attacked only by corrosive media capable of reacting chemically with them under specific conditions. Chemical corrosion occurs in the absence of a liquid phase or above the dew point of the environment and is also called *dry corrosion.* Such corrosion is usually caused by various chemicals in gaseous or vapor form, often at high temperature.

Metals usually exist in nature in combined forms such as oxides, hydroxides, carbonates, sulfides, sulfates, and silicates. A great deal of energy must be expended to extract and refine the metal to its pure form. Thus, a pure metal is in a much higher energy state than its unrefined ore. The metallurgist's definition of corrosion of a metal is extractive metallurgy in reverse, i.e., the

## 240  Introduction to Offshore Structures

natural tendency of the material to seek the lower energy level contained in its combined state.[1]

The dark reddish-brown coating which forms on iron or steel in a moist or wet environment is called *rust*. It consists of a hydrated form of ferric oxide ($Fe_2O_3$). The common usage is to reserve the term "rust" for steel and iron corrosion; when other metals form oxides on their surfaces, they are said simply to corrode.

### Corrosion Mechanism

The mechanism of electrochemical corrosion of metals in aqueous electrolytes is associated with the flow of electric current between anodic and cathodic areas within the system. The *anodic region* is where the metal atoms are transformed into positive ions and free electrons. The *cathodic region* is where the free electrons unite with ions of the electrolyte to produce either hydrogen or oxygen. Corrosion occurs in the anodic region, i.e., where the dissolution of the metal causes metal ions to go into solution.

Corrosion of iron (Fe) in an aqueous electrolyte can be described as[2]:

1. In the absence of oxygen

$$Fe + 2H_2O \rightarrow Fe(OH)_2 + H_2$$
solid  liquid  solid  gas

2. In the presence of oxygen

$$Fe + \tfrac{1}{2}(O_2) + H_2O \rightarrow Fe(OH)_2$$
solid  gas  liquid  solid

### Hydrogen Evolution

The type of reaction described in 1 is called *hydrogen evolution*. The corroding metal simply supplies ions to replace the hydrogen ions in the electrolyte. This type of reaction usually occurs in acid environments. The anodes are usually large areas while the cathodes are small areas. The electrons freed by the dissolution of the corroding metal flow through the metal to the small cathodic areas, and there the hydrogen gas is given off.

### Oxygen Absorption

The type of reaction described in 2 is called *oxygen absorption*. In a neutral (pH = 7) aqueous electrolyte, or one that is slightly alkaline, iron corrodes in the presence of oxygen. In this type of attack the electrons freed by the dissolution of the metal are intercepted by the oxygen according to

$2e + \frac{1}{2}O_2 + H_2O \rightarrow 2OH^-$

Notice that the free electrons are converted into hydroxyl ions. In this case the anodic areas on the surface of the metal are due to cracks in the oxide film coating the metal. Consequently, the anodic areas are very small, whereas almost the entire surrounding surface of the metal constitutes the cathode. The corrosion current is concentrated within a very small area; this concentration causes strong, localized attack.

Since seawater, which is one of the best electrolytes, contains sodium and chlorine ions in solution, the cathodic product of this form of corrosion is sodium hydroxide and the anodic product is ferrous chloride. Both products are readily soluble and will diffuse away from their respective electrodes. As they meet in solution, ferrous hydroxide is precipitated. With plenty of oxygen present, ferrous hydroxide oxidizes to ferric hydroxide, which precipitates even more quickly. Since the products at both electrodes have tendencies to diffuse away, the corrosion will proceed as long as there is a fresh supply of oxygen. In situations where there is agitation of the solution, as with waves and currents offshore, the rate of corrosion increases as the aeration of the solution increases.

## Concentration Cell Corrosion

Concentration cell corrosion occurs because of differential aeration, that is, when one part of a metal body is exposed to a different air concentration than another part. A difference in electrical potential is created between the differently aerated areas. The surface of that part of the body exposed to electrolyte of low-oxygen concentration is anodic; the surface area exposed to electrolyte with a high-oxygen concentration is cathodic. The flow of electrons between the two areas is called a *differential aeration current.*

This form of corrosion occurs just below the waterline for metals that are partially immersed in seawater. Dissolution of the metal just below the waterline releases electrons that flow through the metal body to the highly aerated region just above the waterline. The electrons are said to flow through the metal circuit, or the external circuit. In the cathodic area oxygen combines with the free electrons to form hydroxyl ions. The internal circuit of the corrosion cell consists of the migration of the metal ions through the electrolyte. The amount of oxygen available controls the magnitude of the corrosion current.

A drop of water on a steel or iron surface constitutes a simple example of concentration cell corrosion. The central portion of the surface covered by the drop is farthest from the oxygen of the air and becomes anodic, while the outer periphery of the surface covered by the drop has ready access to oxygen and becomes cathodic. The spot of rust forms at the center of the drop of water. In a moist environment this form of corrosion occurs where some object rests on a metal surface and screens that portion of the surface from oxygen access. For example, steel plate will rust under a block of wood that is left lying for some time. Steel or iron equipment exposed to the weather for a long time will rust because of the stagnant films of water left in recesses on the metal surface. Corrosion from differential aeration leads to pitting of the surface.

## Galvanic Corrosion

Galvanic corrosion is electrochemical corrosion that occurs when two dissimilar metals are in contact with one another in an electrolyte. A potential difference naturally occurs between two dissimilar metals when they are immersed in an electrolyte. If the two metals are physically touching, or if they are otherwise electrically connected, the potential difference causes an electron flow between them. The less corrosion resistant metal becomes anodic and corrodes, and the more corrosion resistant metal becomes cathodic. The cathodic metal corrodes very little or not at all in this situation.

The chemical reaction at the cathode may be either hydrogen evolution or oxygen absorption depending on the nature of the corrosive environment. The free electrons resulting from the anodic metal transforming into ions and going into solution flow through the electrical junction between the two metals and are intercepted by positive ions from the solution at the cathode.

Chemists have developed values of standard electrode potentials for metals completely free of oxide films on their surfaces. In practical situations most metals are covered with oxide films. This oxide film tends to make their electrical potentials more positive, that is, their surfaces have a tendency to be more protected from attack in corrosive environments.

In actual corrosion problems galvanic corrosion occurs between two corroding metals. Most engineering materials are alloys—not pure metals—and only rarely are these alloys in equilibrium with their ions in solution. Consequently, to account for environmental conditions met in practical applications, a series was devised of the more often used industrial metals and alloys listing them from the material most easily corroded to that least likely to corrode. This listing, called the *galvanic series,* is given in Table 12-1.[1,2] *Noble* denotes the materials least likely to corrode (gold and platinum).

In general, the relative positions of the materials in the galvanic series agree with the positions of their constituent elements in the series of standard electrode potentials. When those metals and alloys listed in Table 12-1 that are adjacent to one another (or only slightly separated in the listing) are coupled in a practical situation, there is little danger of galvanic corrosion. This situation results because the relative electrical potential generated between the two materials is not great. The farther apart the two materials in the couple are in the galvanic series, the greater the potential generated by electrically connecting them; hence, the more rapid the corrosion.

**Table 12-1**
**Galvanic Series of Metals and Alloys in Seawater**

**Anodic, least noble, most readily corroded**

Magnesium and magnesium alloys
Zinc
Commercially pure aluminum
Cadmium
Aluminum alloy (4.5% Cu, 1.5% Mg, 0.6% Mn)
Steel or iron
Cast iron
Chromium stainless steel, 13% Cr (active)
High nickel cast iron
18% Cr, 8% Ni steel (active)
18% Cr, 8% Ni, 3% Mo steel (active)
Hastelloy C (active) (62% Ni, 18% Cr, 15% Mo)
Lead-tin solders
Lead
Tin
Nickel (active)
Inconel (active) (80% Ni, 13% Cr, 7% Fe)
Hastelloy A (60% Ni, 20% Mo, 20% Fe)
Hastelloy B (65% Ni, 30% Mo, 5% Fe)
Chorimet 2 (66% Ni, 32% Mo, 1% Fe)
Brasses (Cu-Sn)
Copper
Bronzes (Cu-Sn)
Copper-nickel alloys
Monel (70% Ni, 30% Cu)
Silver solder
Nickel (passive)
Inconel (passive) (80% Ni, 13% Cr, 7% Fe)
Chromium stainless steel, 13% Cr (passive)
18% Cr, 8% Ni steel (passive)
18% Cr, 8% Ni, 3% Mo steel (passive)
Hastelloy C (passive) (62% Ni, 18% Cr, 15% Mo)
Chlorimet 3 (62% Ni, 18% Cr, 18% Mo)
Silver
Titanium
Graphite
Gold and Platinum

**Cathodic, most noble, least likely to corrode**

While the galvanic series gives a better representation of actual galvanic corrosion characteristics than the series of standard electrode potentials, the corrosion processes tend to change with time. The products of corrosion may accummulate at either the anode or the cathode, or both, and reduce the rate of the process. Whenever possible, corrosion tests should be conducted to explore the factors that affect the rate in a particular situation.[3, 4]

### Atmospheric Corrosion of Steel

Atmospheric corrosion is primarily due to the moisture and oxygen in the air. Atmospheres may be classified as industrial, marine, and rural. Industrial and marine atmospheres frequently contain such gaseous products as sulfur dioxide, hydrogen sulfide, and sodium chloride. These products tend to increase corrosion rates, especially in the presence of moisture.

Atmospheric corrosion involves both film formation and film breakdown. The oxidizing action of the air on the exposed metal surface brings about film formation, and electrochemical action causes the breakdown of film. This process occurs when there is moisture in the air.[5]

Most oxide films absorb moisture, so the presence of an electrolyte for electrochemical action is assured. The film may be highly protective of the metal surface, and corrosion will not take place. Should the film develop cracks, or be destroyed by some mechanical impact or abrasion, fresh surfaces of metal are exposed and concentration cells are developed.

Rain, besides providing moisture to the air, washes away part of the oxide film unless the particular oxide is exceptionally adherent. Rain also removes the accumulated corrosion products that help impede the rate of corrosion. Consequently, the atmospheric corrosion rate is increased for a time following a rain.

The gaseous sulfur compounds of the atmosphere in the presence of moisture produce acids which increase the conductivity of the liquid layer on the metal surface. Sodium chloride in marine atmospheres also increases conductivity of the liquid on the metal surface. Whereas with corrosion of metals immersed in solution the corrosion rate is controlled by the amount of available oxygen, in atmospheric corrosion the rate is controlled by the supply of moisture to the metal surface.

In the corrosion of iron and steel the primary product of corrosion is ferrous hydroxide, which oxidizes to hydrated ferric oxide (red rust). This product does not form an adherent film; it falls away or is easily blown from the surface. The resistance of steel to atmospheric corrosion can be increased by adding small amounts of copper (tenths of 1%) to the composition. Small amounts of nickel and chromium also improve corrosion resistance. This protection results because these alloys cause the steel to form tighter, more protective rust films. Nickel and copper form insoluble sulfates during atmospheric corrosion that do not wash off the metal surface. The stainless steels offer the best resistance to atmospheric corrosion. Galvanizing of steel appurtenances like handrails, stairs, walkways, gratings, etc., is a common way of protecting these items from atmospheric corrosion.

## Principles of Cathodic Protection

This method of preventing or reducing electrochemical metal corrosion makes the metal to be protected the cathode of the corrosion process. The electronic current supplied to the metal to be protected may come from either of two sources: sacrificial galvanic anodes, or impressed current through insoluble electrodes or expendable auxiliary anodes. Cathodic protection offers the most effective means for corrosion control. It is used universally in the offshore industry.[6]

## Sacrificial Galvanic Anodes

These anodes are bars of metal welded or bolted to the metal structure to be protected. The anodes are made of magnesium, zinc, aluminum, or alloys of these metals, and have a negative potential below that of the metal to be protected. A continuous and adequate current of electrons flows from the sacrificial metal through the welded or bolted joint into the metal to be protected, thereby assuring that it will be the cathode of the corrosion cell. A good sacrificial anode will develop insignificant anodic polarization during its service life; it will also have a low rate of dissolution.[7]

The electrical current produced by one sacrificial anode amounts to only a few hundred milliamperes; therefore, many anodes are required to protect a large structure. In a typical offshore structure there may be several anodes on a single brace. (See Figure 11-4.)

Figure 12-1 shows some of the various sizes and shapes of sacrificial anodes. Different sizes and shapes are available for every requirement in length-to-weight and surface-area-to-weight ratios. Common weights are 150, 200, 250, and 300 lbs (68, 91, 114, 136 kg). There are many methods for attaching anodes to the structure to be protected. All anode types can be installed or replaced underwater. During fabrication of the structure, welding is the most practical method of anode installation.

Figure 12-2, 12-3a, and 12-3b show typical installations of modern GALVALUM anodes. GALVALUM-III is an Al-In-Zn-Si alloy; to the aluminum base metal is added nominally 0.15% indium, 3.0% zinc, and 0.1% silicon. It contains no mercury, which has become an environmentally controversial element. The anode core for the 725-pound size is 2½-inch diameter steel pipe. Instead of gusset plates, some fabricators weld rectangular face plates to the anode core ends and then weld these doubler plates to the jacket brace.

**246  Introduction to Offshore Structures**

**BOLT-ON TYPES**

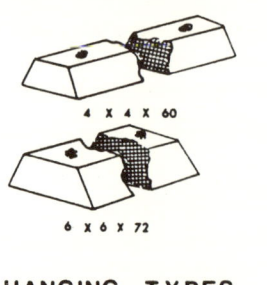

4 X 4 X 60

6 X 6 X 72

**HANGING TYPES**

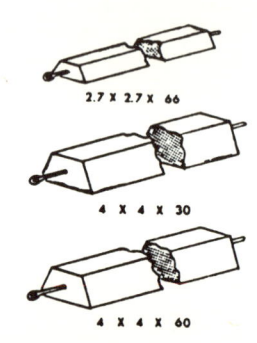

2.7 X 2.7 X 66

4 X 4 X 30

4 X 4 X 60

**WELD-ON AND CLAMP-ON**

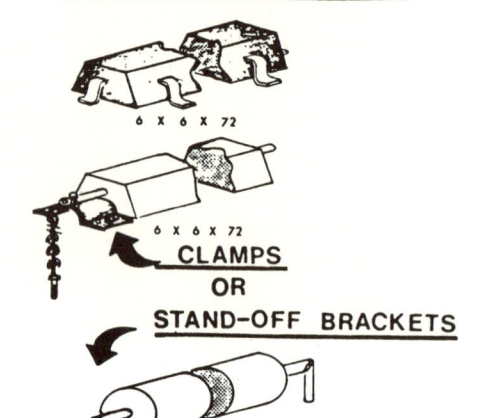

6 X 6 X 72

6 X 6 X 72

**CLAMPS**
**OR**
**STAND-OFF BRACKETS**

*Figure 12-1. Various anode shapes. (Courtesy National Assn. of Corrosion Engineers.)*

The 725-pound GALVALUM anode has a design output of 5.3 amperes and can protect approximately 900 ft$^2$ (84 m$^2$) of jacket structural area in Gulf of Mexico waters. The cathodic protection current density in the Gulf of Mexico is about 6 milliamperes/ft$^2$ (65 miliamperes/m$^2$); in the North Sea the cathodic protection current density approaches 12 milliamperes/ft$^2$ (130 milliamperes/m$^2$).

Figure 12-1 shows some of the various sizes and shapes of sacrificial anodes. Different sizes and shapes are available for every requirement in length-to-weight and surface-area-to-weight ratios. Common weights are 150, 200, 250, and 300 lbs (68, 91, 114, 136 kg). Figures 12-2, 12-3a, and 12-3b show typical installations of modern GALVALUM anodes.

## Impressed Current Cathodic Protection

This system can best be explained with a simple example. Suppose a steel structure to be protected and an expendable auxiliary electrode are immersed in an electrolyte. The two are wired or connected electrically through a battery or dc rectifier such that the negative side of the battery is fastened to the steel structure, and the positive side is fastened to the auxiliary electrode. Electrons

Corrosion 247

*Figure 12-2. GALVALUM-III sacrificial anode fastened to a jacket tubular member employing gusset plates at the anode core-to-structure junction. (Courtesy Cathodic Protection Services, Inc., a member of the CORRINTEC group of companies.)*

flow from the auxiliary electrode through the battery to the steel structure, making it cathodic or negative. The positive ions released into the electrolyte at the auxiliary electrode travel through the electrolyte to the protected structure and complete the electrical circuit.

Before the application of any external electrical power, the protective flow of electrons is limited to the difference in local electrode potentials divided by the sum of the electrical resistances in both the cathodic body and the auxiliary anode. To prevent corrosion, external electrical power is supplied through the battery or dc rectifier to boost the flow of electrons to the protected structure. Figure 12-4 shows the schematic arrangement of an impressed current system.[4]

**248** Introduction to Offshore Structures

**Figure 12-3a.** *A 725-pound GALVALUM-III sacrificial anode (for a 20-year life) welded in place on a jacket brace. (Courtesy National Assn. of Corrosion Engineers.)*

**Figure 12-3b.** *725-pound GALVALUM-III anodes installed between braces of a jacket. (Courtesy Harco Corporation.)*

Corrosion 249

Notice that the phrase "flow of electrons" has been used; this phrase is contrary to the conventional designation that positive electricity flows from positive to negative. Conventionally speaking, the magnitude of the positive protective current flowing from the auxiliary electrode to the protected structure through the electrolyte depends on the difference in the polarized potentials of the anode and cathode and on the magnitudes of their resistances. The magnitude of current per unit area of the protected metal surface expressed in amperes or milliamperes per square foot (mA/sq ft) is called *current density*. For minimum protection of submerged steel, the desired electrical potential measured at the surface of the cathode should be -0.85 volts referred to either a silver-silver chloride or a copper-copper sulfate reference electrode. The optimum or maximum potential depends on two factors: the equilibrium potential of the structure, and the degree of protection required. For galvanic anode systems with 5-10-year lifetimes, the potential range should be between -0.90 and -1.00 volts.

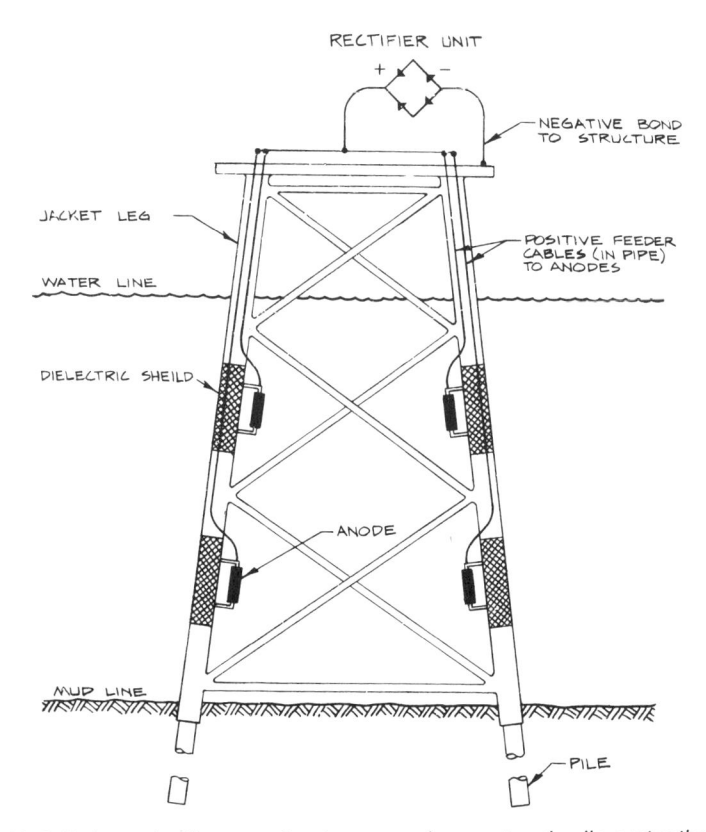

**Figure 12-4.** Schematic diagram of an impressed current cathodic protection system. (Copyright ©1969 Offshore Technology Conference.)

## 250    Introduction to Offshore Structures

For structures in seawater over 50-ft (15-m) deep, a rule-of-thumb provides that the protective current density should be 5 mA/sq ft (55 mA/sq m) for all steel surfaces from the water line to the mudline. The same general rule provides that there should be 3 mA/sq ft (33 mA/sq m) for all the surface area of the piling below the mudline. These numbers become greater as velocity of the water past the structure increases and as the water becomes colder. Seawater normally contains about eight ppm of dissolved oxygen. The dissolved oxygen content increases rapidly as the water gets colder. In Cook Inlet, Alaska, for example, the combination of cold water and ocean current requires a protective current density of as much as 20 mA/sq ft (218 mA/sq m).[7]

Auxiliary electrodes for the impressed current system in the form of expendable anodes are made from the same materials as sacrificial anodes for galvanic corrosion protection. Many materials have been used for inert impressed current electrodes in offshore applications. Graphite and high-silicon iron are widely used; steel and lead-silver-antimony alloy are also used. Platinized titanium and platinum-clad tantalum are the most expensive but, because of a long lifetime, are widely used. Platinized titanium electrodes have lifetimes in the range of five to seven years. Platinum-clad tantalum has a lifetime of approximately 20 years in seawater.

## Offshore Structure Corrosion Zones

The magnitude of the monetary investment required to place an offshore platform in the ocean makes control of corrosion mandatory. The overall corrosion protection system can represent as much as 10% of the total platform cost.

There are three corrosion zones: the immersed, the splash, and the atmospheric. Figure 12-5 illustrates the three zones and indicates the nature of the protection system in each zone.

**Immersed Zone**

Galvanic corrosion predominates in this zone. For galvanic corrosion to occur, the following conditions must prevail:

1. The various regions on the surface of the structure must have different electric potentials.
2. There must be a conducting path between the regions of different potential.
3. There must be an electrolyte on the surface of the structure.

All of these conditions are met by having a steel jacket submerged in seawater.

## Corrosion 251

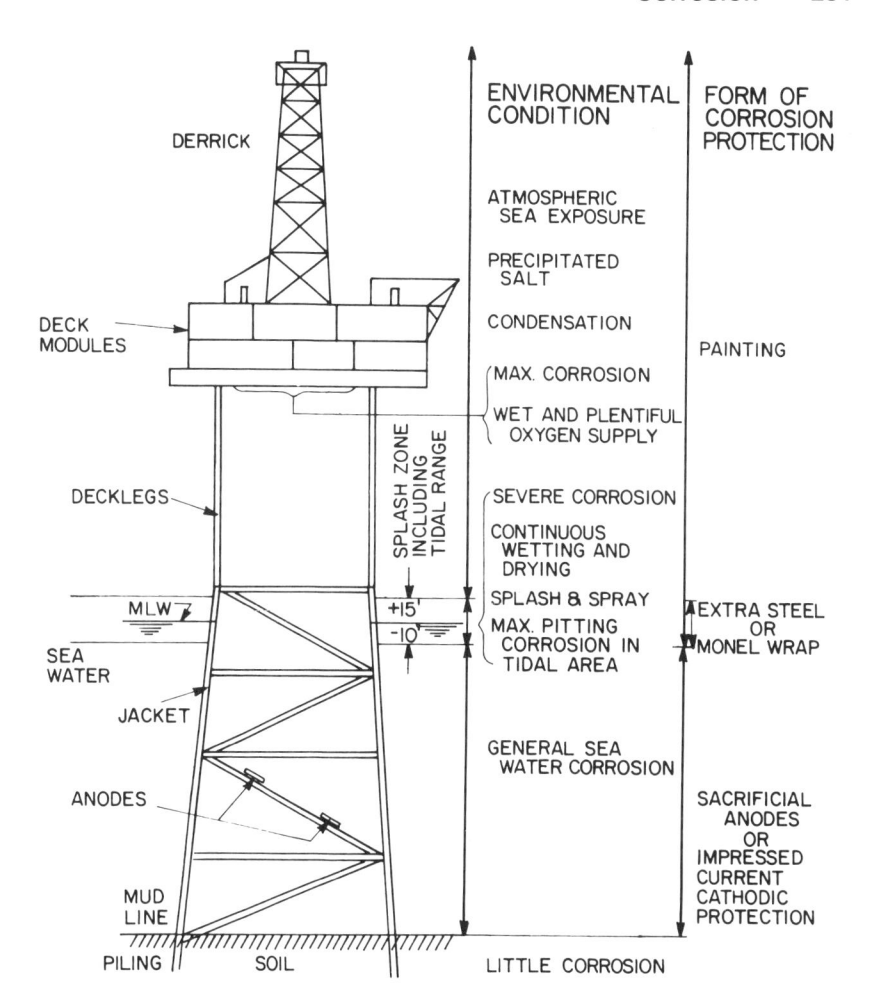

**Figure 12-5.** Corrosion zones on fixed offshore steel structures.

The rate of corrosion depends on these factors:

1. The magnitude of the potential difference between cathodic and anodic regions on the metal surface.
2. The electrical conductivity and temperature of the seawater.
3. The relative sizes of anodic surface areas to cathodic surface areas.
4. The local conditions at the particular anodic region.
5. The amount of oxygen dissolved in the seawater (the extent of aeration).

## 252　Introduction to Offshore Structures

Factor four pertains to such conditions as the impurities in the steel, extent of surface roughness, state of stress in the metal, presence of weld seams, nearness to some more noble metal, and motion of the seawater in the immediate vicinity.

The immersed zone is protected cathodically. Either a system of sacrificial galvanic anodes or an impressed current generator-rectifier with inert anodes is used. Protection is directly influenced by the location of the electric current source, the total area to be protected, water salinity, water temperature, water velocity, tide and current, and the shape of the surface.

Galvanic anodes are special alloys of aluminum and mercury, magnesium with small quantities of zinc and aluminum, or zinc with small quantities of aluminum and cadmium. Magnesium anodes deteriorate rapidly and may require replacement in two-three years. Zinc anodes last the longest, often up to 10-15 years. The mercury alloy of aluminum is considered the best compromise material considering initial cost and life to replacement.

Impressed electric current systems are more economical than galvanic systems, but they require a reliable source of power, and maintenance problems are associated with generators, cables, anodes, and rectifiers. The initial cost of a galvanic system is high. The difference in cost between a galvanic system and an impressed current system increases with increased ocean depth. A lead-silver-antimony alloy is the primary anode material for offshore impressed current systems. Graphite and cast iron deteriorate because of the high rates of electric current required.

### Splash Zone

Air dissolved in seawater increases the corrosion rate significantly. Dissolved air causes the metal surface areas near the sea surface to corrode more severely than surface areas at greater depths.

The splash zone extends from some distance below mean low water (MLW) to about 1.5-2.0 times this distance above MLW. The range depends on local conditions of tide, nominal wave height, and (in cold climates) ice abrasion. There are places in the Gulf of Mexico where the splash zone extends from -3.5 to +7 feet (-1 to 2 m); however, in general the range is larger, perhaps from -10 to +15 feet (-3 to 4.5 m).

It is desirable to avoid horizontal bracing in the splash zone. Since the splash zone is the most active corrosion zone, it is desirable to minimize the steel in this area. It is the most difficult area to sandblast and paint and, in cold climates, ice formations on braces in this zone have caused many failures.

The splash zone is the most critical area of protection because of alternate submergence and aeration. Noncorrosive coverings are used, sometimes in conjunction with increased thickness of steel members. The increased thickness of steel to offset deterioration over a long period of time is called a *corrosion allowance.* In the splash zone this allowance frequently means increasing the member-wall thicknesses by as much as ½ of an inch (12.7 mm).

In addition to the corrosion allowance, the surfaces in the splash zone are sandblasted and painted with about three mils of inorganic zinc paint followed by about nine mils of epoxy, phenolic, neoprene, or vinyl coating. These coatings are usually applied by spraying, or with brushes in those areas which cannot be reached with spray. Application is rarely attempted when the humidity exceeds 85% or the temperature dips below 50° F. The purpose of the coatings is to keep the seawater away from the metal surface. Moisture leaking through the coatings causes them to crack. The corrosion caused by the moisture eventually leads to peeling of the paint. Often, paint and other coatings are damaged by abrasion with objects. Where there are cracks in the paint, very rapid pitting corrosion occurs due to concentration of the corrosion process at the crack or gap.

In cold climates where ice floes impinge against jacket legs in the splash zone additional protection has been achieved with sheet metal leg wraps. Leg wraps are also widely used for splash zone corrosion protection. Stainless steel and Monel metal are most often used. These wraps in thicknesses approaching ⅜ of an inch (10 mm) are formed to the proper shapes and welded onto the members. However, use of leg wraps can create a secondary corrosion problem. The wrap material is more noble than the steel it protects. Those portions of the steel members just below the bottom edge of the wrap become anodic with respect to the wrap; this effect must be accounted for in the design of the overall corrosion protection system.

Monel metal is easily damaged. Floating objects will sometimes penetrate the sheeting due to impact; ice abrasion can be detrimental.

## Atmospheric Zone

The atmospheric zone, while being the area of least corrosion rate, is the most expensive to maintain on the basis of total expenditure over a period of time. Only coatings can be used for this portion of the structure. Usually, heavy-bodied coatings of epoxy mastic over zinc paint are used. The zinc paint contains 85-95% zinc by weight and may be based on organic or inorganic solvents. Inorganic zinc silicate coating is widely used.[8]

The coatings are preferably applied during construction yard fabrication. For optimum adhesion and durability, the steel surfaces are first sandblasted to the white-metal condition. The surface is painted immediately. If the coatings are applied after the platform is installed at the offshore site, the cost may be 50-60% more than construction yard application, and the quality will be much poorer.

Frequent inspection of coatings is required, since full corrosion protection is achieved only when the coating is uninterrupted or continuous over all the exposed metal area. One advantage of zinc paint is that the zinc corrosion products are voluminous and, if a break occurs in the epoxy coating, the galvanic action of the zinc layer beneath produces corrosion products that seal the gap in the coating.

In the offshore environment prompt repair of any damaged areas of paint or coating is necessary to avoid a more severe problem, perhaps even early failure of some superstructure component. After installation of the drilling platform and all the wells have been drilled, the corrosion protection coatings are renewed as the platform is fitted-out for its next function. Repainting is then required about every five to eight years thereafter.

In summary, adequate protection in the atmospheric zone consists of proper surface preparation, proper coating selection, and proper application. Figure 12-6 should be helpful in the initial selection of coating types.[9] The first four inner circles of the wheel give recommended applications (floors, tank linings, maintenance, and high temperature). The next circle gives three basic classifications of coatings: thermosetting, thermoplastic, and elastomer. The outside circle gives the generic types of coatings within the three main types. For example, the following coatings are recommended for high-temperature application (starting at the seven o'clock position and reading clockwise): inorganic zinc, silicone pigmented, silicone alkyd, and acrylic.

## Biological Corrosion

Although not strickly a type of corrosion, this deterioration of metal by various corrosion processes is a result of the metabolic activity of living microorganisms. These microorganisms are found in fluidized mediums with pH values from 6 to 11 (seawater pH is 8), in temperatures from 30 to 180°F (-1.1 to 82°C), and under pressures up to 15,000 psi (98 x $10^6$Pa).

The normal metabolic activities of microorganisms affect corrosion in seawater by:

1. Altering anodic and cathodic reactions
2. Changing surface films
3. Creating corrosion conditions (differential aeration)

There are two classes of microorganisms: aerobic, which require oxygen for their metabolic processes; and anaerobic, which live and grow in environments containing little or no oxygen.[10]

The three microorganism types of most concern in sea water corrosion are the sulfate-reducing anerobic, the thiosulfate-oxidizing aerobic, and the iron

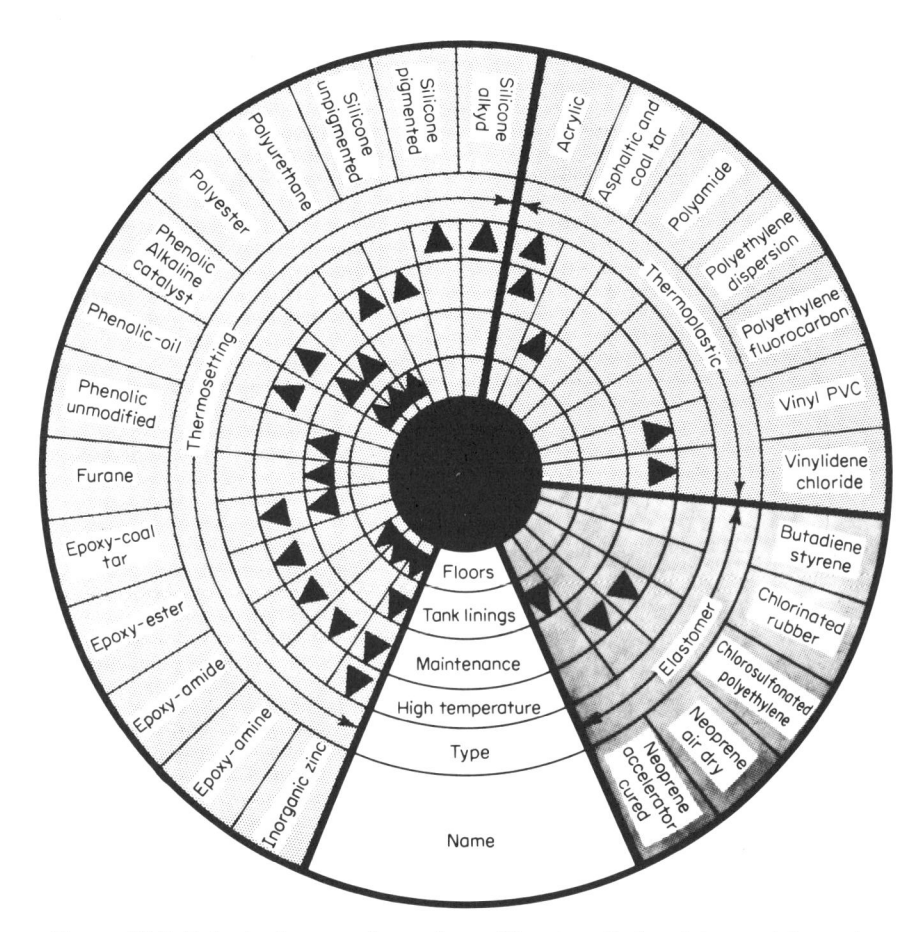

**Figure 12-6.** Selector for organic coatings. (Courtesy National Assn. of Corrosion Engineers and Fontana, M.B. and Green, N.D., Corrosion Engineering, McGraw-Hill Book Co., 1967, p. 220.

aerobic. Sulfate-reducing bacteria are by far the most troublesome. These bacteria have an accelerating influence on the corrosion behavior of buried pipelines, piles below the mudline, and deeply submerged steel components. In a low-oxygen environment the bacteria can utilize the hydrogen cathodically formed at the steel surface to reduce sulfate from the electrolyte and increase the local corrosivity of the environment. The corrosion result is the formation of pits on the steel surface filled with black iron sulfide and ferrous hydroxide products. Marine growth on piles is not a function of pile diameter.

Thiosulfate-oxidizing aerobic bacteria thrive in near neutral pH fluids containing dissolved oxygen, and increase the corrosive conditions by

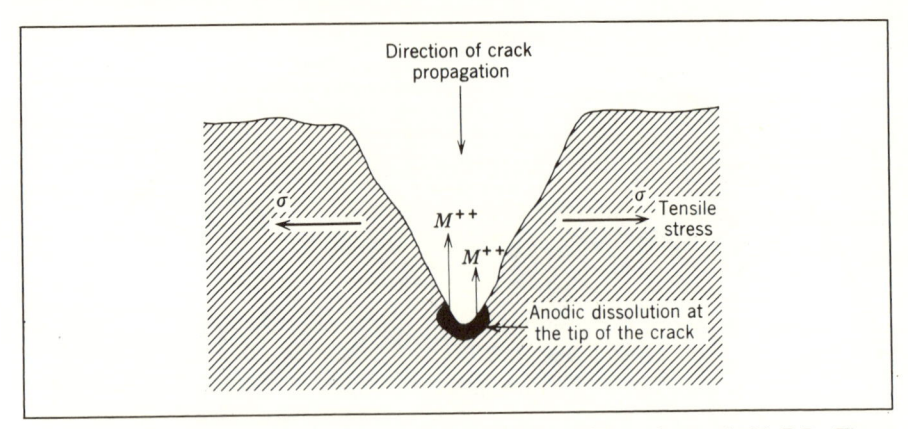

*Figure 12-7. Schematic of stress corrosion cracking. (From Jastrzebski, Z.D., The Nature and Properties of Engineering Materials, 2nd edition, John Wiley & Sons, 1976.)*

increasing the local sulfuric acid concentration. Iron aerobic bacteria assimilate ferrous iron from the electrolyte and precipitate ferrous hydroxide as tubercles or nodules on the steel surface. This accumulation tends to produce crevice corrosion.

Fouling organisms such as barnacles enhance localized corrosion. Differential aeration cells are created; sometimes the excrements of the organism itself cause increased corrosion.

## Stress Corrosion

The term *stress corrosion* is defined as the cracking of a metal under the combined effect of tensile stresses and a corrosive environment. Stress corrosion cracking is distinct and different from cracking caused by hydrogen embrittlement. In the latter case atomic hydrogen, which evolves at the metal surface in a hydrogen sulfide-rich solution, diffuses into the metal, collects in vacancies and dislocations, and causes lowering of metal strength and ductility.

Stress corrosion cracking is shown schematically in Figure 12-7. Cracking progresses in a direction perpendicular to the existing tensile stress. The source of the stress is immaterial; it could be applied, residual from forming or welding, or thermal. Almost no physical evidence of the formation of stress corrosion cracking will be visible, since the cracks are very fine until in the latter stages.

A corrosion pit or gouge on the metal surface acts as a stress raiser. Corrosion at the tip of the notch causes the stress corrosion crack to start there. The corrosion is highly localized, and the failure looks like a brittle fracture. Stess corrosion cracking is prevented by cathodic protection.

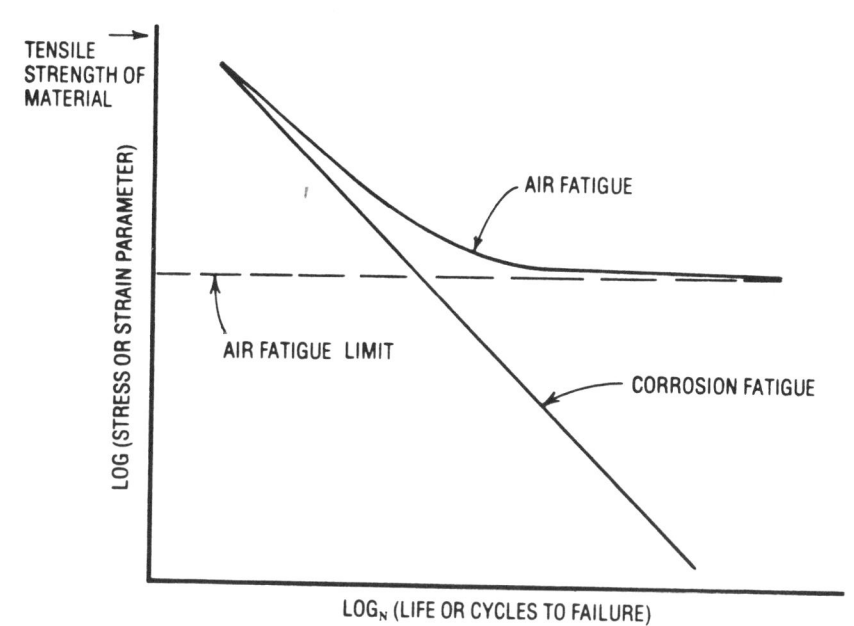

**Figure 12-8.** S-N diagram for corrosion fatigue. (Courtesy Harcourt Brace Jovanovich.) Also, see reference 12.

## Corrosion Fatigue

Corrosion fatigue is simply the reduction of the metal fatigue resistance caused by the presence of a corrosive medium. Fatigue cracks propagate faster in the presence of cyclic tensile stresses that expand the lattice structure of the metal. The frequency of application of the stress cycles has a pronounced effect on corrosion fatigue. The effect is most significant at low-stress frequencies. At low-stress frequencies, the metal lattice in expanded or strained condition has longer contact time with the corrosive environment.

As with stress corrosion, corrosion fatigue is most likely to occur in solutions or electrolytes that cause pitting attack. The corrosion pits serve as stress raisers and initiate cracks. Ferrous alloys submerged in seawater tend to lose the fatigue limits characteristic of their behavior for stress-cycling in air. The term *fatigue limit* indicates that cyclic stress level applied to the metal below which the metal will cycle an infinite number of times without fracture. In seawater steels appear to show continually decreasing ability to withstand cyclic stresses as the number of cycles of stress application increases. Figure 12-8 shows schematically the relationship of metal corrosion fatigue behavior to in-air fatigue behavior.[11]

Chapter 10 includes a more detailed discussion of fatigue behavior.

## References

1. Fontana, M.B. and Greene, N.D., *Corrosion Engineering*, McGraw-Hill Book Co., 1967.
2. Jastrzebski, Z.D., *The Nature and Properties of Engineering Materials*, 2nd Ed., John Wiley and Sons, 1976, Ch. 15.
3. Andrews, J.L., "Cathodic Protection Design," *Materials Protection*, Vol. 6, Jan. 1967, pp. 49-52.
4. Lehmann, J.A., "Cathodic Protection of Offshore Structures," 1969 OTC Preprints, Vol. 1, Paper #1041.
5. Hanson, H.R. and Hurst, D.C., "Corrosion Control: Offshore Platforms," 1969 OTC Preprints, Vol. 1, Paper #1042.
6. "Offshore Platform Design: Corrosion Prevention Begins with Initial Design," *Materials Protection*, Vol. 6, Jan. 1967, pp. 25-28.
7. Sansonetti, S.J., "To Prevent Corrosion at Sea: Think Aluminum Anodes," 1969 OTC Preprints, Vol. 1, Paper #1039.
8. Cook, A.R., "Zinc and the Protection of Offshore Structures," 1969 OTC Preprints, Vol. 1, Paper #1040.
9. Garrett, R.M., "How to Choose the Right Protective Coating," *Material Protection*, Vol. 3, Mar. 1964, pp. 8-13.
10. Cleary, H.J., "On the Mechanism of Corrosion of Steel Immersed in Saline Water," 1969 OTC Preprints, Vol. 1, Paper #1038.
11. Peterson, M.L., "Corrosion Fatigue of Offshore Welded Structures," *Offshore Drilling and Producing Technology*, 1976, pp. 103-107.
12. Peterson, M.L., "Corrosion Fatigue of Offshore Welded Steel Structures," *Ocean Engineering*, November 15, 1975.

# Part II

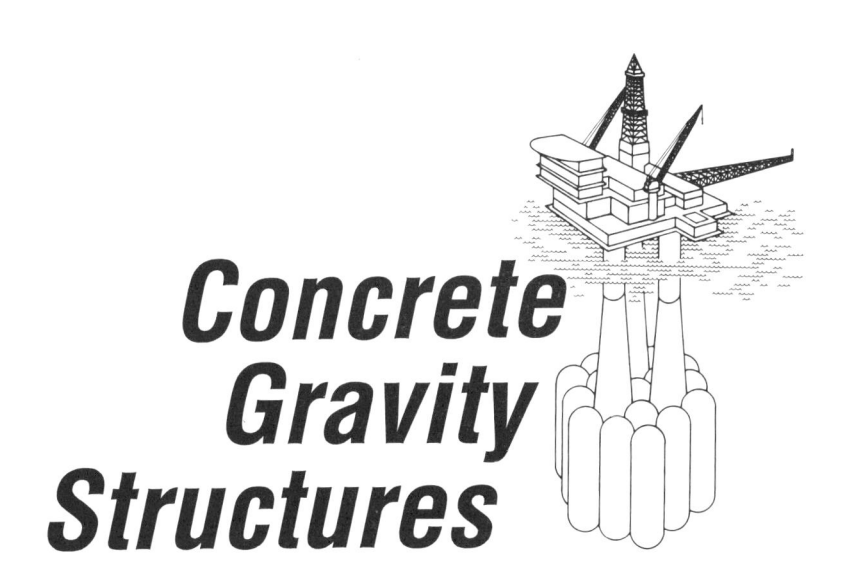

# Concrete Gravity Structures

# Chapter 13

# *The Gravity Platform*

As mentioned in Chapter 1, drilling for oil offshore began in the Gulf of Mexico in the 1930s with wells placed in swamps and marsh areas of Louisiana. Wooden decks were constructed on timber piles driven into the muddy soil.

The building of offshore structures grew gradually and steadily as the search for oil and gas progressed farther and farther offshore. Wooden piling gave way to steel tubular piling, and individual unsupported piles gave way to laterally braced systems of support as water depths increased. By the time of the oil crisis in 1973-74, several thousand steel jacket offshore structures were in service all over the world in such places as Alaska, Australia, Brazil, Indonesia, New Zealand, the Persian Gulf, and Zaire, to mention just a few.

The first well drilled in the North Sea was off the coast of Holland in 1960. The first major offshore gas field in the British sector of the North Sea was discovered in 1965. The Ekofisk oil discovery occurred in 1969; this event marked the beginning of rapid development in the engineering and technology of offshore structures in one of the most hostile environments yet encountered, and in water depths in excess of those previously faced by the industry.

Concrete gravity platforms are a relatively recent development in the offshore industry. The Ekofisk oilfield was the first in the North Sea to be produced commercially. During the last half of 1971 the initial wells in the Ekofisk field were connected through subsea pipelines to a jack-up drilling rig converted into a temporary production platform for separating the oil and gas, flaring the gas, and as an off-loading point for tankers. In June 1973 the first concrete offshore oil platform—the Ekofisk storage tank(*Ekofisk* 1)— was installed. This concrete gravity platform was constructed by the C.G. Doris Company for the Phillips Petroleum Company. If not the largest, the Ekofisk storage tank remains one of the largest offshore gravity structures.

## The Gravity Platform

The original purpose of the Ekofisk tank, which has a capacity of $5.6 \times 10^6 ft^3$ (160,000 m³) and stays in position by virtue of the gravitational pull of its large mass, was to store oil during periods when it was impossible to load tankers because of severe weather. Since 1973, a remarkable increase has occurred in the design, construction, and use of offshore concrete gravity platforms.

By the end of 1979, there were 14 concrete platforms either installed or under construction in the North Sea alone. Three concrete platforms have been installed in waters off Brazil. The principle of holding offshore structures in position by gravity has been employed in several places around the world. Table 13-1 gives information of principal interest about the 17 fixed offshore concrete gravity platforms installed or under construction as of 1979.

### General Features

Concrete gravity structures rest directly on the ocean floor by virtue of their own weight. These structures offer an attractive alternative to piled steel template platforms in hostile waters like the North Sea. The advantages are that the structures can be constructed onshore or in sheltered waters, towed semisubmerged to the offshore location, and installed in a short time by flooding with seawater (ballasting). Other advantages are the elimination of steel piling, the use of traditional civil engineering labor and methods, less dependence on imported building materials, a structure tolerant to overloading, and the high durability of the materials.

Many designs for gravity structures have been proposed; however, only a few of these designs have actually been constructed.[1] Figures 13-1a and 13-1b depict the concrete platform configurations presently employed. Some designs have a cellular concrete foundation, called a *caisson,* with two, three, or four hollow-concrete shafts or towers to support the deck structure. Other concrete designs consist essentially of an arrangement of vertical concentric circular or oblate shells joined by radial walls to form a large-diameter manifold structure. The deck structure is supported on columns extending up from the outer walls and the innermost cylinder. Figure 13-2 shows a platform of the first type as finally installed; Figure 13-3 shows the same platform after deck-mating and ready for towing to the offshore site. Figure 13-4 shows a platform of the manifold type under tow. Note the concrete beam substructure for the steel decks and the temporary booms for deck erection later at the offshore site.

## Table 13-1
### Survey of Fixed Offshore Concrete Platforms Either Installed or Under Construction
### Fall 1979

| | Type of design location | Main function | Design wave height (m) | Depth (m) | Approx. concrete volume (m$^3$) | Base diameter (m) | Storage capacity (mill. barr.) | Installation year |
|---|---|---|---|---|---|---|---|---|
| 1* | Doris Ekofisk 1 (N) | Storage | 24.0 | 70 | 90,000 | 92 | 1.0 | 1973 |
| 2* | Condeep Beryl A (UK) | Drilling, production, storage | 29.5 | 120 | 55,000 | 100 | 0.93 | 1975 |
| 3* | Condeep Brent B (UK) | Drilling, production, storage | 30.5 | 142 | 65,000 | 100 | 1.0 | 1975 |
| 4 | Doris Frigg CDP 1 (UK) | Drilling, compression, production | 29.0 | 96 | 60,000 | 101 | | 1975 |
| 5* | Sea Tank Brent C (UK) | Drilling, production, storage | 30.5 | 142 | 105,000 | 100 | 0.65 | 1978 |
| 6* | Sea Tank Frigg TP1 (UK) | Production | 29.0 | 104 | 70,000 | 72 | | 1976 |
| 7* | Sea Tank Cormorant A (UK) | Drilling, production, storage | 30.5 | 152 | 115,000 | 100 | 1.0 | 1978 |
| 8* | Condeep Brent D (UK) | Drilling, production, storage | 30.5 | 142 | 65,000 | 100 | 1.0 | 1976 |

| | | | | | | | |
|---|---|---|---|---|---|---|---|
| 9* | *Andoc* Dunlin A (UK) | Drilling, production, storage | 30.5 | 152 | 89,000 | 104 | 0.85 | 1977 |
| 10* | *Condeep* Statfjord A (N) | Drilling, production, storage | 30.5 | 149 | 88,000 | 110 | 1.3 | 1977 |
| 11* | *Condeep* Frigg TCP2 (N) | Treatment, compression, production | 29.0 | 104 | 50,000 | 100 | | 1977 |
| 12 | *Doris* Frigg MP2 (UK) | Compression station | 29.0 | 94 | 60,000 | 101 | | 1976 |
| 13 | *Doris* Ninian (UK) | Drilling and production | 31.2 | 139 | 142,000 | 140 | | 1978 |
| 14* | *Pub 3* Petrobras | Drilling, production, storage | 11.0 | 15 | 15,000 | 52 | 0.125 | 1977 |
| 15* | *Pub 2* Petrobras | Drilling, production, storage | 11.0 | 15 | 15,000 | 52 | 0.125 | 1978 |
| 16* | *Pag 2* Petrobras | Drilling, produciton, storage | 11.0 | 15 | 15,000 | 52 | 0.125 | 1978 |
| 17* | *Condeep* Statfjord B (N) | Drilling, production, storage | 30.5 | 149 | 130,000 | 169 | 1.5 | 1981 |

* Supervised by Det Norske Veritas

(Courtesy of Det Norske Veritas, Furnes, O. and Loset, O., "Shell Structures in Offshore Platform Application and Design.")

**264** Introduction to Offshore Structures

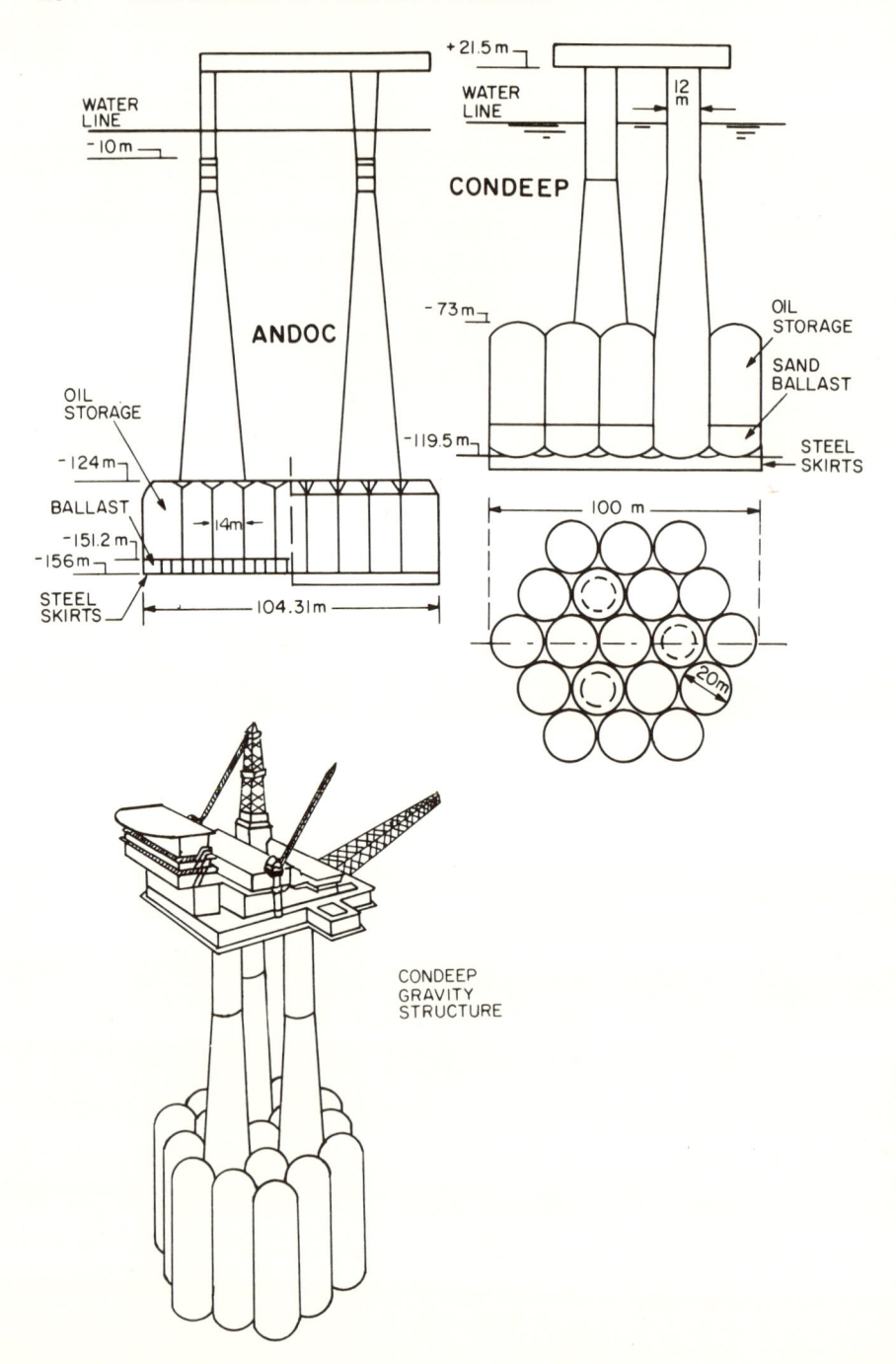

**Figure 13-1a.** Condeep *and* Andoc *gravity structures.*

**The Gravity Platform** 265

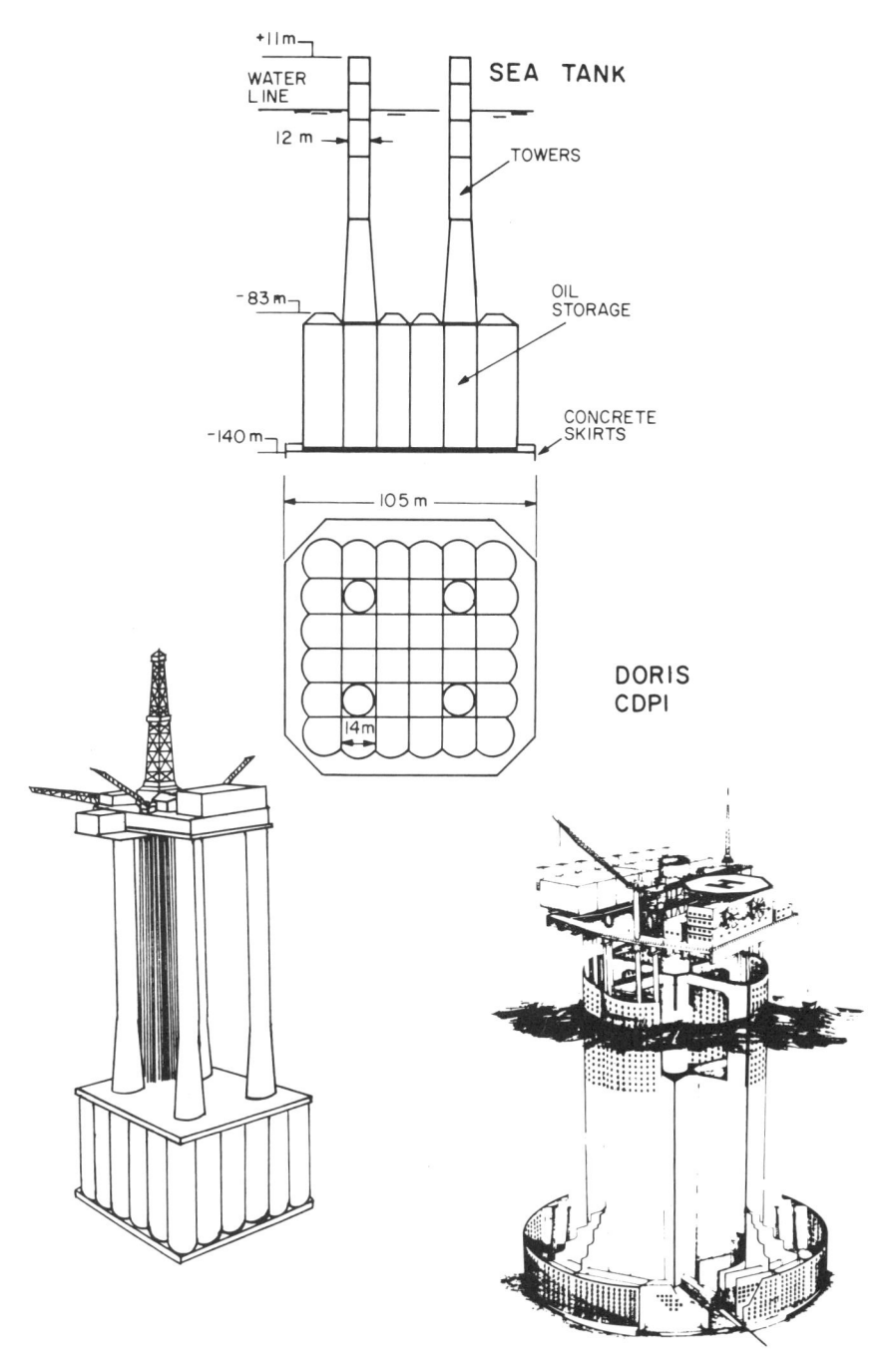

*Figure 13-1b.* Seatank *gravity structure and* Doris CDP 1 *platform.*

**266    Introduction to Offshore Structures**

*Figure 13-2. Mobil* Beryl A *as finally installed in the North Sea. (*BOSS '76, Volume I.*)*

The caisson of the multitower configuration generally extends up about one-third the height of the platform and is on the order of 328 ft (100 m) across or along the side at the bottom. The base of the manifold type, called a *base raft* because of the method of construction, has a diameter on the order of 492 ft (150 m).

The four concrete platform structures shown in Figures 13-1a and 13-1b are constructed almost entirely of various shells of large dimensions. The first *Condeep* platforms had caissons with 19 cells; the *Statfjord B* has a caisson with 24 cells. Each cell has an outer-shell diameter of 66 ft (20 m) and is more than 164 ft (50 m) in height. These vertical cylindrical shells, which are used for oil storage and ballast, are capped at each end with spherical domes. The shafts or towers are extensions of the caisson cells and consist of conical shells for the lower part and cylinders at the top.

The *Andoc* and *Sea Tank* structures are alike in many respects. Both designs use four towers and have caissons built-up of square cells for oil storage. The outer-cell walls of the caissons are constructed as circular cylindrical shell panels, and the caisson cell roofs consist of a combination of conical shells and flat slabs of variable thicknesses. One major difference between the *Andoc* and *Sea Tank* platforms is the mounting of the deck structure. The *Condeep* and *Sea Tank* platforms have the steel deck structure fastened to the tops of the concrete towers by a specially designed steel

*Figure 13-3. Completed Mobil Beryl A platform floating at proper draft for towing. (Copyright © 1976 Offshore Technology Conference.)*

transition joint. The towers on the *Andoc* platform change from concrete shells to steel shells just below the lower astronomical tide water level. The upper steel part of the tower is an internally stiffened cylinder 98 ft (30 m) long and 26-33 ft (8-10 m) in diameter. The steel deck structure mounts on the upper ends of the four steel cylinders.

All of the concrete gravity platforms constructed so far have had concrete skirts beneath their base slabs, or combinations of concrete skirts with steel skirts attached to them. These skirts, which are usually in the form of short cylindrical shells, serve several purposes. By their penetration into the ocean floor, they assure sufficient resistance against horizontal sliding. They also serve as a means of scour protection and form enclosures to facilitate grouting the foundation after platform installation.

**268    Introduction to Offshore Structures**

*Figure 13-4.* Frigg CDP 1 *platform being towed. (Courtesy Elf Aquitaine A/S.)*

Gravity platforms are said to have certain advantages over steel template platforms. Among the more commonly alleged advantages are:

1. Greater safety for personnel and facilities.
2. Towing to offshore site with essentially all deck equipment already installed thus minimizing installation time and cost.
3. Low maintenance cost of the submerged portion of the structure once it is on the site due to the nature of the basic material.
4. Adjustable crude oil storage capacity.
5. Protection against corrosion and shock to the steel risers by placing them inside concrete towers, tunnels, or the central cylindrical shaft.
6. Capability of supporting larger deck areas.
7. Possible access to the sea floor from inside one or more of the compartments in the cellular foundation.

The general design procedure for a gravity platform is shown in Figure 13-5.

## The Gravity Platform  269

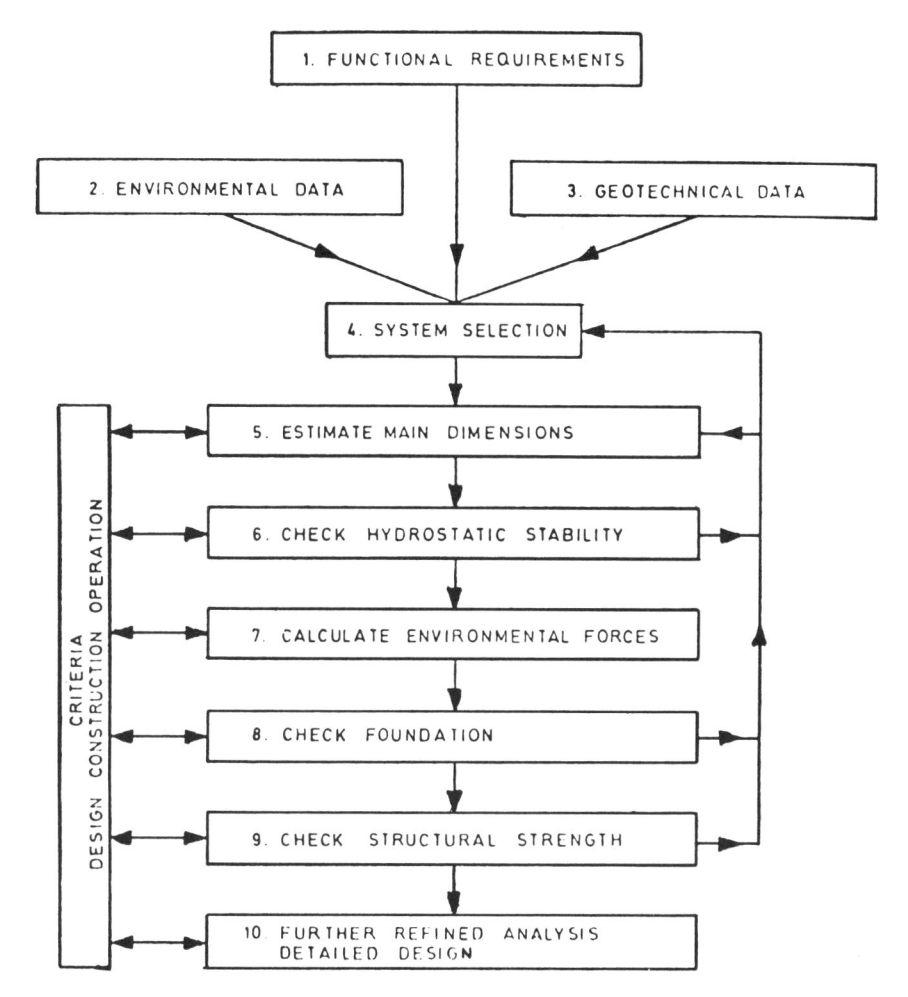

**Figure 13-5.** *General design procedure for a gravity platform. (From Svein Fjeld, "Concrete Structures," Second WEGEMT Graduate School on Advanced Aspects of Offshore Engineering, 5-16 March 1979, Technical University Aachen; Ship Model Basin Wageningen.)*

## References

1. "North Sea Oil Gravity Platforms: Who Is Proposing What," *New Civil Engineer Special Review,* London, England, May 1974, pp. 39-44.

# Chapter 14

# *Environmental Loads*

As in the case of steel template offshore platforms, waves and wind constitute the principal environmental loads on concrete gravity platforms. Water current, especially near the ocean bottom, is also of major concern. The ocean waves combine with the current as dynamic loads on the submerged parts of the platform. Wind acts on the superstructure, the modules, exposed equipment, derricks, and deck substructure. The distribution of wave heights and periods is of great importance.

The environmental loads under the design storm conditions constitute the major part of the total overturning moment on the platform and the transverse base shear at the mudline. Live loads on the platform are composed of stored materials, operating equipment, cranes, helicopters, etc. Live loads may shift in magnitude and location. Permanent loads are primarily from structure weight, ballast, and external water pressure.

Reliable environmental data are of utmost importance for adequate design of any offshore structure. The environmental loads used for design are usually based on extreme conditions with a recurrence period of 100 years. The normal design lifetime of the platform is 20-30 years.

## Wave Loads

On a concrete gravity platform, wave loads account for the major part of the environmental loading. Two basic methods of evaluating wave loads are the design wave method and the spectral analysis method.

**Design Wave Method**

This method has been used often in the past. The loads are derived from the passage of a single regular wave of given height and length past the structure.

The design load is determined by calculations based on regular waves and verified or supported by model tests in a wave tank. The most commonly used wave period for the North Sea is 15 seconds. The period used should cause the worst loading on the structure. This method works fine on piled steel template structures where nonlinear drag force (from Morison's equation) usually dominates. For large-volume structures like those shown in Figures 13-1a and 13-1b, the calculated wave force sometimes increases continuously with increasing wave period. In this situation the difficulties of establishing a design criterion have not been entirely resolved, but the placing of a 20-second limit on a wave period is usually adopted.

The usual design wave for the North Sea has been a wave of 100 ft (30 m) height with a 14-16-second period.

## Spectral Analsyis Method

Spectral analysis has frequently been used to evaluate wave loads as an alternative to the deterministic design wave method. This method has been found appropriate for gravity structures. Spectral methods are useful where the nonlinear drag loads are small compared to the linear inertial loads. Hence, a linear relationship exists between wave heights and wave forces for a given period.

If in the spectral analysis method only a few wave spectra are specified, the problem of choosing a peak period for the spectrum is much the same as in the design wave method. Thus, for short-term analysis, careful consideration must be given to the location of the wave spectrum along the frequency axis. In the long-term prediction of wave distributions and wave-induced loads on ships and other floating structures, the problem of choice of periods is largely avoided. The long-term prediction is obtained by summing up a great number of short-term statistically stationary conditions, each combined with a specified probability of occurrence. When determining the transfer function for loads on a large-volume structure, this influence of the structure on the flow field must be taken into account, e.g., with diffraction theory as mentioned in the next section.

### Importance of Morison's Equation

Concrete platforms of the tower type may be separated into two parts for determining the effects of environmental forces on the structure: the towers and superstructure, and the caisson. For tall, slender shafts or towers, use of the Morison equation for drag and inertia components of the wave force is reasonable. The Morison equation is applicable when the structural member diameter is small in comparison to the incident wavelength. When the tower

## 272   Introduction to Offshore Structures

diameter/wavelength ratio is 1:10 or less, Morison's equation is considered to apply satisfactorily. When the diamater of the towers appears large compared to the wavelength of the incident wave, a large dynamic amplification must be applied to the wave force determined by the Morison equation.

For large manifold type structures like the Ekofisk tank or a tower type platform caisson, the wave is scattered and reradiated by the structure. This effect tends to increase the wave particle velocities, and Morison's equation is not directly applicable. The horizontal wave force and overturning moment on such a large, submerged structure should be determined by diffraction theory.[1,2,3] Diffraction theory (linear three-dimensional potential flow theory) employs distributed sources in the flow field to evaluate the pressure distribution and resulting forces and moments acting on the large-volume structure. Use of diffraction theory is not so readily understandable as is application of Morison's equation with appropriate force coefficients. Diffraction theory also requires the use of a large digital computer.

In general, for concrete gravity platforms, the inertial forces predominate because of the size of the submerged structure relative to the wave dimensions. That is, the force from diffraction (which includes inertial effects) becomes large compared with drag forces. Calculating wave forces with diffraction theory is complicated.* In some situations the acoustic theory of scattering of plane waves by circular cylinders has been applied to ocean waves. The wave load on an arbitrarily shaped structure should be determined by a combination of theoretical analysis and model tests in a wave tank. Model tests are well suited for the determination of inertia forces.

The main dynamic motions of a concrete gravity platform are rocking and sliding. Sliding can be eliminated by the proper use of skirts and steel pipe dowels. Consequently, dynamic analysis for possible rocking motion is most important. Dynamic analysis is also important from the standpoint of resonance effects. Maximum load at resonance occurs when the natural period of the structure is such that it is excited by a wave of wavelength equal to an integer multiple of the distance between the towers.

### Wind Loads

The force of the wind on an offshore structure is a function of the wind velocity, the orientation of the structure, and the aerodynamic characteristics of the structure and its members. The forces exerted on a structure by the wind are expressed as

---

* For more information refer to *Rules for the Design, Construction and Inspection of Fixed Offshore Structures, Second Edition,* Det Norske Veritas, Oslo, Norway, 1977.

## Environmental Loads

$F_D = C_D \, 1/2 \, p V_z^2 A$ (force parallel to wind)
$F_L = C_L \, 1/2 \, p V_z^2 A$ (force perpendicular to wind)

where:

$C_D$ = drag coefficient
$C_L$ = lift coefficient
$p$ = density of the air
$V_z$ = wind velocity at height $z$
$A$ = area perpendicular to wind velocity

Values for $C_D$ and $C_L$ may be found in Reference 4. In preliminary calculations the wind force perpendicular to the wind direction is often neglected.

Because of the shear forces with the earth's surface, the wind velocity is not constant but is zero at the surface and increases exponentially to a limiting maximum speed known as the *gradient wind*. Over water, the wind speed at any elevation is represented as

$$V_z = V_{Ref} \left[ \frac{z}{Ref} \right]^{1/7}$$

where $V_{Ref}$ represents the wind speed at a height of 30 ft (10 m), which is the customary elevation for such measurements; $z$ is the desired elevation, either in ft or m; and Ref is the reference height, either 30 ft or 10 m, depending on whether $z$ is in ft or m. This equation is called the One-Seventh Power Law.

The wind effects on all parts of the above-water structure should be calculated. Two kinds of wind speeds are normally considered:

1. *Sustained wind speed.* This is the average wind speed during a time interval of one minute.
2. *Gust wind speed.* This is the average wind speed during a time interval of three seconds. Wind is quite gusty. The gust factor is that multiplier which must be used on the sustained wind speed to obtain the gust speed, or the fastest-mile velocity. The average gust factor $F_{10}$ at the 10 m elevation is in the range of 1.35-1.45. The variation of gust factor with height is negligible.

Structures and components loaded primarily by wind (as oppposed to wind and wave), such as pieces of equipment or modules on the platform decks, should be designed for the fastest-mile velocity with a period of recurrence at a given site of 100 years. For computation of wind forces in conjunction with

## 274   Introduction to Offshore Structures

maximum wave force, the 100-year sustained wind velocity should be used. As a large variation of wind direction may occur during a storm, all orientations of the structure should be analyzed to assure a safe design.

The design wind gust speed often used for the North Sea is 120 knots.

### Current Forces

There are two major components of the current to be considered: tidal current and wind-driven current. The tidal current is selected from available statistics. The wind-driven current at the still water surface is generally taken as 1% of the sustained wind at the reference level of 30 ft (10 m) above the still water level.

The presence of current in the water produces three distinct effects. The current velocity should be added vectorially to the wave particle velocity before computing the drag force. Because drag depends on the square of the horizontal particle velocity, and because the current velocity decreases slowly with depth, a comparatively small current can increase drag significantly. The second effect is a steepening of the wave profile from changing the wave celerity. This effect is very small and may be neglected. The third effect makes the structure itself generate waves which in turn create diffraction forces. However, these diffraction forces are negligible for realistic values of current.

Water current may be important for the design of auxiliary equipment such as exposed risers and well conductors. Such components should be investigated for the possiblity of vortex shedding, which is a result of large current velocity.

### References

1. Garrison, C.J.; Torum, A.; Iversen, C.; Lievseth, S.; and Ebbesmeyer, C.C.; "Wave Forces on Large-Volume Structures: A Comparison Between Theory and Model Tests," 1974 Offshore Technology Conference Preprints, Paper 2137.
2. Torum, A.; Larsen, P.K.; and Hafskjold, P.S.; "Offshore Concrete Structures: Hydraulic Aspects," 1974 Offshore Technology Conference Preprints, Paper 1947.
3. Loken, A.E.; and Olsen, O.A.; "Diffraction Theory and Statistical Methods to Predict Wave-Induced Motions and Loads for Large Structures," *1976 Offshore Technology Conference Proceedings,* Paper 2502.
4. Myers, J.J.; Holm, C.H.; and McAllister, R.F.; "Wind and Wave Loads," Section 12 of *The Handbook of Ocean and Underwater Engineering,* McGraw-Hill Book Co., 1969.

# Chapter 15

# *Geotechnical Design*

### General Aspects

Concrete gravity platforms depend on the bearing pressure of large concrete slabs, often with downward projections underneath called *skirts*, that rest on the unprepared ocean floor to provide foundation support against the maximum environmental loads imposed on the structure. Thus, marine soil mechanics is of utmost importance to the foundation design of such platforms. Conversely, confidence in the foundation design can be no better than the confidence in the soil mechanics data on which it is based.

The foundation has to resist extremely large overturning moments and large horizontal forces resulting from waves and wind. These effects tend to cause the gravity platform to topple over, slide along the ocean bottom, or rock back and forth until a bearing capacity failure of the soil occurs. Scouring of the ocean floor around and adjacent to the foundation is a problem during storms. The magnitudes of the wave-induced forces resulting from design waves acting on several platforms are given in Table 15-1.

Soil conditions at the site must be mapped carefully to allow safe installation of the platform and ensure permanent positioning throughout the platform lifetime.

The major problems to be considered in foundation design are:

1. Type and extent of contact between the platform and the seabed.
2. Stability of the platform against sliding and overturning.
3. Installation (dowel and skirt penetration, grouting process).
4. Settlement of the platform (initial and long-term).
5. Effects of cyclic loading on the soil.

**276** Introduction to Offshore Structures

Table 15-1
Wave-Induced Forces

| Site | Water depth (m) | Extreme wave height (m) | Maximum horizontal force (tonnes) | Maximum overturning moment (tonne-meters) |
|---|---|---|---|---|
| *Ekofisk* | 70 | 24.0 | 78,600 | $3.35 \times 10^6$ |
| *Frigg CDP 1* | 100 | 29.0 | 69,130 | $3.33 \times 10^6$ |
| *Frigg MP 2* | 97 | 29.0 | 69,150 | $3.13 \times 10^6$ |
| *Ninian Central* | 139 | 31.2 | 102,900 | $4.18 \times 10^6$ |

Note: 3.28 ft = 1m
0.907 tonne = 1 ton
1 tonne m = $7.2 \times 10^3$ lb ft

6. Modeling of the soil for dynamic analysis of the water-soil-structure interaction behavior.
7. Instrumentation for performance measurements during and after installation on the ocean floor.

After construction work has begun on a gravity platform, modifications in design can be accommodated to only a small extent. Thus, it is important for the ocean bottom soil survey to be for the exact site and in detail. The soil investigation must be completed early while the analytical models for analysis are being formulated.

The soil investigation should include a geophysical survey with soundings, core samples for laboratory testing, and *in situ* testing.[1] The geophysical survey is to map the seabed typography and to determine the depth of the different soil layers. A geophysical survey does not give any information about the soil itself, only changes in the soil properties are registered.

A site investigation for a gravity structure includes soil borings accompanied by measurements with *in situ* vane and cone penetration equipment. Cone penetration tests are widely used for *in situ* testing of the ocean bottom. Such tests give a continuous profile and also allow detection of thin layers with different material properties. Cone penetrations depend on the denseness of the soil, but typically will range from 20 to 80 ft (6 to 24 m). In addition, a detailed bathymetric survey is necessary to determine the general relief of the bottom topography. A radar sidescan survey provides information regarding obstructions such as boulders and debris that may interfere with the installation of the gravity structure.

Gravity platforms are continuously subjected to repeated force reversals because of wave action. Sands and clays in the ocean floor behave differently when subjected to cyclic loading. The foundation stability for a given wave must therefore take into account the previous loading history of the soil. Cyclic loading by ordinary waves may tend to reduce the ability of the foundation to resist large design forces. The stress level is important in determining the effects of cyclic loading. The wave distribution for the maximum design condition must be defined, and the relationship between wave height and soil shear stress must be established.[2, 3]

To design risers and conductor tubes, the settlements of the foundation must be estimated. There is an initial settlement which comes immediately with platform installation. Other settlements from consolidation and secondary creep occur gradually during the platform lifetime. These latter settlements depend on the wave action effects and the preconsolidation pressure in the soil. The soil deformations introduce significant stresses in the conductor tubes under the platform as well as in the risers. Axial stresses arise from negative skin friction along the conductors, and bending stresses develop from horizontal and shear deformations within the soil.

The analyses performed for the design of the conductor tubes are similar to the ones performed for laterally loaded piles. The local soil foundation stiffness increases around conductors. This increase may cause local concentration of contact stresses on the platform base.

## Foundation Stability

The foundation design must be finalized before construction begins. The foundation is the first structural element of the gravity platform to be completed.

For stability, a gravity structure requires a large base area and sufficient weight for stability.[4, 5, 6] One of the primary conditions to be met by the foundation is that it be nonlifting. This condition implies that the stabilizing moment from the large weight must increase simultaneously with the applied overturning moment. The required weight is usually provided for by constructing the caisson or foundation as a cellular unit which can be filled with sand, water, or oil. Often, the oil storage cells are surrounded by water storage cells so that any oil leak can be detected by monitoring the water going in and out of the outer cells.

Ordinarily, as the oil storage cells are filled with crude oil, the seawater is automatically forced out. Likewise, as the oil is pumped out, seawater is automatically pumped back in. The storage cells are filled at all times. In some gravity platforms which do not have storage capability the cellular foundation is permanently ballasted with sand and seawater.

# 278     Introduction to Offshore Structures

Certain limiting conditions control the analysis for the stability of the foundation. The assumptions are:

1. The soil/structure contact is always maintained.
2. The motions of the foundation (slipping and oscillation) remain within permissible limits.
3. No liquefaction risk exists in the sand layers.

Overall gravity platform foundation stability must be examined for the situation of a long storm followed by a wave of magnitude equal to the design wave. In such a situation the foundation has reduced strength and stiffness from the previous cyclic loading. Because of the long wave period associated with the design wave, the stability analysis is performed as a static analysis. To get the total wave force, the static force is multiplied by the dynamic amplification factor.

Because of storm magnitude wave forces, the gravity platform experiences cyclic horizontal displacements and rotations as well as a gradual increase in vertical settlement—even if there is no dissipation of the excess pore water pressure. A redistribution of stresses under the platform takes place as the soil weakens locally.[7,8] During a storm there may be significant dissipation of excess pore pressures in sand. However, for a clay foundation, no significant drainage occurs even during a series of storms.

For a platform resting on sand or normally consolidated clay, the cyclic displacements will decrease, and the stability will improve with time. But for a structure on overconsolidated, dilating clay, the long-term effect may be increased cyclic displacements and reduced stability. The ocean floor of the North Sea generally consists of clay strata covered by a relatively thick layer of dense sand.

## Skirts

For foundations resting on clay layers, the surface layers frequently show insufficient slipping strength. Therefore, the more resistant soil layers must be reached with skirts installed beneath the base raft. These skirts must be thin enough to penetrate the ground and yet be capable of transmitting the vertical and horizontal forces imposed on the foundation.

Skirts serve three purposes:

1. They penetrate the weaker near-surface soils
2. Transmit the applied loads to the stronger soil structures beneath
3. Protect the foundation from scour.

Skirts are designed (width, length, and spacing) so that the platform's structural weight will be sufficient to produce full skirt penetration during

# Geotechnical Design 279

installation. They must also provide adequate resistance against sliding of the platform under the action of the large horizontal wave load.

The presence of an uneven ocean floor topography or the presence of boulders or other obstructions on the sea bottom may prevent full skirt penetration. In some cases boulders are removed prior to installation. Steel pipe dowels project below the skirts to penetrate the ocean floor first and prevent aquaplaning or skidding at the moment the skirt touches the bottom. The cavities created inside the skirts after they penetrate the soil are filled with concrete grout to provide firm contact pressure at all points within the base.

Figures 15-1 and 15-2 show diagrammatically the dowels and skirts for a *Condeep* gravity platform. The location of the steel dowels in plan view between the caisson cells is shown in Figure 15-1. The general vertical arrangement of dowels and skirts is shown in Figure 15-2. The heavy lines in Figure 15-1 outline the several skirt compartments formed beneath the caisson by the steel skirts.

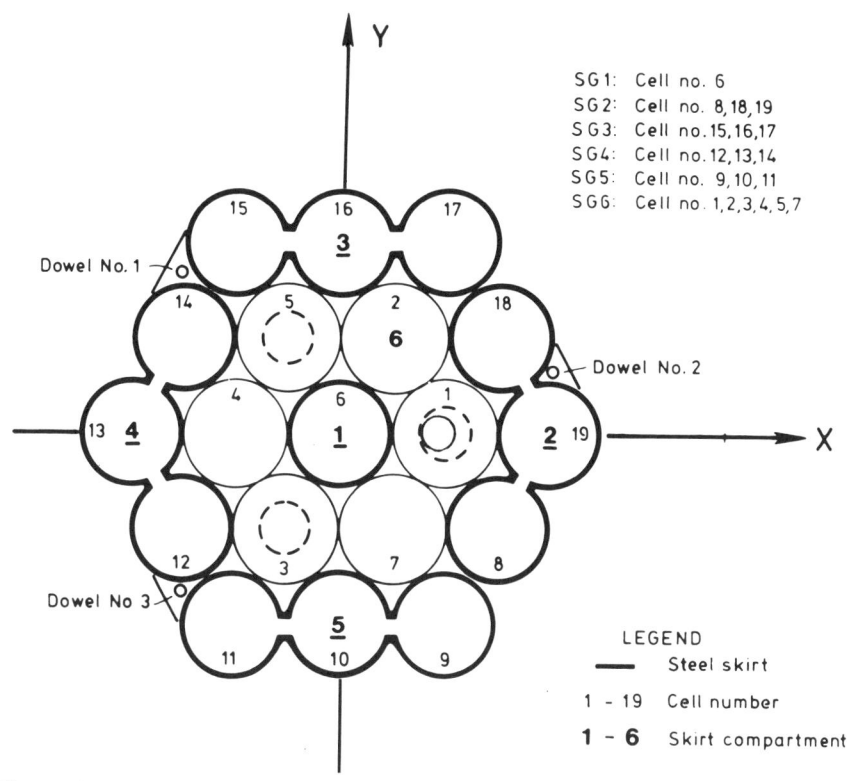

**Figure 15-1.** *Cell and skirt compartment number and dowel location. Note: SG stands for skirt compartment. (Copyright © 1976 Offshore Technology Conference.)*

**280** Introduction to Offshore Structures

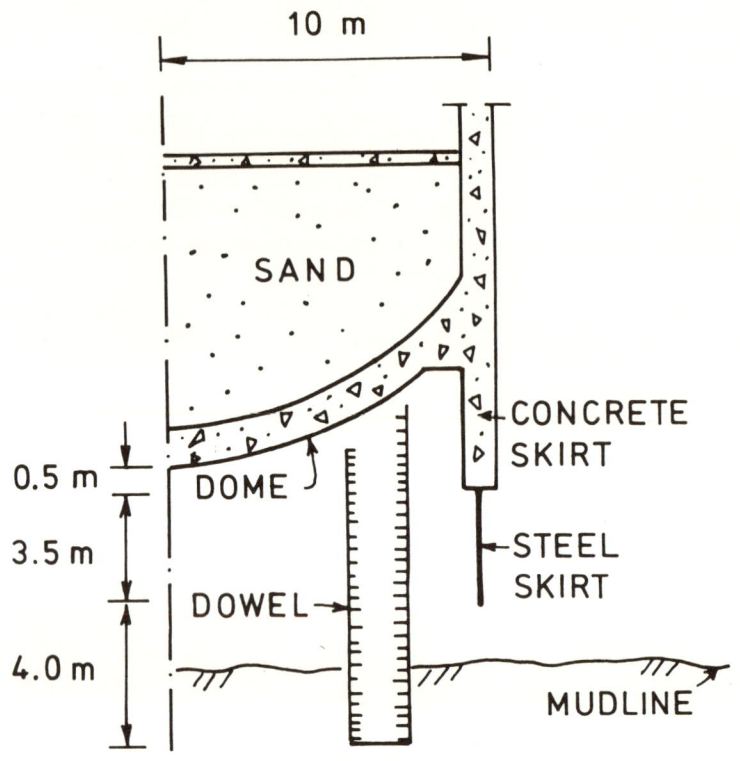

*Figure 15-2. Detail of the base of the structure. (Copyright © 1976 Offshore Technology Conference.)*

## Foundation Failure Modes

The vertical loads on a concrete gravity platform are primarily dead loads which can be estimated with reasonable accuracy. A greater uncertainty is involved in the horizontal loads. In general, horizontal loads vary from about 15-35% of the vertical loads.

The distributed wind and wave loads on the platform can be resolved into a horizontal force on the foundation at the mudline and an overturning moment about the mudline. When the foundation is subjected to a horizontal force, at least three potential modes of sliding failure exist. Which particular failure mode develops depends on many factors, including soil strength characteristics of both the near-surface and the underlying layers, skirt height, orientation and spacing of skirt elements, and the net vertical load on the foundation.

Sliding is the simplest foundation failure mode. Figure 15-3 shows schematically six variations of this failure mode for a typical cylindrical caisson with concentric steel skirts.[9,10] The simplest type of sliding failure

**Geotechnical Design** 281

*(a) PASSIVE WEDGE FAILURE*

*(b) DEEP PASSIVE FAILURE*

*(c) SLIDING BASE FAILURE*

*(d) SLIDING FAILURE IN SHALLOW WEAK ZONE WITH WIDELY SPACED SKIRTS*

*(e) SLIDING FAILURE IN SHALLOW WEAK ZONE AVOIDED WITH CLOSELY SPACED SKIRTS*

*(f) SLIDING FAILURE IN DEEP WEAK ZONE*

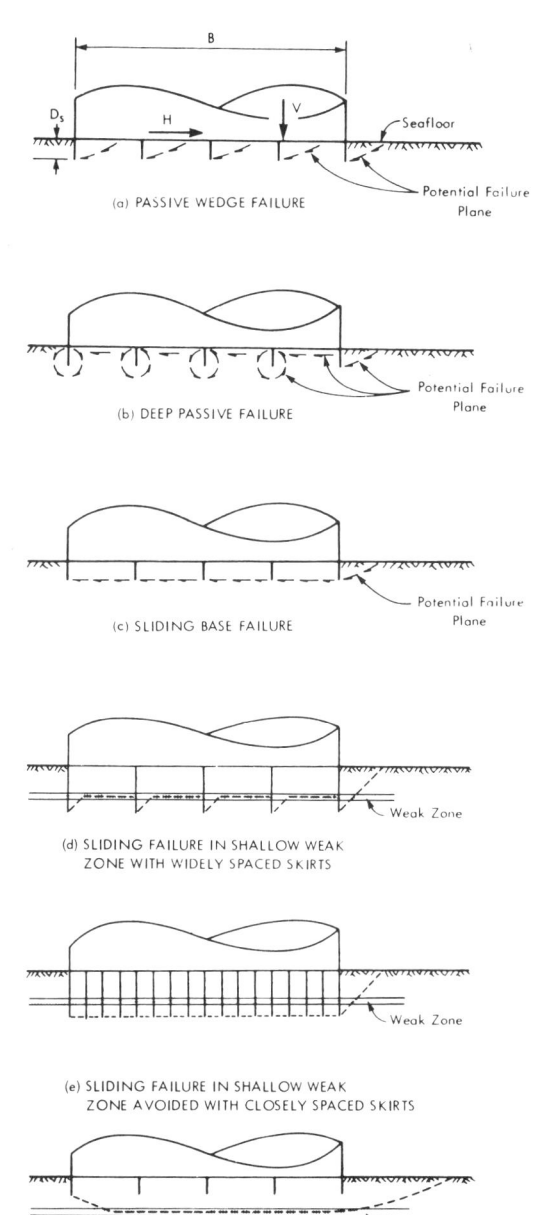

*Figure 15-3.* Schematic illustration of some possible failure modes for sliding resistance. Note: H and V are the horizontal force and vertical force, $D_s$ is the depth of skirt, and B is the diameter of the caisson. (Copyright © 1975 Offshore Technology Conference.)

occurs when the shear resistance of the soil at the soil-structure interface is smaller than the shear resistance in the greater soil mass. Failure wedges develop beneath the slab, cause the foundation to move upward, and destroy the sliding resistance. (See Figure 15-3a.) If the net vertical load is larger than some critical value (which varies with the skirt and soil conditions), upward movement of the foundation from the development of a failure wedge is not possible. In this second failure mode (Figure 15-3b) sliding failure may develop within almost all of the soil foundation area and extend along and beneath each skirt. A third possible type of sliding failure may be thought of as a horizontal failure plane in the soil passing through the skirt tips as shown in Figure 15-3c. The sliding resistance is effectively the soil shear strength along this plane. Three other possible types of sliding failure are shown in Figures 15-3d, 15-3e, and 15-3f. Skirts which penetrate to the deeper, stronger soil layers normally prevent foundation failure by sliding.

While the sliding failure mode is thought to be the most critical because it depends on the properties of the weakest soil layer near the soil-structure interface, there are other possible failure modes. The bearing capacity of the soil primarily depends on its average strength over a considerable depth, a distance about equal to the foundation radius. Bearing capacity failures are deep-seated and occur when the shear strength of the soil is exceeded. (See Figure 15-4.) This type of failure is typical for foundations resting on clay. The effect of repeated loading on such a foundation may be critical, as it causes a reduction in soil shear strength.

Whether or not a large foundation resting on sand will have a bearing capacity failure from transient wave loadings depends on the undrained shear strength of the soil. Large-foundation gravity platforms resting on sand are not very likely to experience a bearing capacity failure. However, the high shear stresses in the soil may lead to large deformations. In combination with high hydraulic gradients around the periphery of the foundation, repeated loading and large deformations may lead to soil softening and a progressively growing rocking motion. A rocking type failure could develop. (See Figure 15-4c.)

Liquefaction may be considered an important but separate type of failure for large foundations resting on sand. (See Figure 15-4d.) In a storm the alteration of the soil stress/strain properties and the build-up of pore water pressure under cyclic loading significantly affect foundation movements. Not only may the weakening of the soil affect the foundation movements during the storm, but the dissipation of the excess pore water pressure may cause further foundation movements afterward. Liquefaction failure is a possibility if the pore water pressure in the sand should increase so much during a storm that the effective stress in the soil approaches zero. The sand loses its strength, behaves like a heavy liquid, and the platform may sink into the fluidized soil. Eccentric loading on the deck may cause the platform to tilt at the same time.

a) SLIDING

b) BEARING CAPACITY FAILURE

c) ROCKING

d) LIQUEFACTION

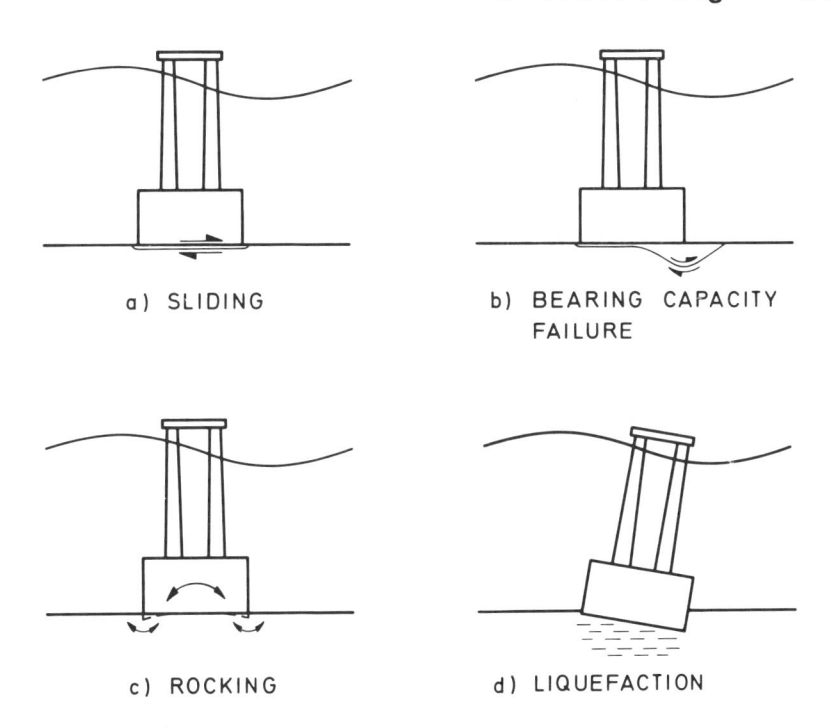

**Figure 15-4.** *Possible failure modes for the foundation of a typical gravity platform. (Copyright © 1974 Offshore Technology Conference.)*

## References

1. Hitchings, G.A.; Bradshaw, H.; and Labiosa, T.D.; "The Planning and Execution of Offshore Site Investigations for a North Sea Gravity Platform," *1976 Offshore Technology Conference Proceedings,* Paper 2430.

2. Hoeg, K.; "Foundation Engineering for Fixed Offshore Structures," *Proceedings of the First International Conference on the Behavior of Offshore Structures, BOSS-76, Vol. I,* pp. 39-69.

3. Roren, E.M. and Furnes, O.; "Behavior of Structures and Structural Design," *Proceedings of the First International Conference on the Behavior of Offshore Structures, BOSS-76, Vol. I,* pp. 70-111.

4. Janbu, N.; Grande, L.; and Eggereide, K.; "Effective Stress Stability Analysis for Gravity Structures," *Proceedings of the First International Conference on the Behavior of Offshore Structures, BOSS-76, Vol. I,* pp. 449-466.

5. Vaughan, P.R.; El Ghamrawy, M.K.; and Hight, D.W., "Stability Analysis of Large Gravity Structures," *Proceedings of the First International Conference on the Behavior of Offshore Structures, BOSS-76, Vol. I,* pp. 467-487.

6. Lauritzen, R. and Schjetne, K.; "Stability Calculations for Offshore Gravity Structures," *1976 Offshore Technology Conference Proceedings,* Paper 2431.
7. Dibiagio, E.; Myrvoll, F.; and Hansen, S.B.; "Instrumentation of Gravity Platforms for Performance Observations," *Proceedings of the First International Conference on the Behavior of Offshore Structures, BOSS-76, Vol. I,* pp. 516-527.
8. Foss, I.; "Instrumentation for Operation Surveillance of Gravity Structures," *Proceedings of the First International Conference on the Behavior of Offshore Structures, BOSS-76, Vol. I,* pp. 545-556.
9. Young, A.G.; Kraft, L.M.; and Focht, J.A.; "Geotechnical Considerations in Foundation Design of Offshore Gravity Structures," *1975 Offshore Technology Conference Proceedings,* Paper 2371.
10. Hove, K. and Foss, I.; "Quality Assurance for Offshore Concrete Gravity Structures," 1974 Offshore Technology Conference Preprints, Paper 2113.

# Chapter 16

# *Structural Design*

There are four principal differences between concrete structures onshore and those offshore. These differences are: (1) safety requirements, (2) functional requirements, (3) environmental loads, and (4) structure size.

Offshore concrete gravity structures are huge by any standard. Their size, coupled with the large environmental forces, cause design problems without precedence. The safety as well as the functional requirements must be determined with great care when considering the environment in which such platforms are placed and the grave consequences of a possible failure.

The structural design requirements on a concrete gravity structure may be divided into three categories: (1) strength, (2) material quality, and (3) serviceability. It is because of the properties of reinforced and prestressed concrete that gravity platforms have been made possible. Concrete in general, and prestressed concrete in particular, offer good resistance to marine corrosion and fatigue.

Gravity platforms are installed by controlled ballasting (flooding) with sea water. The installation operation (immersion and placing on the ocean floor) usually requires less than 24 hours. Often, the caisson cells are not pressurized; consequently, they must be made strong enough to resist the external hydrostatic pressure developed during immersion. The shafts or towers that support the deck structure and the operating equipment must be designed to offer minimal resistance to wave and current forces. These forces are at maximum on or near the ocean surface; therefore, the diameters of the towers should be minimal in this region.

This basic principle followed throughout the design phase is to keep the structure simple. Temporary loading conditions, such as immersion of the caisson and towers for deck-mating, may govern the structural design. A detailed description of the operations required in the different project phases is prepared to determine construction and installation loads and ensure adequate design of all structural parts.

## 286 Introduction to Offshore Structures

Because of the size of a concrete gravity structure, it is not possible to analyze all of the design problems with one analytical model of the structure. Loading conditions vary from design storm loads and small amplitude wave loads requiring dynamic analyses to temporary immersion loads for deckmating. The overall water/soil/structure system is divided into many analytical models for analysis. The results of these separate studies are combined with appropriate interaction effects to determine the overall structural response. For example, a crude structural model is sufficient for making soil deformation calculations, since soil response depends mainly on structural size and weight. Figure 16-1 is a simple model for soil analysis. Likewise, to determine dynamic structural response, a structural model of perhaps 100,000 degrees of freedom may be required whereas the soil characteristics might be included in the analysis by relatively few parameters. Figure 16-2 shows the nodes and elements of a typical finite element method (FEM) model for structural analysis.

Most concrete gravity platforms are designed for several functions: combined drilling, production, and oil storage. The designs in Figure 13-1 divide into two basic types. To be consistent with their functional requirements, each of the basic types is designed to offer the least possible resistance to environmental loads while providing suitable support for the particular deck structure required.

The subject of structural design can best be presented by describing the two basic types of platforms: one designated as the tower concept, or *Condeep* design, which includes in general the *Andoc* and *Sea Tank* configurations, and

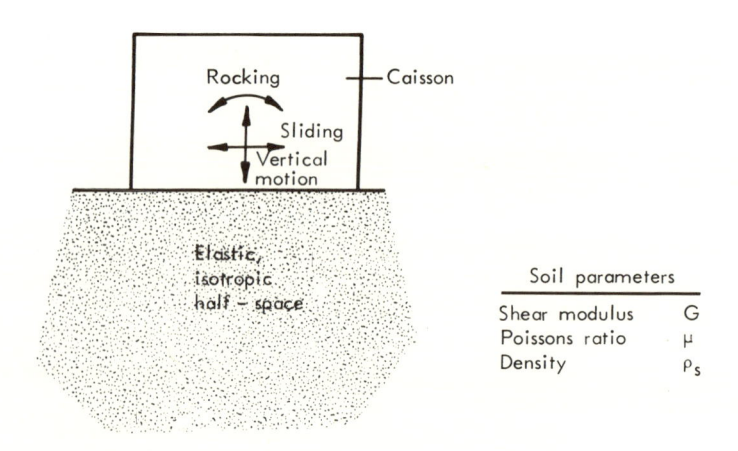

**Figure 16-1.** Soil model. (Copyright © 1975 Offshore Technology Conference.)

Structural Design    287

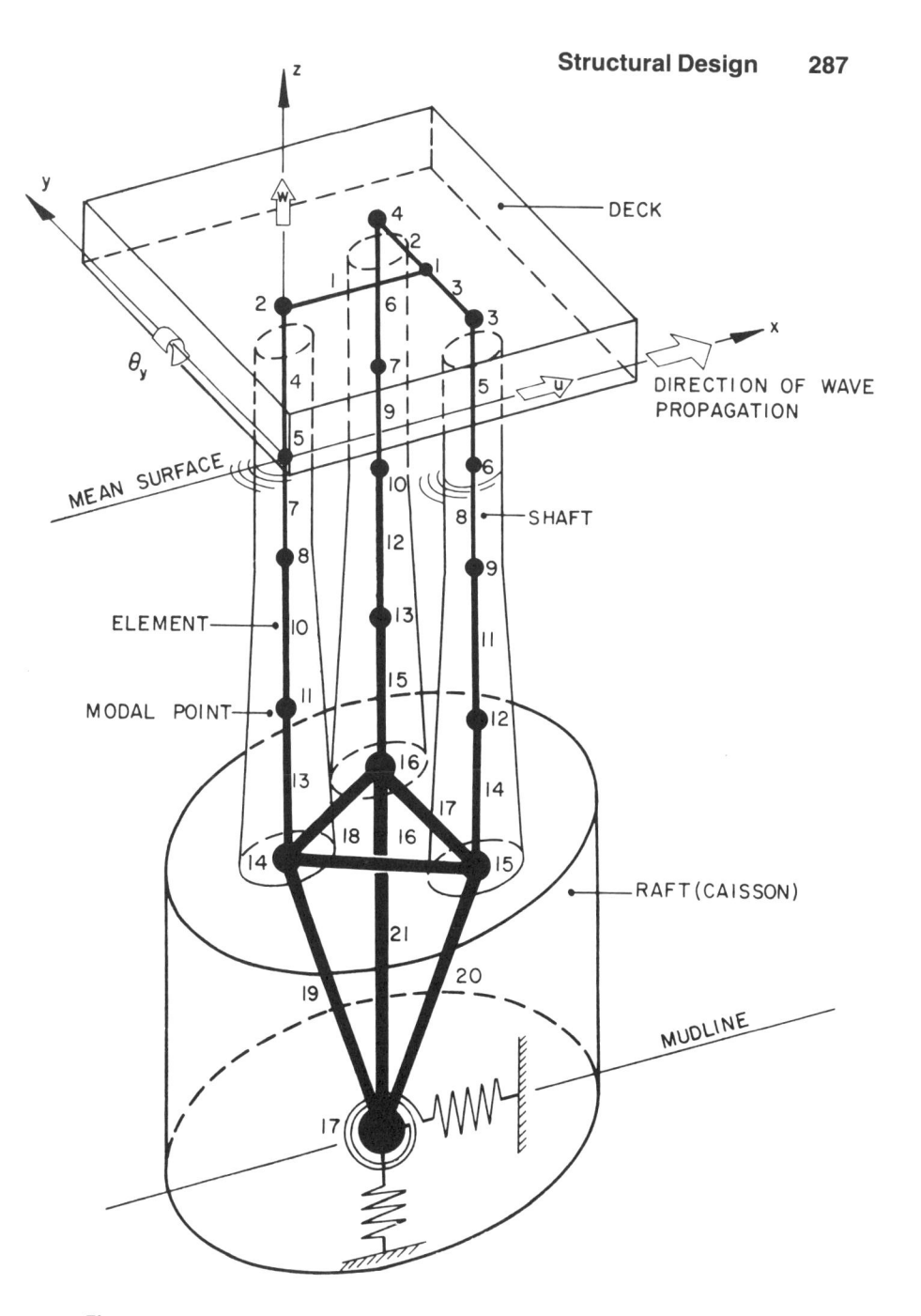

*Figure 16-2. Typical structure and F.E.M. model. (BOSS '76, Vol. I, p. 848.)*

the manifold, or the *Doris* concept, as exemplified by the *Ekofisk storage tank,* the *Frigg CDP 1, Frigg MP 2,* and *Ninian* platforms.

Briefly, the tower structure consists of a wide cellular base forming a caisson or reservoir surmounted by several (two, three, or four) slender, tapered towers that extend up through the water surface and on which the deck structure is mounted. The manifold platform consists of concentric cylindrical or lobate vertical shells connected radially by prestressed walls. In the Ekofisk structure the upper portion of the outermost cylindrical shell (302 ft or 92-m diameter) is perforated to form a Jarlan breakwater wall. The *Frigg CDP 1* and *Ninian* structures consist of a pyramid-stepped cylinder with sea level diameters of 203 ft (62 m) and 148 ft (45 m), respectively. Surrounding the base raft of both of these structures is another perforated wall for antiscour protection. Figure 13-1b shows the *Frigg CDP 1,* and Figure 16-3 shows the *Ninian Central* platform.

## Limit State Design

The possibility for error always exists in the determination of applied loads from the interpretation of environmental conditions. This possibility warrants the use of safety factors applied to the loads. These are introduced into the design as load factors in accordance with the principles of limit state design. The limit state method is believed to produce a more reliable design with a more uniform level of safety throughout the structure than what could be obtained with the allowable stress method of structural design.

An alternate name for limit state design is semi-probabilistic design. From the beginning of the development of concrete gravity platforms, the designs have been based on this method. The crux of the method depends on the proper definitions for critical relevant limit states and the associated partial load and material safety factors used.

A limit state is reached when a structure or structural member just ceases to fulfill the resistance requirements of structural performance for which it was designed.[1] The various limit states used in design are given in Table 16-1.[2]

The principal parameters governing structural design are listed in Table 16-2. These are basically random variables. Usually, full statistical information is not available, so the main uncertainties are included as partial load and material safety factors. Characteristic load effects $Q_c$ and characteristic resistances $R_c$ are defined as certain percentiles of the distribution functions $P(Q)$ and $P(R)$ for load and resistance, respectively. (See Figure 16-4.) $Q_c$ is the mean of the characteristic load effects distribution plus $k_q$ (e.g., one or two) standard deviations. $R_c$ is the mean of the characteristic structural resistance distribution minus $k_r$ (e.g., one or two) standard deviations.

**Structural Design** 289

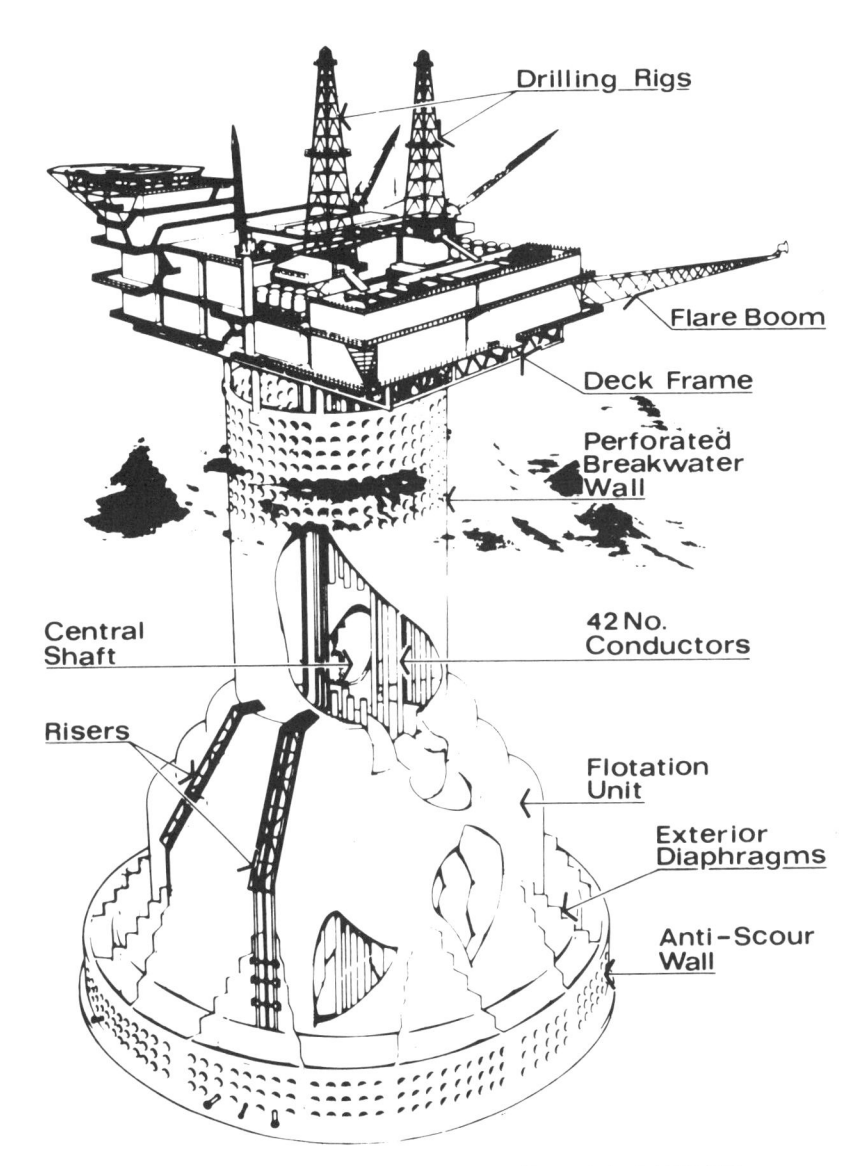

*Figure 16-3. The* Ninian Central *platform. (Copyright © 1979 Offshore Technology Conference.)*

# Introduction to Offshore Structures

**Table 16-1**
**Design Limit States**

| Limit states of unfitness | Main characteristics |
|---|---|
| ultimate limit state (ULS) | Ultimate load-carrying capacity |
| | Rupture or yielding of sections |
| | Collapse or instability of single member or structure |
| | Transformation into mechanisms |
| | Loss of equilibrium, etc. |
| progressive collapse limit state (PLS) | Accidental loss or overloading of single members that may render the structure or major parts thereof into a condition where progressive failure may take place |
| fatigue limit state (FLS) | Accumulated effects caused by cyclic or repeated stresses during service life |
| | Disintegration caused by accumulated fatigue damage |
| | Insufficient residual strength |
| serviceability limit state (SLS) | Specifications as regards serviceability or durability |
| | Excessive deformations (vibrations) without loss of equilibrium |
| | Damage caused by corrosion |
| | Aspects of durability in the general sense, including unforeseen amount of maintenance and repair |

## Table 16-2
## Design Against Structural Failure

*Design modes (limit states)*
  Serviceability
  Fatigue
  Buckling and ultimate collapse
  Progressive collapse

*Loads (In various conditions: construction immersion, tow-out installation, operation)*
  *Permanent Loads*
    Weight of structure, permanent ballast
    hydrostatical pressure (in permanent location)

  *Live loads*
    Stored bulk material, equipment
    Operation of cranes
    Fendering and mooring of vessels

  *Deformation loads*
    Prestressing forces
    Thermal loads, creep, and shrinkage
    Differential settlements

  *Environmental loads*
    Wind, waves, currents, ice,
    earthquake

  *Accidental loads*
    Collision loads and dropped objects
    Explosion and fire

*Capacities*
  Effect of production tolerances on capacity

*Structural analysis*
*(determination of load effects and capacities of components)*
  Static vs. dynamic, linear vs. nonlinear
  Codes, criteria, safety factors (partial coefficients)

**292  Introduction to Offshore Structures**

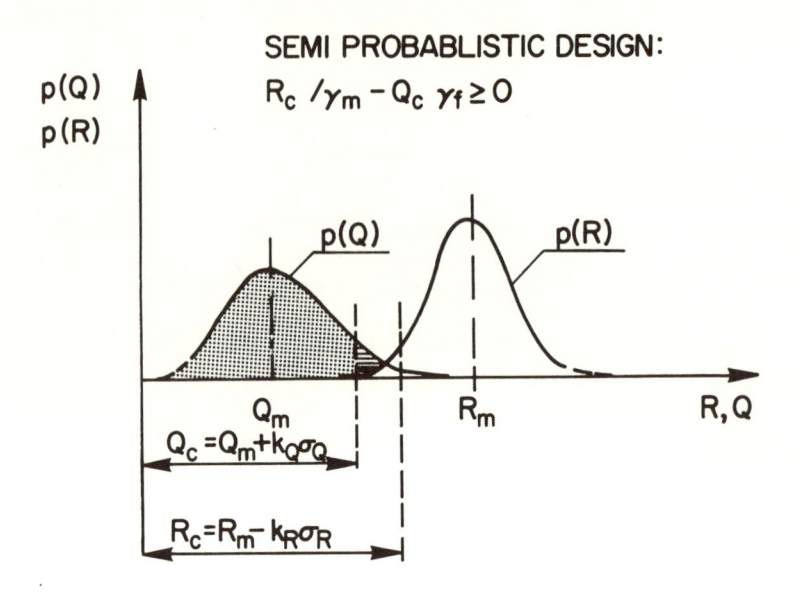

**SEMI PROBABLISTIC DESIGN:**

$$R_c/\gamma_m - Q_c \gamma_f \geq 0$$

**Figure 16-4.** Characteristic load effect $Q_c$ and resistance $R_c$ defined as certain percentiles of the distribution functions p(Q) and p(R), respectively. (From O. Furnes, "Overview of Offshore Oil Industry with Emphasis on the North Sea," Lectures on Offshore Engineering, 1978, Institute of Building Technology and Structural Engineering, Aalborg University Centre, Aalborg, Denmark.)

In general, the design load effect $Q_d$ is the most unfavorable combination of a specified set of loads and associated partial load factors $h_f$

$$Q_d = \text{effect of } \Sigma\, h_f\, Q_c$$

The design resistance $R_d$ is the most unfavorable combination of relevant characteristic and substitutional resistance parameters $R_c$ and associated partial material factors $h_m$

$$R_d = \text{combination of } \Sigma\, K\, \frac{R_c}{h_m}$$

where $K$ represents constants that define the geometry and composition of member sections. The verification of safety against any limit state is finally reached when

$$R_d - Q_d \geq 0$$

## Structural Design

The magnitudes of the partial load and material factors as recommended by Det Norske Veritas are given in Table 16-3.[1] Hove and Foss present these factors in slightly different form.[3]

Concrete gravity structures constructed thus far have been built according to the loading conditions shown in Table 16-4. The external hydrostatic pressure in phase 3 (immersion for deck installation), and the water pressure underneath the base slab during phase 5 (installation), often govern the dimensions of a major part of the caisson. Sometimes compressed air is used within the caisson to reduce the external overpressure. In general, the ultimate limit state (ULS) governs the overall dimensions of the structure.

Oil storage is usually one of the principal functional requirements of the platform. After installation in the ocean, repair of the oil storage part of the structure can be done at best only under extreme difficulty. Thus, the establishment of realistic serviceability limit state (SLS) criteria is very important. Structural components containing oil should be designed completely tight, which means only very limited cracking of the concrete may be allowed. If the internal pressure of the stored oil is always kept below the external pressure of the surrounding water, then somewhat less severe requirements may be used. Sometimes a separate water-containing chamber is provided between the stored oil chamber and the outside ocean water.

### Prestressing

Prestressing is essential to the construction of a concrete gravity platform, since it permits the concrete to be kept always in compression. Compression precludes cracks and corrosion and provides durability against fatigue.

The term "prestress" includes both pretensioning of cables or tendons before the concrete is poured, and post-tensioning after the concrete has been poured and hardened. Although there usually are many precast pretensioned concrete members in a platform, this discussion of prestressing will deal only with the post-tensioning method, since it must be done at the platform construction site and in schedule with the slipforming.

Several words tend to become confused and are sometimes used interchangeably. These are: wire, strand, cable, and tendon. A *wire* is a long, slender solid rod, literally a cold-drawn steel wire. A *strand* is madeup of several wires twisted around a center wire. The number of wires in a strand is important in relation to flexibility. In general, the greater the number of wires for a given diameter of strand, the more flexible the strand. Commonly, a strand consists of seven wires, a straight center wire and six wires of slightly smaller diameter winding helically around and gripping it. A *cable* is composed of a certain number of strands encased in a flexible metal or plastic duct. Whereas in a wire rope the strands are twisted around a hemp center, in a

## Table 16-3
### Partial Load and Material Factors

| Loading conditions | | Load factors $h_f$ | | | | Material factor $h_m$ | |
|---|---|---|---|---|---|---|---|
| | Permanent loads (P) | Live loads (L) | Deformation loads (D) | Environmental loads (E) | Accidental loads (A) | Concrete | Steel(3) |
| ULS | Ordinary (1) | 1.3 | 1.3 | 1.0 | 0.7 | | 1.5 | 1.15 |
| ULS | Extreme | 1.0 | 1.0 | 1.0 | 1.3 | | | |
| PLS, FLS, SLS | | 1.0 | 1.0 | 1.0 | 1.0 | 1.0(2) | 1.0 | 1.0 |

(1) For well-controlled loads, $h_f = 1.2$ may be used for the loads P and L when temporary.
(2) Only accounting for accidental loads in PLS.
(3) Value depends on mode of failure and how the resistance is evaluated.

### Table 16-4
### Construction Phases and Load Conditions
### for Concrete Structures

| Phase | Load conditions/Construction sequence |
|---|---|
| 1 | Construction in drydock |
| 2 | Structure as launched depends on ballasting |
| 3 | Immersion for deck installation Maximum hydrostatic pressure |
| 4 | Towing |
| 5 | Installation Maximum pressures on bottom slab |
| 6 | Operation Maximum deck and tower forces shock and collision forces |

prestress cable the strands lie parallel to one another inside the duct. The generic term for a prestress cable is *tendon;* a tendon may be simply a single wire or several parallel wires (cable) in a duct.

Post-tensioning is frequently used for shaped cast-in-place members and long members like towers or beams. In post-tensioning the tendons are tensioned and permanently fastened to anchorages embedded in the hardened concrete. The anchorages and ducts, either with or without tendons, are carefully positioned (along with the reinforcing steel) prior to pouring the concrete. Special attention is given to the reinforcement in the vicinity of the anchorages to leave access for proper compaction of the concrete and to provide adequate strength to accommodate the prestressing later.

The tendons are tensioned after the concrete has attained sufficient strength. This tensioning is done with a specially designed double-acting hydraulic jack. (See Figures 16-5 and 16-6.) Figures 16-7, 16-8, and 16-9 show typical anchorages. Figure 16-7 shows a tendon composed of wires, and Figure 16-8 shows a tendon (cable) of 12 strands. The purpose of the duct or sheath is to provide a passageway and prevent bonding of the tendons with the concrete. Tendons are normally shop-fabricated, cut to length, and coiled for transport to the construction site. Sometimes the assemblies are made complete with flexible sheathing and anchorages attached.

**296** Introduction to Offshore Structures

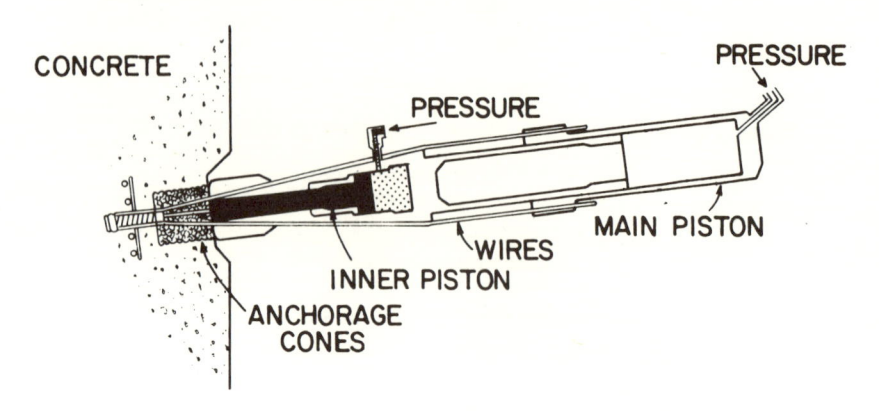

*Figure 16-5.* The Freyssinet double-acting jack. (Courtesy of John Wiley and Sons, Inc., T. Y. Lin, Design of Prestressed Concrete Structures, 2nd edition.)

*Figure 16-6.* The Freyssinet K-1000 jack. (Courtesy Freyssinet International and the Prescon Corporation.)

**Structural Design** 297

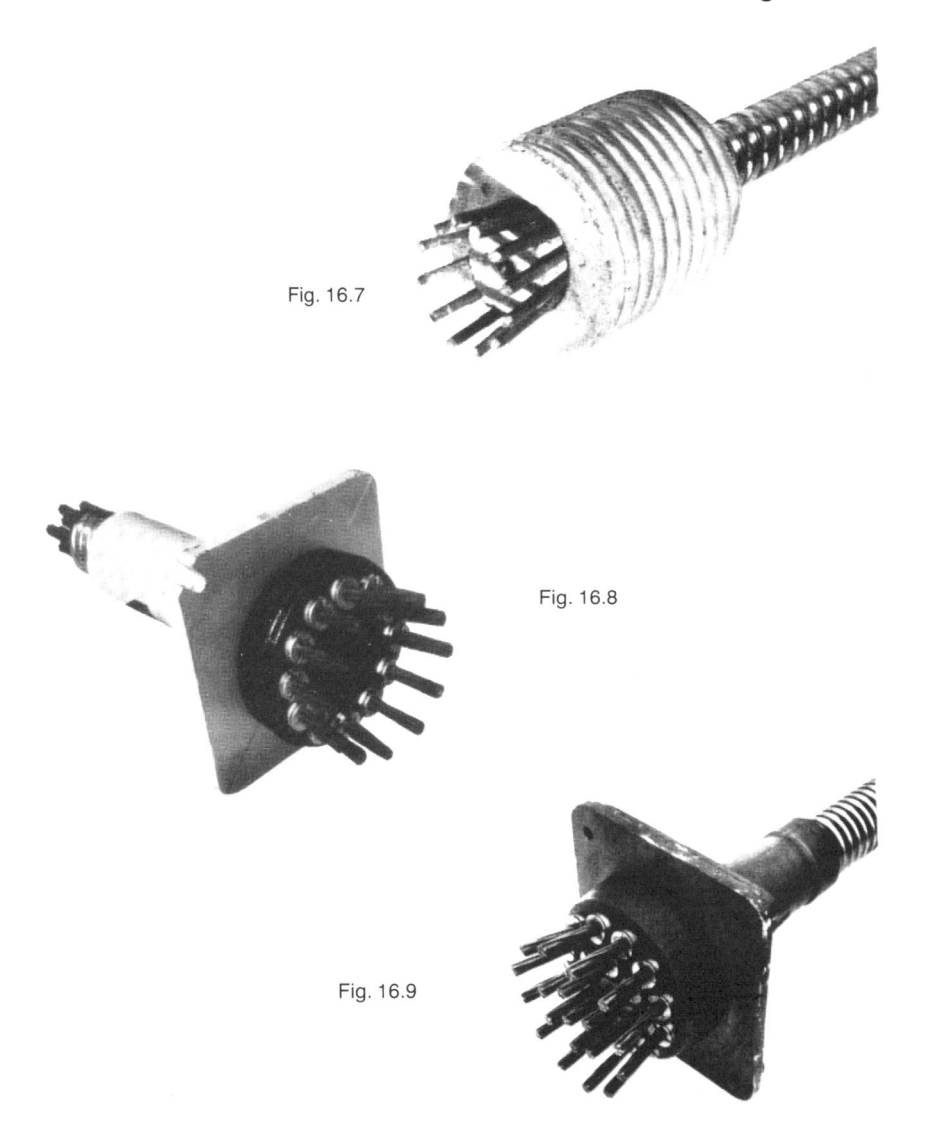

Fig. 16.7

Fig. 16.8

Fig. 16.9

***Figure 16-7.*** *Typical Freyssinet multiwire anchorage assembly. The cable consists of 12 wires laid parallel to each other; such a cable is very simply and economically threaded into a sheath before or after concreting. The whole cable is tensioned in one operation as each wire is individually attached to the jack.* ***Figure 16-8.*** *Freyssinet multistrand anchorage of the 12 V 13 type, trumplate version.* ***Figure 16-9.*** *Anchorage of the 19 K 15 type. (All courtesy Freyssinet International and the Prescon Corporation.)*

In some parts of the structure precast concrete elements with compressive strengths greater than 6000 psi (42 N/mm²) are placed side-by-side so that space is provided in-between for vertical and horizontal networks of prestressing ducts to tie together the cast-in-place and the precast members. Following installation of the prestressing cables in both directions, the cable passageways are filled with concrete. This procedure forms, with the precast parts, one monolithic structure.

Tendons may be draped in a U-shape in a vertical plane to develop upward and downward prestress forces. These tendons are tensioned by jacking simultaneously against the embedded anchorages at both ends.

Usually, grout is pumped into the duct after tensioning to establish bond with the concrete and protect the tendons against corrosion. The grout is injected under a pressure of about 100 psi (0.7 N/mm²). Grout usually is made of cement mortar using ordinary Portland cement proportioned to produce good fluidity, low bleeding, and low expansion.

In the *Condeep* design the lower and upper domes of the caisson cells and the shafts or towers are post-tensioned. For the *Mobil Beryl A* and *Shell Brent B* platforms, the Freyssinet system with 12 T 13 tendons and anchorages were used. Table 16-5 lists the types of tendons used on the first nine North Sea concrete gravity platforms. The designation 12 T 13 or 12 V 13 means 12 strands in the tendon, each strand being 0.5 in (13 mm) in diameter. The T or V indicates the anchorage type. Figure 16-8 shows the anchorage for a 12 V 13 tendon. The nominal ultimate strength of a 12 T 13 or 12 V 13 tendon is 493 kips (2.2 x $10^6$ N).

For post-tensioning, high-strength concrete is generally used because it develops higher bond stresses with the tendons, greater bearing strength to withstand the pressure of anchorages, and a higher modulus of elasticity. Concrete with a 28-day compressive strength of 5100 psi or more, measured on six-inch diameter cylinders (corresponding to a strength of approximately 45 N/mm² measured on 10-cm cubes) is generally used in offshore structures. The concrete shell elements of the structure are prestressed in two directions.

Figures 16-10 and 16-11 show the prestressing cable scheme for the *Ekofisk* storage tank. Figures 16-12, 16-13, and Figure 16-14 show prestressing for the *Condeep* platform. In the manifold platform configuration the main types of prestressing cables are: (1) curved (circular arc) horizontal cables in the vertical cylindrical or lobate shell walls, overlapping at the radial diaphragm walls; (2) straight horizontal cables in the radial diaphragms; (3) "U"-shaped cables in the vertical walls where the heights are not too great; (4) "J"-shaped cables in the vertical walls, which were too high for use of the "U" tendons. In the tower type structure the three main types of prestressing cables are: (1) curved horizontal cables in the cell domes of the caisson and at certain critical elevations in the towers, (2) vertical "U"-shaped cables in the cell walls of the

## Structural Design    299

**CABLE LAYOUT
TYPICAL HORIZONTAL SECTION**

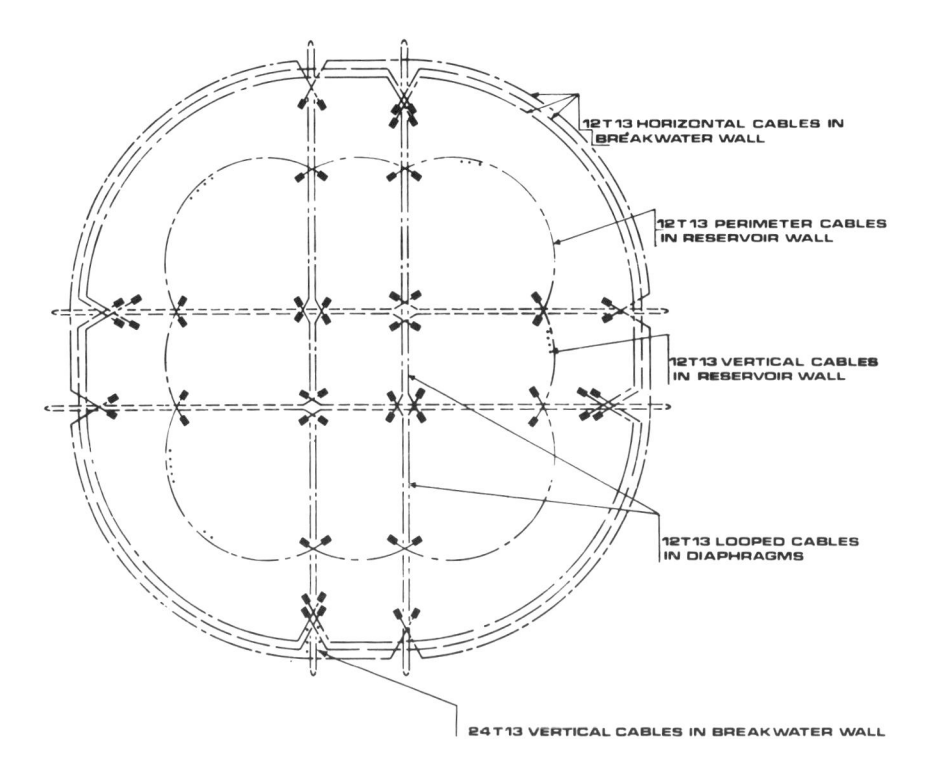

- 12T13 HORIZONTAL CABLES IN BREAKWATER WALL
- 12T13 PERIMETER CABLES IN RESERVOIR WALL
- 12T13 VERTICAL CABLES IN RESERVOIR WALL
- 12T13 LOOPED CABLES IN DIAPHRAGMS
- 24T13 VERTICAL CABLES IN BREAKWATER WALL

*Figure 16-10. Horizontal prestressing cable scheme for the Ekofisk oil storage tank now called the* Ekofisk Center. *(Courtesy Phillips Petroleum Company Norway.)*

caisson, and (3) vertical "J"-shaped cables in the towers. The longest vertical prestressing cables in the *Ekofisk* storage tank are 262 ft (80 m), while the longest cables in a *Condeep* design are 394 ft (120 m).

All horizontal prestress cables must be installed, stressed, and grouted in a rhythmic pattern matching the slipforming rise rate of the walls.[4] This matching is for the practical reason of gaining access to the anchorages. Special multistory scaffolds are suspended from the framework of the slipforms in the locations where the cables are anchored. Figure 16-15 shows how the scaffolds were arranged for prestressing the horizontal cables in the *Ekofisk* storage tank. Figure 16-16 shows in general the slipform arrangement for constructing vertical walls.

Table 16-5 gives the proportions of prestressing and reinforcing steel in typical concrete platforms.

300     Introduction to Offshore Structures

**CABLE LAYOUT
PARTIAL ELEVATION ON BREAKWATER WALL**

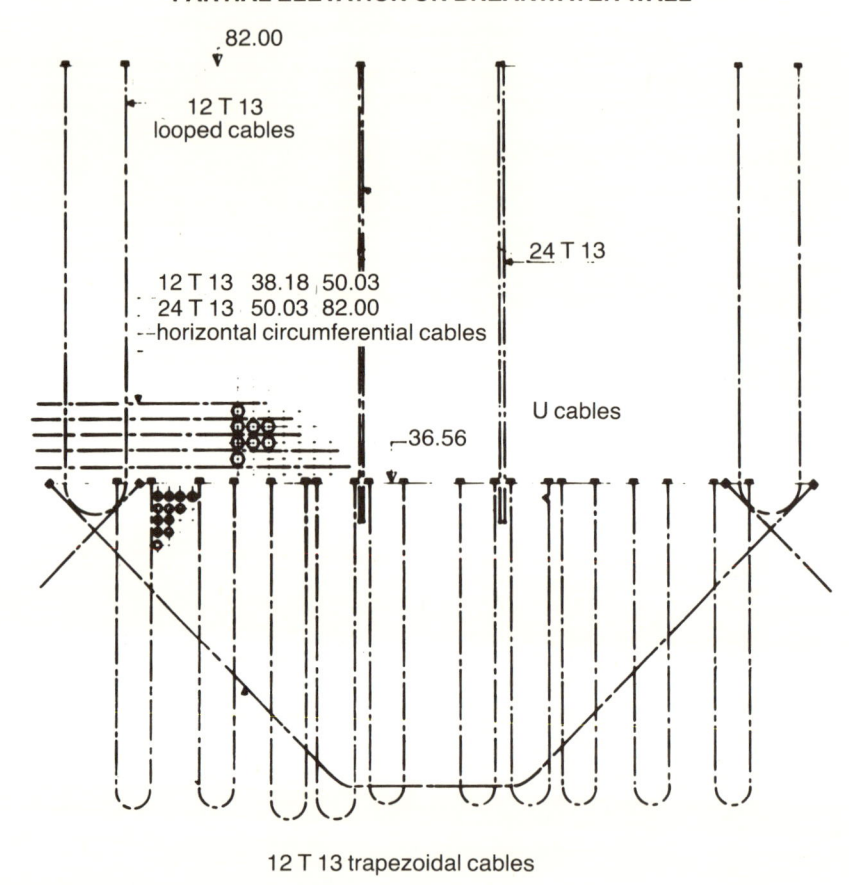

82.00

12 T 13
looped cables

24 T 13

12 T 13   38.18  50.03
24 T 13   50.03  82.00
horizontal circumferential cables

U cables

36.56

12 T 13 trapezoidal cables

*Figure 16-11.* Vertical prestressing cable scheme for the Ekofisk oil storage tank now called the Ekofisk Center. *(Courtesy Phillips Petroleum Company Norway.)*

**Structural Design** 301

(Figure 16-11 continued)

**PARTIAL ELEVATION ON THE RESERVOIR WALL**

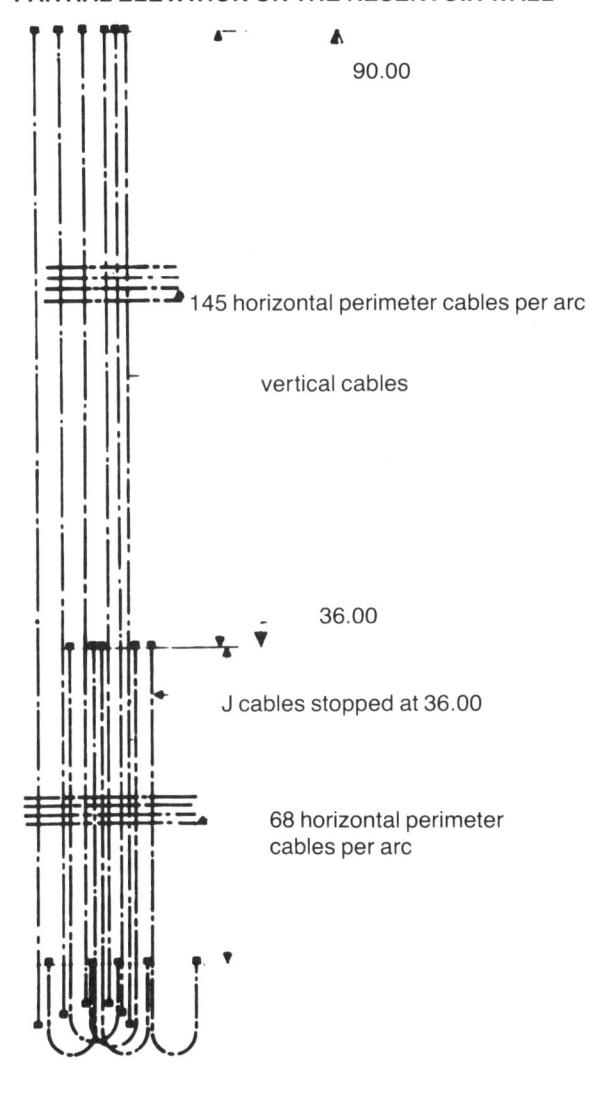

90.00

145 horizontal perimeter cables per arc

vertical cables

36.00

J cables stopped at 36.00

68 horizontal perimeter cables per arc

**302**  Introduction to Offshore Structures

*Figure 16-12.* Prestressing cables for cell domes of Condeep caisson. *(Courtesy Mobil North Sea, Ltd.)*

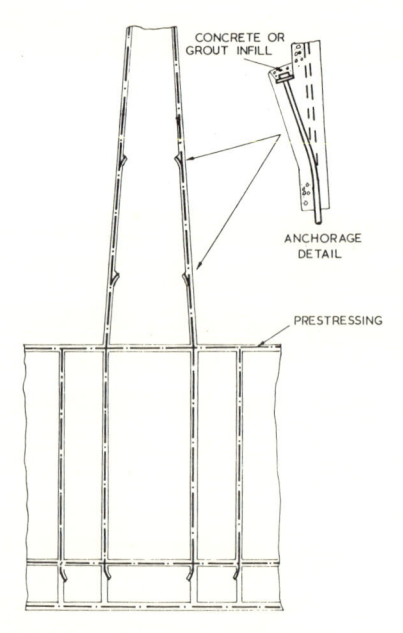

*Figure 16-13.* Schematic diagram of prestressing system for columns in a tower type concrete gravity platform. *(Copyright © 1977 Offshore Technology Conference.)*

**Structural Design** 303

*Figure 16-14. Inspection of the reinforced bottom dome of one of the* Condeep *caisson cells. (Courtesy Mobil North Sea, Ltd.)*

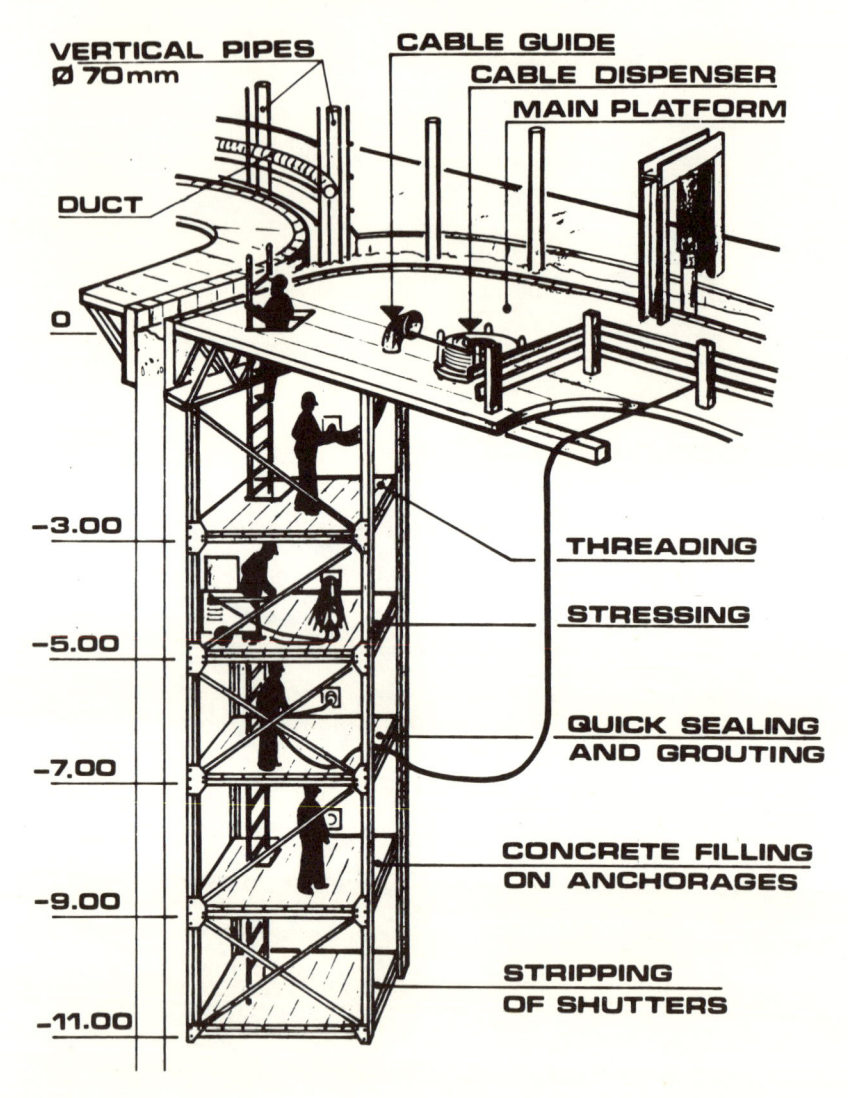

**Figure 16-15a.** Arrangement of the prestressing platforms under the slipforms for the Ekofisk storage tank. *(Courtesy Phillips Petroleum Company Norway.)*

Structural Design 305

*Figure 16-15b.* Threading of a vertical prestressing cable. (Courtesy Phillips Petroleum Company Norway.)

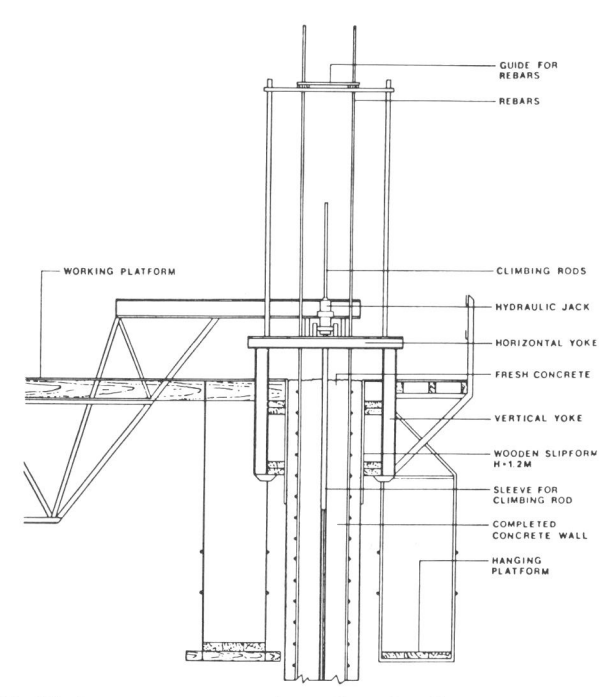

*Figure 16-16.* Slipform arrangement for cell walls. (Copyright © 1975 Offshore Technology Conference.)

## Table 16-5
## Proportions of Prestressing and Reinforcing Steel in Typical Concrete Platforms

| | Ekofisk storage tank | Frigg CDPI | Frigg MP-2 | Mobil Beryl A | Shell Brent B | Condeep TCP-2 | Andoc Dunlin A | Ninian | Petrobras |
|---|---|---|---|---|---|---|---|---|---|
| Volume of Concrete (m³) | 90,000 | 60,000 | 60,000 | 55,000 | 65,000 | 50,000 | 90,000 | 145,000 | 9,500 |
| Prestressing steel (tonnes) | 3,300 | 2,600 | 2,600 | 820 | 1,270 | 600 | 2,200 | 4,000 | 270 |
| Reinforcing steel (tonnes) | 9,000 | 8,000 | 8,000 | 8,000 | 9,800 | 14,000 | 9,000 | 19,000 | |
| Prestressing tendon type | 12 T 13 24 T 13 | 12 T 15 | 12 T 15 24 T 15 | 12 T 13 | 12 T 13 | 12 T 15 | 12 T 13 | 12 T 15 | 12 T 13 |
| Tank Capacity (m³) | 160,000 | | | 149,000 | 160,000 | 160,000 | 220,000 | | 19,500 |
| Water depth (m) | 70 | 96 | 94 | 120 | 142 | 104 | 152 | 139 | 15 |
| Diameter of base raft (m) | 92 | 98 | 98 | 100 | 100 | 100 | 104 | 140 | 52 |

Source: *Freyssinet Offshore Booklet.*

Note: 1 m = 3.28 ft
1 m³ = 35 ft³
1 tonne = 1.103 tons

## References

1. Furnes, O.; "Overview of Offshore Oil Industry with Emphasis on the North Sea," *1978 Lectures on Offshore Engineering* (combined proceedings of a one-day conference plus eight weekly seminars) Institute of Building Technology and Structural Engineering, Aalborg University Centre, Aalborg, Denmark.
2. Moan, T. and Graff, W.J.; "Structural Analysis and Design of Steel and Concrete Fixed Offshore Structures," *1978 Lectures on Offshore Engineering,* Institute of Building Technology and Structural Engineering, Aalborg University Centre, Aalborg, Denmark.
3. Hove, K. and Foss, I.; "Quality Assurance for Offshore Concrete Gravity Structures," 1974 Offshore Technology Conference Preprints, Paper 2113.
4. Moksnes, J.; "Condeep Platforms for the North Sea: Some Aspects of Concrete Technology," *1975 Offshore Technology Conference Proceedings,* Paper 2369.

# Chapter 17

# *Manifold Platforms*

Several platforms using the manifold, or C.G. Doris, concept have been constructed. Because it is typical, the *Frigg CDP 1* platform will be described here. CDP stands for concrete drilling and production platform. Significant features of other manifold type platforms will be mentioned in this discussion.

### General Description

The *CDP 1* platform rests on a 331 ft (101 m) diameter base raft. The bottom slab of the raft is from 2.3 to 2.6 ft (0.7 to 0.8 m) thick and is stiffened by a system of radial and circumferential walls 49 ft (15 m) high. The two outer circumferential walls are perforated and serve as antiscour walls.

When there are currents at the ocean floor, the presence of large-diameter vertical cylinders (walls) disturb the fluid flow and generate local fluid accelerations capable of putting sand from the bottom in suspension. These accelerations produce scour as the sand is carried away in the disturbed flow. The low, perforated wall built around the base raft periphery effectively reduces the scour problem. Gravel piled against the base of the perforated wall eliminates any other scour.

As shown in Figure 13-1b, the main body of the platform is approximately a cylinder made of six vertical cylindrical shell panels the radii of which are less than the radius of the overall body so that the panels form a lobed shape as shown in Figure 17-1. This lobed cylinder has a 203 ft (62 m) outside diameter. For a height of 223 ft (68 m) from the base slab, the lobed cylinder is stiffened by six radial diaphragms 1.8 ft (0.55 m) thick. The upper part of this lobed cylinder is a 4 ft (1.2 m) thick breakwater wall stiffened at the top (elevation 351 ft or 107 m) by radial precast prestressed beams 13 ft (4 m) deep. The intermost cylinder, or shaft, has an inside diameter of 30 ft (9 m) and encloses the gas pipelines. Twenty-four vertical tubes are installed through the ballast

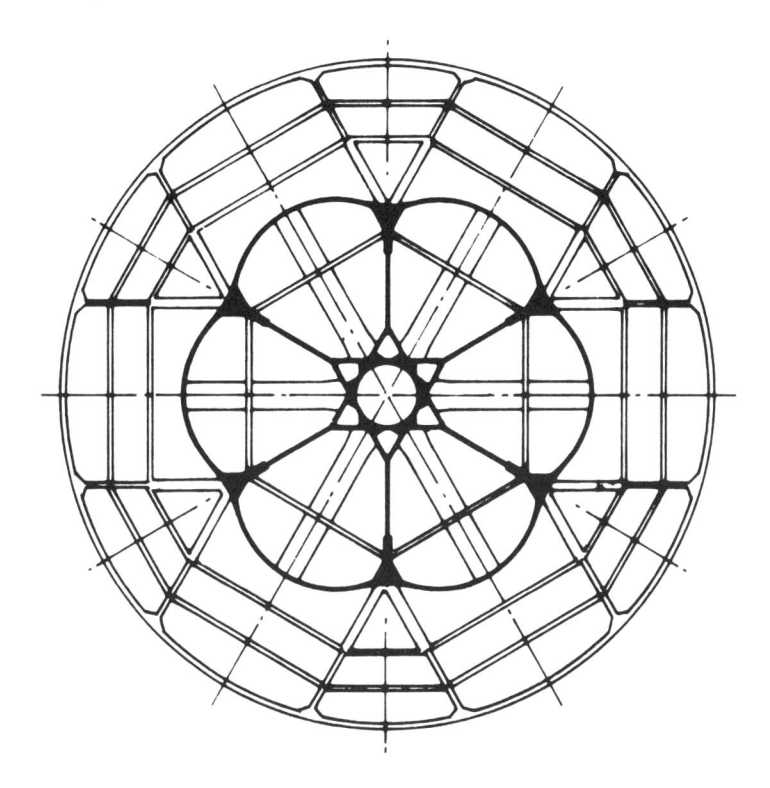

**Figure 17-1.** Horizontal cross section of Frigg CDP 1 platform. (Courtesy Elf Aquitaine A/S.)

inside the lobed wall for well drilling and oil production. The deck structure is supported by 14 6.5-ft (2-m) diameter steel columns filled with concrete; these bear upon the precast, prestressed radial concrete girders at the top of the shell walls. The deck structure is mounted on a system of 45 parallel deck beams outlining an area 210 ft x 207 ft (64 m x 63 m) and located 403 ft (123 m) above the ocean floor. The beams are between 13 and 16 ft (4 and 5 m) deep and weigh about 303 tons (275 tonnes) each. The total deck area is $10.8 \times 10^4$ ft$^2$ (10,000 m$^2$).

Figure 17-2 shows a cut-away of the *Frigg MP-2* platform. The structural features of this platform are the same as for the *CDP 1* platform except that the height of the perforated breakwater wall is 6.5 ft (2 m) lower because the water depth is 6.5 ft (2 m) less, and there is only one deck. Figure 17-3 shows the *Frigg CDP 1* in an early stage of dry dock construction. Figure 17-4 shows construction of the *Frigg CDP 1* at the sheltered deepwater site.

**310** **Introduction to Offshore Structures**

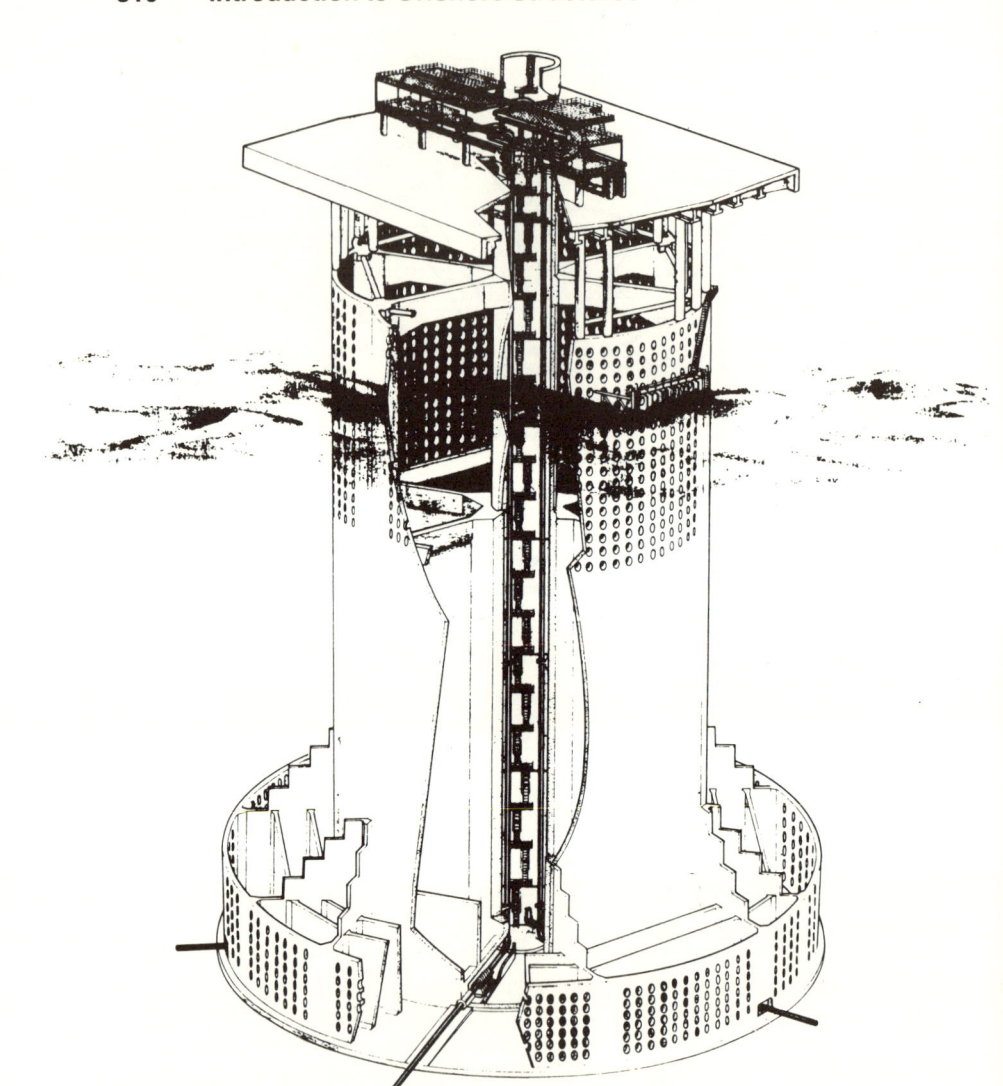

*Figure 17-2.* Schematic of Frigg MCP-01 platform. *(Courtesy Total Marine, Ltd.)*

**Manifold Platforms** 311

*Figure 17-3.* View of dry dock construction of Frigg CDP 1 platform. *(Courtesy Elf Aquitaine A/S.)*

**312    Introduction to Offshore Structures**

**Figure 17-4.** *Construction of the* Frigg CDP 1 *platform in the Andalsnes fjord. (Courtesy Elf Aquitaine A/S.)*

The base slab is divided into compartments by a series of walls 49 ft (15 m) high. The base is in the form of an annulus with the central section 52 ft (16 m) in diameter being raised approximately 1.6 ft (0.5 m). The large annular area of the base slab was cast on corrugated steel sheet to produce an unevenness to enhance the sliding resistance. The annulus ID was 26 ft (8 m) while the OD was 166 ft (50.5 m).

The design water depth for the *Frigg CDP 1* platform was 326 ft (99.4 m). The design storm wave had a height of 95 ft (29 m) and a wave period of 16 seconds. The ordinary operating wave had a height of 59 ft (18 m) and a wave period of 12.5 seconds. Water current at the surface was 4.9 ft/sec (1.5 m/sec); at the ocean floor it was 0.82 ft/sec (0.25 m/sec). The hourly mean wind was 118 ft/sec (36 m/sec). These environmental criteria resulted in the following forces and moments on the platform with a submerged weight of 226,000 tons (205,000 tonnes):

*Horizontal force:*
  storm wave 72,765 tons (66,000 tonnes)
  operating wave 33,300 tons (30,200 tonnes)

*Overturning moment at the ocean floor:*
storm wave 23.1 x 10$^6$ kip-ft (3.2 x 10$^6$ tonnes-meter)
operating wave 12.3 x 10$^6$ kip-ft (1.7 x 10$^6$ tonnes-meter)

For stability, sand ballast was provided inside the lobed walls to a height of 171 ft (52 m). The loads resulting from the wave action on the upper breakwater walls are transmitted by arch action into the vertical beams at the intersections of the lobed walls. These vertical beams may be classified structurally as propped cantilever beams. They are propped by the radial beams at level 351 ft (107 m) and are fastened at level 223 ft (68 m) to the tops of the radial diaphragm walls.

Prestressing in the walls and slab was sufficient to ensure that no tension stresses would be developed in the concrete.

The base raft diameter was calculated so that the platform would float under the service loads exerted during the various phases of construction, towing, and installation.

## Base Slab Design

For design purposes, the base slab was inverted. It was considered to be supported by the circumferential walls and the radial diaphragm walls and to be loaded by soil pressures ranging from 8.2 to 16.4 kips/ft$^2$ (40 to 80 tonnes/m$^2$). Because of uncertainties in the nature of the soil-slab contact, the slab was designed to support a much higher local pressure, at some places as much as 41 kips/ft$^2$ (200 tonnes/m$^2$).

The structural joints formed where the vertical walls joined the base slab were critical in the design. The walls were embedded into the base slab; the reinforcing systems were designed to reduce local distortions.

## Solid Vertical Wall Design

For a manifold platform, the vertical walls must withstand hydrostatic pressures ranging from 8.2 kips/ft$^2$ (40 tonnes/m$^2$) during construction to as much as 28.7 kips/ft$^2$ (140 tonnes/m$^2$) during installation. Because of the desire to limit draft during tow-out to the offshore site, the vertical walls should be made as thin as practical to reduce the weight to be floated and towed. Only concrete with a 28-day compressive strength of 5700 psi (400 kg/cm$^2$) or more should be used. The concrete is vibrated into place to ensure high tightness. Multidirectional prestressing must be used because of the high stress states existing in the walls. The placing of prestressing ducts and tendons causes numerous difficulties during the slipforming operation. Ordinarily, the slipforms rise at a rate of 3-5 ft/day (1.0-1.5 m/day).

## Perforated Vertical Wall Design

Hydraulic considerations govern the dimensions and detailed shape of the perforated breakwater wall. Its purpose is to reduce the effects of wave forces on the platform. The perforated wall must withstand wave-induced forces which generate bending, ovalization, and shearing moments, especially on the edges of the apertures. This general wall type is commonly called a "Jarlan wall," which is a patented, perforated caisson breakwater.[1,2]

A solid vertical-face wall causes reflection of the nonbreaking wave striking it and creates a *clapotis,* which is a system of stationary waves characterized by amplitudes approaching twice the height of the incident wave. The clapotis can result in a large mass of water inundating the top surface. Use of the top surface is restricted unless the solid vertical wall is built higher than normally required or unless the wall is perforated.

The Jarlan wall consists of a perforated concrete wall having an enclosed hollow wave chamber behind. The design wave size governs the chamber width, the thickness of the perforated wall, and the diameter and spacing of the holes. The Jarlan wall produces dissipation of the wave energy such that the resulting standing waves are appreciably smaller than the amplitude of the clapotis associated with the same size solid vertical wall.

The Jarlan wall breaks the incident waves by reflection. A portion of the waves penetrate into the enclosure. Inside the enclosure a large part of the wave energy is dissipated by creating violent and multiple currents, turbulences, and deflections.

When the Jarlan wall is located next to the ocean floor, it becomes a means of scour protection. The holes in this antiscour wall must be sealed with dished steel plate plugs and neoprene gaskets when the base raft is floated out of dry dock. The breakwater holes must be closed with steel plugs and neoprene seals when the platform is immersed (ballasted) to a deep draft for deck-mating. The plugs also provide buoyancy during the controlled seawater ballasting to touchdown on the ocean floor. They are removed as soon as grouting beneath the base slab is finished. The holes may be of uniform diameter or conical in shape with the larger-diameter holes facing the incident waves.

The dimension and shape of the holes, their respective position, and the thickness of the wall governs the proportion of energy reflected outward, the energy dissipated by turbulence and friction when crossing the wall, and the residual energy which penetrates into the chamber in the annular space. The wave chamber size must take into account any possible resonance, and the desired attenuation rates for the most probable waves to be encountered under ordinary operating conditions, including exceptional waves.

The ratio of void area with respect to the total surface area varies from 0.2 to 0.4. The perforated wall reduces noticeably the horizontal forces and the overturning moment on the platform.

## Manifold Platforms

In Figure 17-5 the partially completed, perforated breakwater wall for the *Ekofisk 1* platform is clearly seen. Figure 17-6 is a photograph of the precasting yard where the breakwater wall quadrants for the *Ninian Central* platform were cast. Erection of the precast quadrants is shown in Figure 17-7. For the *Ninian Central* platform, the breakwater holes were 5.3 ft (1.6 m) in diameter, and the wall was 3 ft (900 mm) thick.

### Petrobras Gravity Platforms

The manifold type platform changes its appearance when adapted for shallow water. Representative of the manifold design for a 50-ft (15-m) water depth are the three concrete gravity structures constructed for Petroleo Brasileiro S.A. (*Petrobras*). The first of these drilling/production/storage platforms was installed May 1, 1977, in the Ubarana oilfield near Macao, Brazil.

**Figure 17-5.** Deepwater construction of the Ekofisk storage tank. (Courtesy Phillips Petroleum Company Norway.)

### 316  Introduction to Offshore Structures

*Figure 17-6.* Construction of breakwater wall quadrant units for the Ninian Central platform at the Kishorn precasting yard. (Copyright © 1979 Offshore Technology Conference.)

The platforms are rectangular in plan, 149 x 172 ft (45.36 x 52.38 m). The central reservoir consists of 16 cells 23 x 23 ft (7 x 7 m) surrounded by ballast cells with an exterior lobed wall. Figures 17-8 and 17-9 show the side and plan views, respectively. The wellhead channel passes across the center of the structure with provision for drilling 8-12 wells. The bottom and top slabs are about 3 ft (1 m) thick. From bottom slab to top slab the structure is 84 ft (25.7 m) high. There are 298 tons (270 tonnes) of prestressing steel in each of the three platforms. The outer walls and the base are prestressed vertically and horizontally by Freyssinet 12 T 13 cables. The reservoir capacity is 688,110 ft$^3$ (19,500 m$^3$).

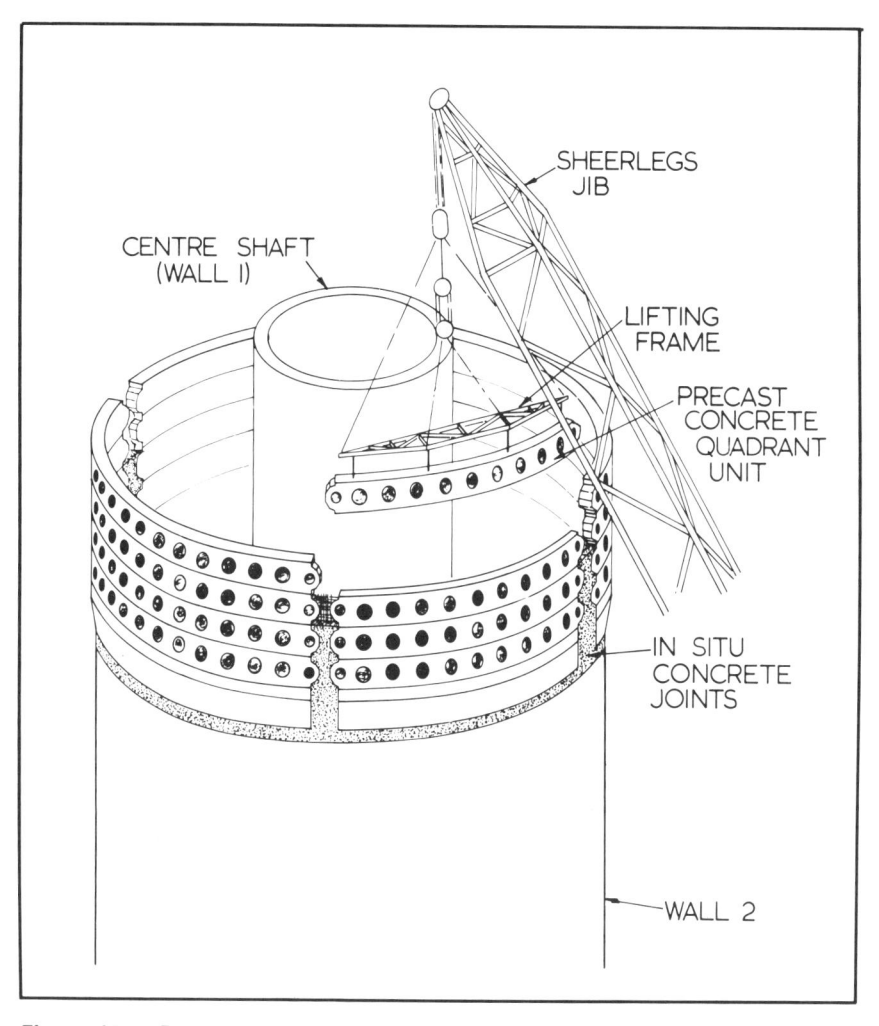

**Figure 17-7.** *Breakwater wall precast unit erection. (Copyright © 1979 Offshore Technology Conference.)*

The *Petrobras* gravity platforms were constructed like other manifold platforms. Concrete for the base slab and walls to a height of 23 ft. (7 m) was poured in a dry dock. This base raft was then towed out of the flooded dock and construction was completed using slipforms.

The platforms were constructed at a site near Salvador and towed 600-800 miles (1000-1300 km) to the oilfield.

**318** Introduction to Offshore Structures

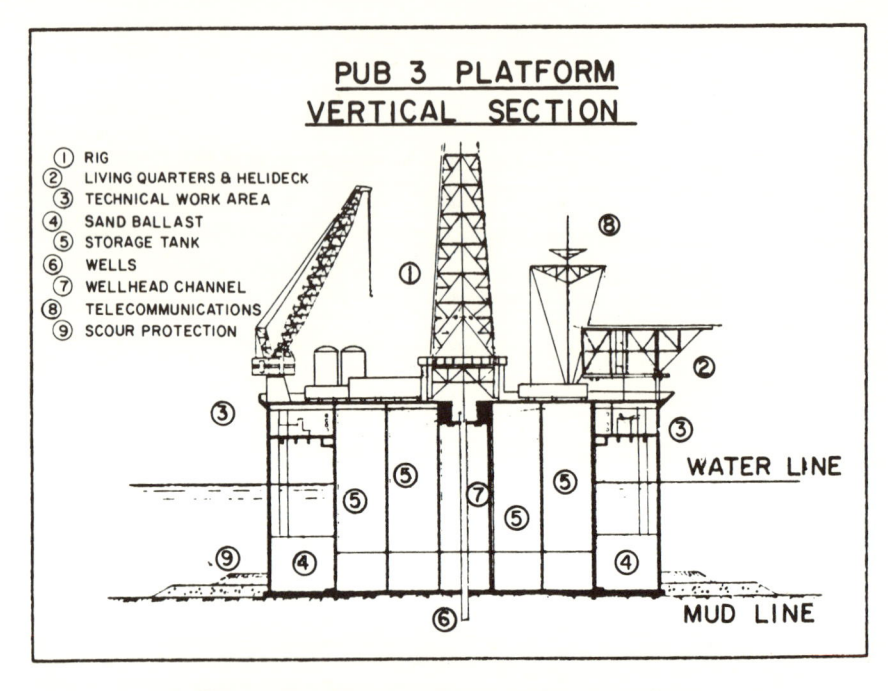

*Figure 17-8.* Petrobras Pub 3 *platform side view.*

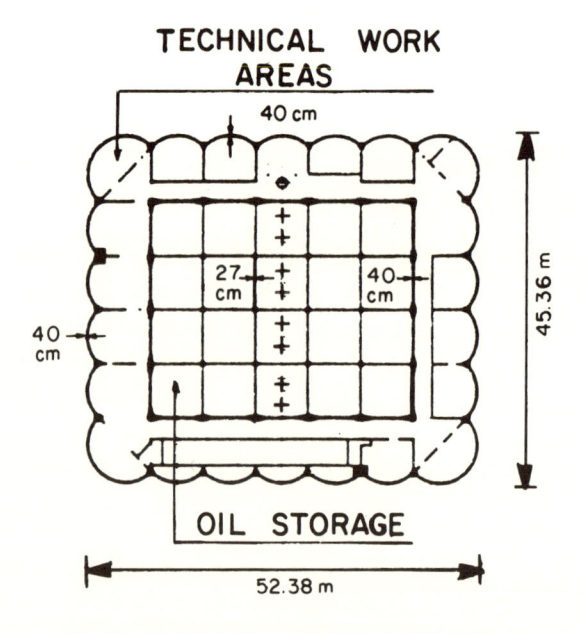

*Figure 17-9.* Petrobras Pub 3 *platform plan view.*

## References

1. Jarlan, G.L.E., "A Perforated Vertical Wall Breakwater," *Dock and Harbour Authority,* Vol. 41, No. 486, Apr. 1961, pp. 394-398.
2. Marks, W. and Jarlan, G.L.E., "Some Unique Characteristics of a Perforated Cylindrical Platform for Deep-sea Operations," *Proceedings of the Offshore Exploration Conference (OECON),* 1966, pp. 349-372.

# Chapter 18

# *Tower Platforms*

The tower design consists essentially of a caisson or cellular base and several (two, three, or four) hollow concrete shafts or towers extending up from the caisson supporting a steel deck structure. The most frequently used foundation shapes are circular, square, hexagonal, or octagonal. Typical of the tower concept is the three-tower Condeep design shown in Figure 13-1a. The first two of this type were the drilling/production platforms *Beryl A* for Mobil and *Brent B* for the Shell Company. (See Table 13-1.)

### General Description

The tower concept is generally known as the Condeep concept, although there are at least two other designations, the Andoc and the Sea Tank. (See Figure 13-1b).[1,2] The Condeep platforms were constructed by Norwegian Contractors. Norwegian Contractors was the name given to a consortium of Norwegian maritime companies.

In the Condeep design the submerged caisson is hexagon-shaped and usually consists of 19 vertical, interconnected cylindrical cells with spherical domes at each end. The cells provide buoyancy during construction and tow-out and later serve as oil storage volume. This oil storage volume is typically large. (See Table 16-5.) As shown in Figure 18-1, three of the caisson cells in the Condeep design are extended into shafts or towers. These are slender, tapered cylinders constructed by slipforming techniques.

Below the spherical dome bottoms of the cells the cylinder walls extend downward as 3 ft (0.95 m) thick concrete skirts to a level of 1.6 ft (0.5 m) below the tip of the domes. Beneath these concrete skirts, corrugated steel skirts extend 11.5 ft (3.5 m) from the 12 outer cells and the center cell. The object of the skirts is to improve the foundation stability, make the structure less vulnerable to erosion, and facilitate the grouting process. The skirts prevent

**Figure 18-1.** *Shipforming of the towers for the* Beryl A *platform at the deepwater site. Note the floating batching plant at the rear. The barracks on the upper domes indicate the size of the structure. (Source: Jan Moksnes, "Concrete in Offshore Structures," Second WEGEMT Graduate School on Advanced Aspects of Offshore Engineering, The Norwegian Institute of Technology, Trondheim, Norway, Jan. 1979.)*

lateral movement of the platform on the ocean floor. They contain and facilitate evacuation of water from beneath the structure during penetration and serve as containment walls for the grout. The skirt area is usually divided into several compartments. (See Figure 18-2.)[3, 4]

To prevent the platform from sliding and causing damage to the steel skirts as touchdown is achieved, three steel pipe dowels are provided. These extend down 13 ft (4 m) below the level of the steel skirts and penetrate the ocean floor first. The dowels are positioned outside the cylindrical cells as shown in Figure 18-2. The acceptable sea-state conditions for touchdown (more specifically, the initial contacting of the ocean floor with the steel dowels) are based on the axial and bending strengths of the dowels.

**322** Introduction to Offshore Structures

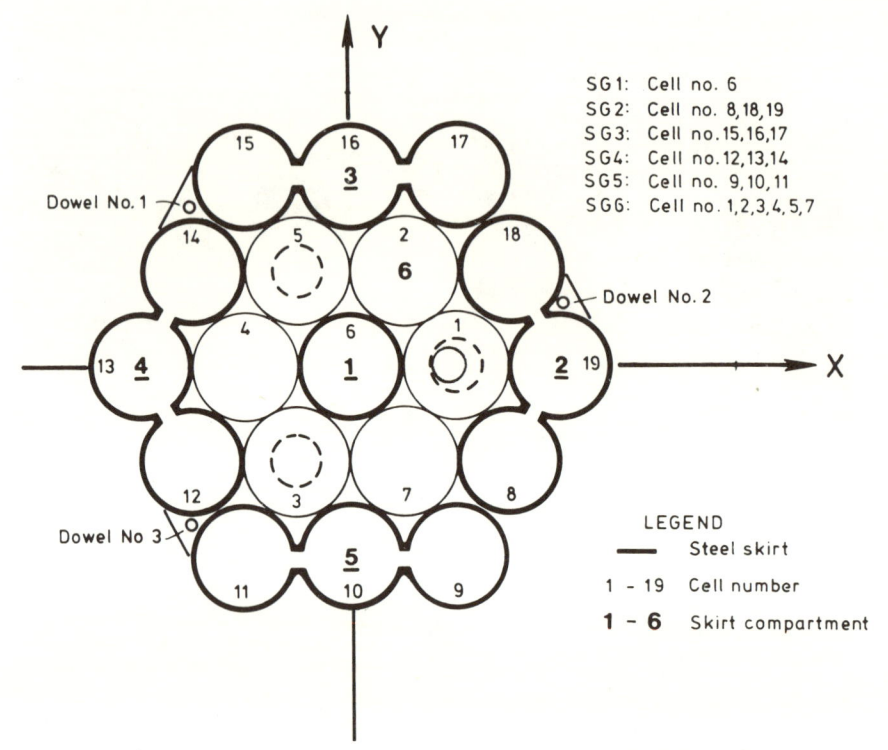

```
SG1:  Cell no. 6
SG2:  Cell no. 8,18,19
SG3:  Cell no.15,16,17
SG4:  Cell no.12,13,14
SG5:  Cell no. 9,10,11
SG6:  Cell no. 1,2,3,4,5,7
```

LEGEND
— Steel skirt
1 - 19  Cell number
**1 - 6**  Skirt compartment

*Figure 18-2. Cell and skirt compartment number and dowel location.*

To cause the skirts to penetrate into the seabed, the downward force must be larger than the frictional resistance of the soil on the dowels and skirts. The penetration force is provided by ballasting, i.e., letting water into the caisson cells. The magnitude of the available penetration force is 165,000-220,000 tons (150,000-200,000 tonnes). Ballast water is used to control platform inclination also. Following penetration into the soil, the pressures within the skirt enclosures can be adjusted to correct the inclination of the platform. After this procedure, the space beneath the caisson is grouted. A permanent platform inclination of less than 0.1° can be achieved.

The platforms are built partly in a dry dock, and partly in a floating position at a deepwater site with sufficient water depth to permit submersion of the structure and installation of the steel deck structure, which is fabricated at another location and transported by barges.

The Condeep platform has its well conductors enclosed in the concrete legs where they are protected against wind and waves. Steel guides with easily drilled plugs are cast into the lower concrete domes so that conductor tubes may be inserted past the plugs with little effort.

## Caisson Walls and Skirts

As shown in Figures 18-3 and 18-4, the cell walls are constructed by slipforming. Figures 18-5, 18-6, and 18-7 show caisson construction at various stages.[5] Figures 18-6 and 18-7 show the same four platforms at two slightly different times. Figure 18-6 shows the base raft of the caisson of the *Statfjord A* being floated in preparation for towing from dry dock to the deepwater site. The two platforms in the distance are the *Mobil Beryl A* and the *Shell Brent B*, both without deck structures. The bottom domes of the caisson for the *Shell Brent D* platform are shown in the foreground. In Figure 18-7 the platforms have changed: the deck structure has been installed on the *Mobil Beryl A*, the *Statfjord A* is on location at the deepwater construction site, and the *Brent D* base raft is being floated into the flooded dry dock where the *Statfjord A* base raft was constructed in preparation for towing through the opened sheet-pile sea wall out into deep water.[6]

In the construction shown in the previous figures, note that the concrete slipforms and working platforms encompass the entire periphery of the caisson. For caisson construction, the total length of the slipform was between 3600 and 4000 ft (1100 and 1200 m). The wooden part of the formwork was 4 ft (1.2 m) high and was supported by steel yokes lifted by hydraulic jacks which climbed notched vertical rods. The height of a lift (one pour of concrete) was about 8 inches (20 cm).

The base slab and the first 20 ft (6 m) of the cell walls were poured in the dry dock. Then, after the base raft had been floated to the sheltered deepwater construction site, the remaining cell walls were poured up to an elevation of 167 ft (51 m) for the *Beryl A* platform and up to 199 ft (60 m) for the *Brent B* platform.

The average rise rate for the slipforms was about 72 inches/day (184 cm/day). This rate required the setting of the concrete to be retarded 8-10 hours to maintain a sufficient depth of plastic concrete in the forms so that each new pour could be revibrated into the old. Again, for the construction of the *Beryl A* and *Brent B* platforms, the horizontal alignment of the slipforms was maintained by a level system based on laser beams in cells 1, 3, and 5. The cells are identified in Figure 18-2.

The concrete skirts were poured in layers of approximately 10 inches (25 cm); each layer or lift was revibrated with the one beneath it to ensure homogeneity of the overall mass. Sacrificial anodes were provided to protect the steel skirts from corrosion.

324  Introduction to Offshore Structures

*Figure 18-3. Slipforming cells of the Shell/Esso Brent B platform in deepwater. Notice the concrete batching plant in the left foreground. (Courtesy Shell Exploration and Production Co. UK, Ltd.)*

Tower Platforms 325

**Figure 18-4.** Casting of cell domes for the Shell/Esso Brent B platform. (Courtesy Shell Exploration and Production Co. UK, Ltd.)

**326**  Introduction to Offshore Structures

*Figure 18-5.* Caisson cell construction in deepwater. (Copyright © 1976 Offshore Technology Conference.)

### Caisson Cell Domes

The concrete pours started in the center of the lower domes and proceeded in concentric lifts outward toward the ring beam at the juncture with the cylinder. Ordinarily, a stiff concrete mix with a slump of about 3 inches (8 cm) was used. But because the array of reinforcing bars and prestressing ducts was so dense in the ring beam, concrete with a slump of about 4.75 inches (12 cm) was used in that vicinity.

For the upper domes, the concrete pour began at the ring beam with a mix having a slump of about 4.75 inches (12 cm) for the same reason as in the lower

**Figure 18-6.** *Four platforms at various stages of construction. Notice that the sea wall bulkhead has been opened to flood one of the construction sites in preparation for floating the base raft out of dry dock. (Copyright © 1976 Offshore Technology Conference.)*

domes. As the lifts approached the center of the domes, a stiffer mix with a slump of about 2.4-3.0 inches (6-8 cm) was used.

As discussed in Chapter 16, in the Condeep design the lower and upper domes of the caisson cells and the shafts or towers are post-tensioned. Circular horizontal overlapping tendons are used in the dome base rings and in the outer reaches of the top dome shells. The towers are post-tensioned by vertical tendons. Figures 16-12, 16-13, and 16-14 show the prestressing arrangements for the tower type platform.

After installation of the *Shell Brent B* platform, it was found that the contact pressures between the domes and the soil beneath ranged from approximately 10 to 33 kips/in$^2$ (50 to 160 tonnes/m$^2$).

**328  Introduction to Offshore Structures**

*Figure 18-7. Four platforms: Mobil Beryl A, Shell Brent B, Statfjord A, and Shell Brent D. (Copyright © 1977 Offshore Technology Conference.)*

**Shafts or Towers**

The shafts or towers have a conical shape for about the lower two-thirds to three-quarters of their height. (See Figure 18-8.) As the shafts are extensions of the cells, the lower shaft diameter is the same as that of the cells, i.e., 66 ft (20 m). The shaft diameter reduces to 39 ft (12 m) at the top. The wall thickness of the conical shafts decreases from 30 to 20 inches (75 to 50 cm) for the *Beryl A* platform. For the *Brent B,* the shaft walls decrease from 41 inches (105 cm) at the bottom to 20 inches (50 cm) at the upper end of the conical section. The slipforming forms used for the shafts were made of overlapping steel plates supported by yokes and lifted by hydraulic jacks. Hydraulic jacks and manual screws enabled changing both diameter and wall thickness of the forms. The actual metal slipforms were about 8 ft (2.5 m) below the working deck. (See Figure 3-26.)

Tower Platforms 329

*Figure 18-8.* Condeep towers being slipformed. (Copyright © 1976 Offshore Technology Conference.)

To ensure homogeneity of the concrete, the new pour is revibrated to increase its bond with the concrete already in the forms. To keep the concrete plastic for a longer period of time, additives are put in the mix. Care must be taken when retarding the setting time to obtain a continuous transition and prevent setting of the upper lifts before the underlying ones.

# 330   Introduction to Offshore Structures

The three sets of slipforms for tower construction were interconnected by steel girders (see Figure 18-8), and the forms were advanced simultaneously. Concrete was poured into the forms by the tremie method.

## Tremie Method of Concrete Placement

A tremie is a downpipe or vertical tube about one foot (0.3 m) or more in diameter through which concrete enters the form. Usually the tube flares slightly at the bottom. The lower end of the tremie is always kept immersed in the concrete just deposited. The tremie is kept full when concrete is being placed. It is raised as the level of concrete rises.

Vibration of the concrete as it is placed in the forms eliminates voids. Vibration produces consolidation which ensures close contact of the concrete with the forms, reinforcement, prestressing ducts, anchorages, and other embedded items. Flexible poker immersion vibrators operate from 7000 to 12,000 rpm. They are manually shifted and positioned at desired locations within the fresh concrete. Some types of vibrators attach rigidly to the outside of the concrete forms.

### Additional Aspects

In construction of several early Condeep type platforms the slipforms were raised about 8 ft (2.5 m) per day. There was a significant heat loss from the freshly poured concrete due to the steel forms and the large amount of reinforcement and prestressing ducts. Thus, to compensate for the heat loss and the related retarding effect on concrete setting, it was necessary to heat the fresh concrete to about 77°F (25°C) and shield the whole assembly with reinforced plastic sheeting. Because of the severe weather conditions during construction, an epoxy mortar was sprayed on the outside faces of the concrete to ensure proper curing. An ordinary curing compound was used on the inside faces.

As with the dome ring beams, a highly workable concrete mix with a slump of about 4.75 in (12 cm) had to be used for the shaft walls because of the congestion of reinforcing steel, prestressing ducts, anchorages, and other embedments. Concrete with greater air entrainment was used for those portions of the cylindrical shafts passing through the splash zone.

In the finished condition of the *Beryl A* and *Brent B* platforms the shafts above caisson cells 3 and 5 are used for drilling purposes; each shaft contains 20 conductor tubes. The shaft above cell 1 is for utilities. It houses the mechanical equipment necessary for platform operations. Inside cell 1 of the caisson is a minicell which contains all the ballasting and grouting facilities.[7,8] Figure 18-9 shows the various levels within the utility shaft of the *Beryl A* platform.

## Andoc Platform

The Shell Company *Dunlin A* platform was constructed by Anglo-Dutch Offshore Concrete (Andoc), a joint venture of seven Dutch and British firms. Designed according to the same general principles as the Condeep platform, the Andoc base structure consists of 81 cells on a 341 x 341 ft (104 x 104 m) pattern. The cells are 105 ft (32 m) high. Inside the caisson the cells are square 36 x 36 ft (11 x 11 m). On the periphery, the cells have flat walls on two or three sides with the remainder consisting of a segment of a cylinder that gives a lobed appearance to the foundation. The base raft, which was 13 ft (4 m) high, included a base slab, rib beams under each cell wall, and steel skirts along the periphery and around the outside of the square formed by the nine center cells.

Four tapered concrete towers 358 ft (109 m) high extend from the caisson upward to support the steel tower extensions which support the decks. The concrete towers taper from 74 ft (22.65 m) at the bottom to 20 ft (6 m) at the top.

Two-dimensional prestressing is used in the base raft and in the vertical cell walls. The towers are prestressed only in the vertical direction.

Construction was begun by placing the steel skirts in trenches and casting the base raft in a dry dock. Cell walls were poured to a height of 23 ft (7 m), and the base raft was floated out of the dry dock to a nearby shallow-water construction site. There the remainder of the cell walls were slipformed. Because the water depth was only 89 ft (27 m), an air cushion was maintained under the raft during slipforming to control the vessel draft. After the tops of the caisson cells were cast, the four concrete towers were constructed by slipforming. Internally stiffened steel columns 30 ft (9 m) long were joined onto the tops of the concrete towers. The steel deck structure, composed of box girder construction and having an area of 48,400 ft$^2$ (4500 m$^2$), was installed by floating it over the tops of the towers when the platform was in deep immersion position. The caisson was deballasted gradually to lift and remove the decks from the barges.

The concrete towers terminated just below the elevation of the lowest astronomical tide so that the steel columns passed through the tide zone. The steel columns were 20 ft (6 m) in diameter at the lower end and 33 ft (10 m) in diameter where they joined the deck structure. Figure 18-10 shows the Andoc design, while Figure 18-11 shows the slipforming construction of the caisson.

**332** Introduction to Offshore Structures

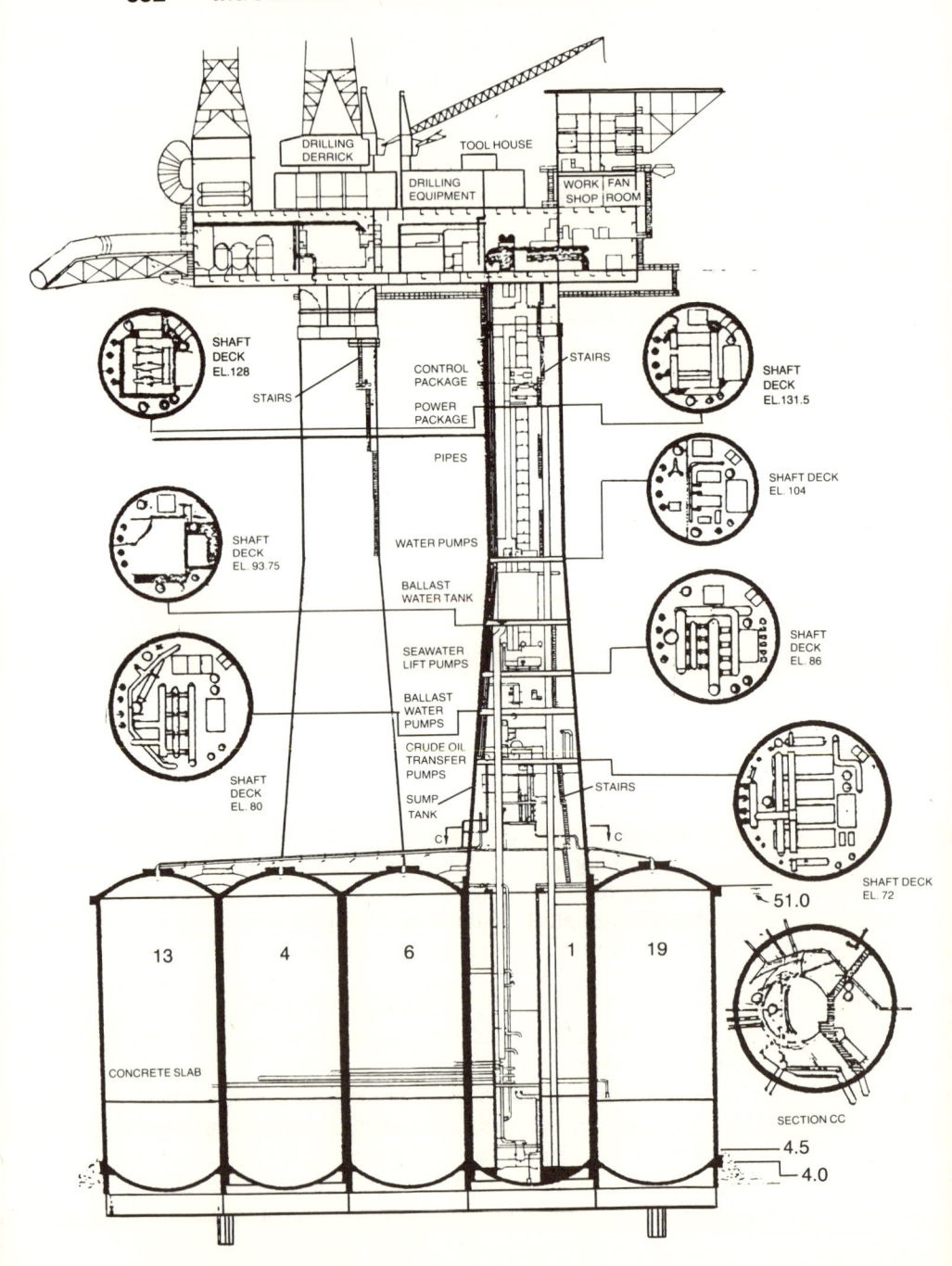

**Figure 18-9.** Equipment at various levels in the Beryl A platform utility shaft. (Copyright © 1976 Offshore Technology Conference.)

Tower Platforms 333

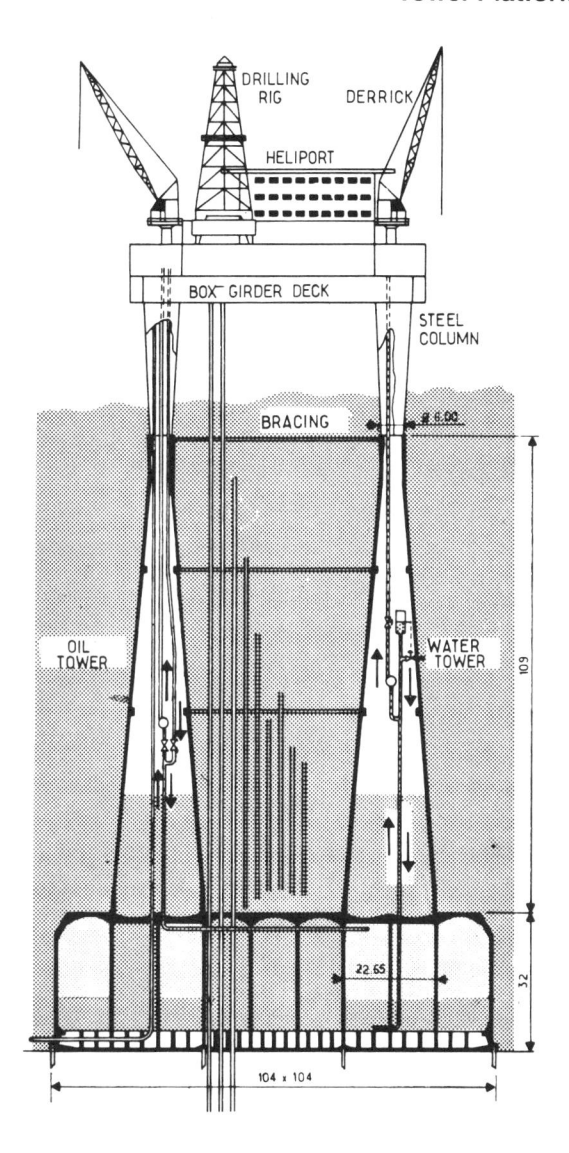

*Figure 18-10. The Andoc design for Shell/Esso Dunlin A platform installed June 1977. (Courtesy Shell Exploration and Production Co. UK, Ltd.)*

**334 Introduction to Offshore Structures**

*Figure 18-11. Slipforming the Shell/Esso Dunlin A base in deepwater. (Courtesy Shell Exploration and Production Co. UK, Ltd.)*

## References

1. Zijp, D.; Van der Pot, B.; Vos, C.; Otto, M.; "Dynamic Analysis of Gravity Type Offshore Platforms: Experience, Development and Practical Application," *1976 Offshore Technology Conference Proceedings,* Paper 2433.
2. LaCroix, R.L. and Pliskin, L.; "Prestressed Concrete Gravity Platforms for Deep Water," 1973 Offshore Technology Conference Preprints, Paper 1888.
3. Dibiagio, E.; Myrvoll, F.; and Hansen, S.B.; "Instrumentation of Gravity Platforms for Performance Observation," *Proceedings of the First International Conference on the Behavior of Offshore Structures, BOSS-76,* Vol. I. pp. 516-527.
4. Eide, O.T. and Larsen, L.G.; "Installation of the Shell/Esso Brent B Condeep Production Platform," *1976 Offshore Technology Conference Proceedings,* Paper 2434.
5. Gausel, E.; Fluge, F.; and Hafskjold, P.S.; "Concrete Offshore Structures, Strains Measured on Full Scale Shells," *1976 Offshore Technology Conference Proceedings,* Paper 2435.
6. Gausel, E. and Hafskjold, P.S.; "Concrete Offshore Structures: Behavior of Full Scale Shells Hydrostatically Loaded," *1977 Offshore Technology Conference Proceedings,* Paper 3007.
7. Holter, E. and Hagen, K.; "Integrated and Semi-Integrated Deck Concepts for Production Platforms With Emphasis on Capital Cost Optimization," *1976 Offshore Technology Conference Proceedings,* Paper 2686.
8. Fjeld, S.; "Concrete Structures," Second WEGEMT graduate school on the advanced aspects of offshore engineering, Technical University of Aachen, Netherlands Ship Model Basin Wageningen, March 5-16, 1979.

# Chapter 19

# Materials, Corrosion, and Fatigue

## Concrete

It is generally accepted that reinforced and prestressed concrete made with hard, dense aggregates combined in a low water/cement ratio are well suited to resist the extreme exposure conditions found in marine environments.[1-4]

> "Based on the experience so far gained from various offshore concrete structures in operation, the following conclusion seems appropriate at this stage: Reinforced and prestressed concrete, when designed and constructed in compliance with relevant criteria, is a suitable, durable and practically maintenance-free construction material for sea structures subject to environmental conditions as in the North Sea. By and large, this is confirmed through the in-service inspections which systematically and regularly are being carried out as part of the platform certification scheme. Valuable feed-back from these inspections is also obtained as concerns relevancy of design criteria and constructional and operational requirements for the structures."[5]

The necessary durability can be ensured only through the careful selection of materials. The quality requirements for good concrete are the physical properties of the mix: compressive strength, permeability, and dimensional stability. Of equal concern are the composition and soundness of the constituent materials. Only high-strength concrete should be used in offshore platforms. With minor variations, one basic concrete mix is usually used throughout. Such concrete should have a minimum 28-day compressive strength of 5100 psi measured on standard six-inch diameter cylinders (corresponding to a strength of approximately 45 N/mm² measured on 10 cm cubes). The modulus of elasticity lies in the range of 4060-4350 kips/in$^2$ (2.8-3.0 x 10$^4$ N/mm$^2$), and Poisson's ratio falls between 0.18 and 0.24. To get this high strength, a high cement content on the order of 25-31 pounds of cement per cubic foot (400-500 kg/m$^3$) of concrete is used. The water-cement ratio is in the range 0.40-0.43.

Aggregates used for concrete platforms built in Norway have been of granite origin. The sand (0-0.39 inches or 0-10 mm in size) is predominantly quartz, and the coarse aggregate is mainly fragments of granular Precambrian rocks such as granite, gneiss, and quartzite. As used, the coarse aggregate consists of a 3:1 blend of two fractions, 0.63-1.25 inches (16-32 mm) and 0.39-0.63 inches (10-16 mm) in size.

Deep immersion of the platform for mating of the deck structure constitutes a severe loading condition. During the deep submersion the caisson walls may have to resist a net hydrostatic pressure in excess of 9 tons/ft$^2$ (100 tonnes/m$^2$). Should failure occur, it would cause implosion of the cell walls of the caisson. Very careful attention must be given to determining the load-carrying capacity of the cellular structure. Besides the nonlinear influence of geometry on deformations, the material nonlinearities aggravate the tendency of the concrete to crack. Sufficient load-carrying capacity must be present in the deformed condition to preclude any inelastic instability.

Instability failure modes in reinforced concrete members should be interpreted as ultimate load-bearing capacity failures rather than critical stress problems as reflected by the classical theory of elastic instability. Reinforced concrete instability failures should be analyzed on the basis of critical strains rather than critical stresses. Many simplifications must be made for a realistic analysis of the nonlinear behavior of a large, complex concrete structure. Understanding of the nonlinear material characteristics of cracking, creep, and other rheological effects is still in the development stage. Predicting the margin of safety of a concrete cellular shell structure may become a difficult task—even when dealing with a well-defined load like hydrostatic pressure.

To produce a durable and impermeable structure, admixtures are used in the concrete. Admixtures retard the setting time, improve workability, and entrain air for the concrete located in the splash zone. Addition of sodium lignosulphonate reduces the water requirement for a given workability by 5-10%. It will also retard the setting time of the concrete by 2-12 hours.

Large, automatic batching plants to provide the concrete are mounted on barges located adjacent to the structure. Ordinary Portland cement is used. The batching plants are capable of producing 44-49 yd$^3$ (40-45 m$^3$) of concrete per hour. The mixing time for the 0.55 yd$^3$ (0.5 m$^3$) batches is in the range of 45-60 seconds after all the constituents have entered the mixer.

**338    Introduction to Offshore Structures**

Because of the stiffness of the mix, concrete pumps are not used. However, all of the concrete is vibrated into the forms for high compaction with high-frequency flexible poker vibrators. These operate in the range of 7000-12,000 rpm and are usually about 2.75 inches (70 mm) in diameter. Occasionally, in locations where the reinforcement is congested vibrators of 1.6 inches (40 mm) diameter are used.

Ordinarily, steel with a yield strength of 58 kips/in$^2$ (400 N/mm$^2$) is used for reinforcement. For prestressing, solid steel rods have yield strengths in the range of 116-180 kips/in$^2$ (800-1250 N/mm$^2$), while cable tendons have strengths of 175-260 kips/in$^2$ (1200-1800 N/mm$^2$). Tendons are tensioned to, typically, 70-80% of their ultimate strengths.

Concrete is highly alkaline, having a pH value around 12.5. Thus the steel reinforcement is normally protected from corrosion by passivation. The ducts through which the tendons pass are usually made of helically wound steel sheet. (See Figure 19-1.) The ducts can be penetrated by seawater under high hydrostatic pressure. Because of the chloride ions in the seawater, the passivation which protects the steel may be destroyed, leaving the steel vulnerable to corrosion.[6] Concrete grout is pumped into the prestressing ducts following tensioning of the tendons to fill the voids around the cables and thus reduce the threat of corrosion. The primary factors influencing the corrosion of steel in concrete are shown in Figure 19-2.

*Figure 19-1. Spirally folded sheath for prestressing tendon. (Courtesy Freyssinet International and the Prescon Corporation.)*

**Materials, Corrosion, and Fatigue** 339

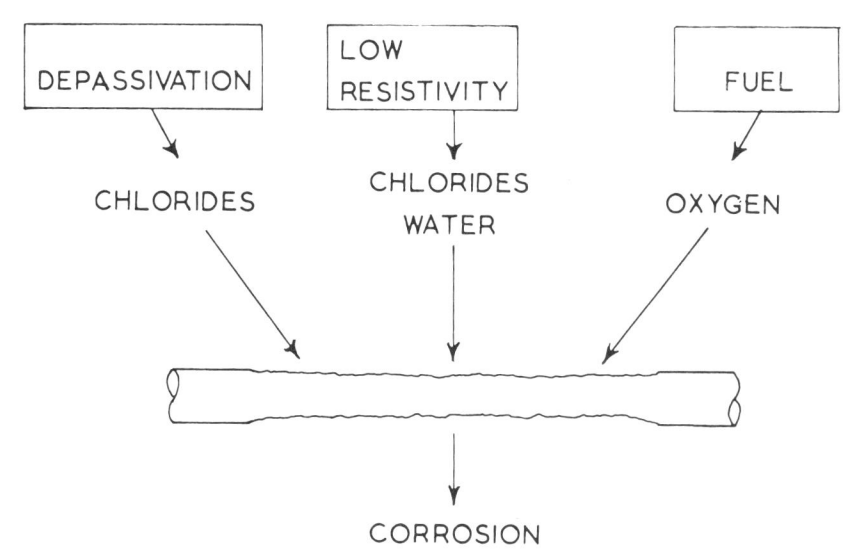

*Figure 19-2. Primary factors influencing the corrosion of steel in concrete. (Copyright © 1977 Offshore Technology Conference.)*

## Corrosion

While concrete has proven to have great durability in marine environments, corrosion of the steel reinforcement may sometimes be a problem. The total amount of steel reinforcement in a concrete gravity platform is about the same as the steel required for an equivalent steel jacket structure. The degradation problems occur, almost exclusively, in the splash zone where extreme steel corrosion may lead to cracking or spalling of the outer concrete surrounding or covering the reinforcement. Fortunately, the vast majority of the structure is permanently submerged.

Reinforcement corrosion can be prevented by using concrete with a low permeability which excludes the dissolved oxygen necessary for the corrosion process. The thickness of concrete cover around the steel reinforcement depends on where the particular component is located within the structure. For bottom, sides, and liquid-tight bulkheads of structures which cannot be dry docked for inspection and repair, the depth to normal reinforcement is 1.5-2.0 inches (40-50 mm). The minimum depth to ducts carrying prestressing tendons is usually about 3 inches (80 mm). For components above water, and for bulkheads that need not be liquid-tight, the minimum depth to normal reinforcement is 1 inch (25 mm); the minimum depth to prestressing ducts is 2 inches (50 mm).

Regardless of cover depth, penetration of seawater is likely to reach the reinforcement in a few years—even in well-compacted concrete. This penetration is well within the design life of the structure. Surveys of existing reinforced concrete structures in marine environments show that no significant degradation occurs over the years in permanently immersed concrete, even when severe corrosion occurs in the splash zone. This condition is a result of the low, uniform levels of chloride present coupled with the low availability of dissolved oxygen.

Because corrosion of steel in the splash zone is a problem, and because the splash zone is accessible only with great difficulty after the platform is installed, several recommendations concerning this zone may be made:

1. Exposed concrete surfaces should be smooth and void of embedded steel parts.
2. Steel risers, utility pipes, and their fixing brackets should be located inside one of the towers or cylindrical cores.
3. Towers or cores should be designed either for access or emptying so that the structure as well as the contents can be inspected periodically.
4. If steel parts are placed externally in the splash zone, they should be isolated from the reinforcement to avoid the possible spreading of corrosion to the interior of the structure.

## Fatigue Behavior

The fatigue strength of reinforced and prestressed concrete structures in submerged ocean service is not well-documented as to the effect of changes in hydrostatic pressure, hydrodynamic pumping, and pore-pressure gradients.[7,8] Until recently, most research related to fatigue of concrete structures has concentrated on the fatigue strength of prestressing tendons, comparisons of free and embedded reinforcing bars, and simple laboratory beam tests.

Experience with structures in marine environments seems to indicate that the fatigue strength of reinforced concrete is adequate as long as the steel reinforcement is protected by a continuous concrete layer. Cracking of the concrete by mechanical, thermal, or other effects may significantly increase the corrosion of the reinforcement, leading to a diminished fatigue strength. Thus, in general, the fatigue behavior of concrete structures is dependent on the type and magnitude of the steel reinforcement, the concrete strength, the bond strength, and the interaction between the concrete and steel reinforcement.

An offshore concrete gravity platform will be subjected to approximately $1 \times 10^8$ cycles of wave loading during a lifetime of 20 years. This effect causes significant fatigue loading in parts of the structure, notably at the base of the

## Materials, Corrosion, and Fatigue 341

columns. The stress levels produced are significant for the fatigue behavior in reinforced concrete (the caisson) but are not important to the stress levels in prestressed members (tall columns). Storm waves are not as important from a fatigue standpoint as are the many smaller, ordinary waves. Ordinary operating waves constitute most of the fatigue loading on a platform. These waves can be quite high; in the North Sea they are estimated to average 65% of the maximum storm wave height. In a 20-year lifetime from 12,000 to 15,000 of these ordinary operating waves will pass the platform. Stresses in the deck structure are principally the result of wave load distributions applied near the tops of the towers where the drag force dominates. This effect is especially true for wave loadings from ordinary operating waves.

Fatigue properties of a material are usually reported by means of an $S$-$N$ diagram, where $S$ (ordinate) is the stress range or the ratio of stress range to yield or ultimate material stress, and $N$ (abscissa) is the number of loading cycles necessary to cause failure at a particular stress level. An $S$-$N$ diagram for plain concrete is shown in Figure 19-3. For concrete tested up to $10^7$ cycles, it was found that the fatigue strength at that time was about 60% of the short-term static strength. The fatigue strength varied very little with concrete strengths of up to 8.7 kips/in² (60 N/mm²).

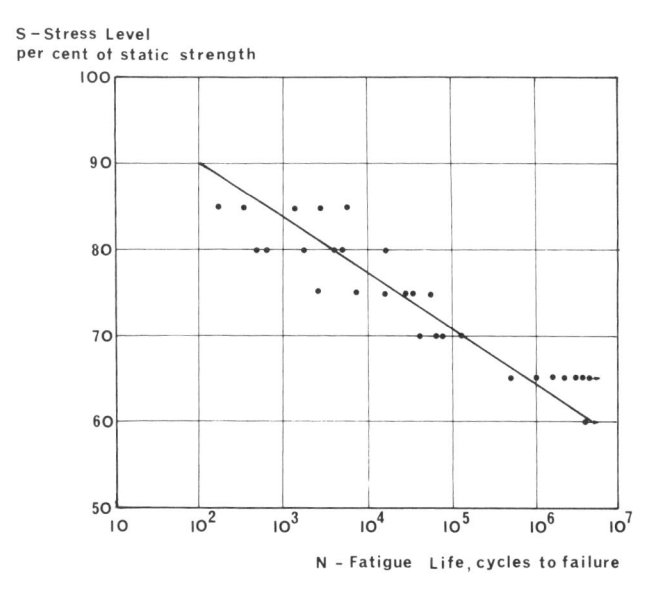

*Figure 19-3.* Fatigue S-N curve for plain concrete. (Copyright © 1974 Offshore Technology Conference.)

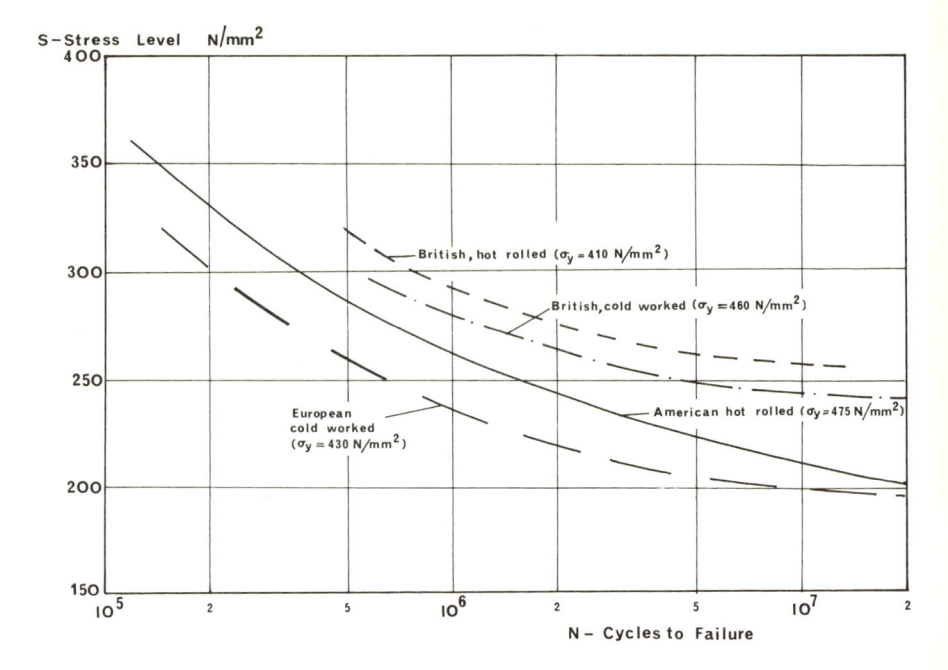

**Figure 19-4.** Fatigue S-N curves for various reinforcing bars. (Copyright © 1974 Offshore Technology Conference.)

Fatigue tests on reinforcing bars have been carried out in uniaxial loading in air or as flexural loading of reinforced concrete beams. The fatigue life of reinforcing steel depends on the range of stress applied and not on the absolute values of minimum and maximum stress, provided the maximum stress does not exceed the yield stress of the material. The work of several investigators is shown in Figure 19-4 for both mild steel and a number of ribbed high-tensile strength steel reinforcing bars. In general, for reinforced members designed to be structurally economic, the reinforcement is likely to fail in fatigue before the concrete. Heavily over-reinforced sections seldom fail before the concrete.

The Palmgren-Miner cumulative damage hypothesis is used for fatigue analysis of reinforced and prestressed concrete but is considered unsafe with a sum of 1.0. As a safety measure, a sum limit of 0.2 is deemed more appropriate.

## References

1. Bury, M.R.; and Domone, P.L.; "The Role of Research in the Design of Concrete Offshore Structures," 1974 Offshore Technology Conference Preprints, Paper 1949.

## Deck Structures 343

2. Moksnes, J.; "Concrete in Offshore Structures," Second WEGEMT graduate school on advanced aspects of offshore engineering, The Norwegian Institute of Technology, Trondheim, Jan. 1979.
3. Hafskjold, P.S.; "Concrete Marine Structures," Second WEGEMT graduate school on advanced aspects of offshore engineering, The Norwegian Institute of Technology, Trondheim, Jan. 1979.
4. Furnes, O. and Sorensen, K.T.; "The Use of Reinforced and Prestressed Concrete in Offshore Structures," paper presented at Institution of Engineers and Shipbuilders in Scotland annual meeting in Glasgow, December 19, 1978.
5. Furnes, O.; "Structural Instability—Problems and Solutions," paper presented at the 25th anniversary convention of the Prestressed Concrete Institute: seminar on precast prestressed concrete, Dallas, Texas, October 17-18, 1979.
6. Browne, R.D.; Domone, P.L.; Geoghegan, M.P.; "Inspection and Monitoring of Concrete Structures for Steel Corrosion," *1977 Offshore Technology Conference Proceedings,* Paper 2802.
7. Holmen, J.O.; "Fatigue of Concrete Structures," Second WEGEMT graduate school on advanced aspects of offshore engineering, The Norwegian Institute of Technology, Trondheim, Jan. 1979.
8. Fjeld, S.; Furnes, O.; Hansvold, C.; Roland, B.; Blaker, B.; and Morley, C.; "Special Problems in Structural Analysis and Design of Offshore Concrete Platforms," *1978 Offshore Technology Conference Proceedings,* Paper 3087.

# Chapter 20

# *Deck Structures*

## General Description

Deck structures for concrete gravity platforms are similar to those of steel template platforms. They support the main drilling equipment such as the derrick, the engine room, the pipe rack, and some of the utility machinery plus the living quarters, the processing equipment, and the heliport.

In principle, all deck structures are made of three-dimensional steel elements. The main load-carrying members may be classified as plate girders, box girders, or trusses. Truss girders are composed either of tubulars or plate sections. Deck types have been categorized into two groups: integrated and modularized. In the integrated deck system the equipment is installed as the structure is fabricated. In the modular system a deck substructure is fabricated, and equipment modules are lifted and placed within or on the substructure. In common use the word "deck" may refer to the entire deck structure or to any one of the many levels or elevations within the overall deck structure.

Concrete platforms have steel decks to minimize the topside structural weight. The topside structural weight is of great significance during tow-out of the platform to its final offshore location. The deck structure is connected to the platform towers using transition pieces designed to spread the applied loads from the steel elements to the concrete columns.

For concrete platforms, the girders are generally of larger dimensions than girders for steel template platforms because of the larger total deck weights for concrete platforms and the smaller number of support columns. Tower concrete platforms have from two to four legs whereas steel template platforms have from four to twelve legs, depending on water depths and wave forces. Concrete platforms of the manifold type support the deck structure on columns extending upward from the outer wall and on the extended core cylinder. Some have concrete deck beams for the lowest deck with steel columns and steel decks above.

Geometrical dimensions of typical steel decks on concrete platforms are listed in Table 20-1.[1] The fractional composition of deck loading for several platforms is given in Table 20-2. The commonly used structural arrangements of girders and trusses are shown in Figure 20-1. Because of the girder or truss size, the webs and flanges are designed as individual stiffened or unstiffened plate elements. The design of these individual elements is complicated by the compactness of the available space and the complexity of installed equipment and piping. Common irregularities are cut-outs, irregular stiffener arrangements, variations of dimensions within a plate, and uneven load distributions. The design is not complete, however, until the three-dimensional effects have been taken into account.

Consideration is given to several types of loading in the design of individual plate elements, namely longitudinal compression, transverse compression, in-plane shear, and lateral load. Interaction effects must be considered because combinations of these load components generally occur.

## Design Loads

For deck design, the inertial component of the passing wave force is predominant, although the drag force is also considered. Ordinarily, the wave period which yields the highest loads on the entire structure is not the one which yields the highest loads on the deck structure. Several wave periods must be investigated to establish the maximum wave loading.

The wind loading is considerably less than the wave loading but is frequently the governing design load for various parts of the deck. Flare towers, telecommunication towers, drilling derrick, and heliport transmit large wind loads through their anchorage points. These points must be carefully analyzed in the design.

Maximum deck loads cause compressive stresses in the concrete columns of the platform. Minimum deck loads combined with wave loads may cause tensile stresses in one or more of the columns. The concrete columns are prestressed by post-tensioning a series of internal vertical cables to keep the stresses in the concrete always compressive. This stress may approach zero in the worst storm condition, but it should not become tension.

**Table 20-1**
**Dimensions of Steel Decks for Concrete Gravity Platforms**

| Platform name | Number of towers | Deck size length x width (m) | Deck area (sq m) | Structural self-weight (tonnes) | Total weight (thousands of tonnes) | Type | Height x Width (m) | Thickness (cm) | Spacing (m) |
|---|---|---|---|---|---|---|---|---|---|
| Andoc | 4 | 82 x 64 | | 6200 | 25 | A | 6 x (6 to 15) | 2-7 | 10-55 |
| Beryl A | 3 | 72 x 70 | 3650 | 6500 | 32 | B | 10 x 2.5 | 8 | 13-24 |
| Brent B | 3 | 70 x 70 | 3650 | 6300 | 30 | B | 6 x 3 | 10 | 11-18 |
| Brent C | 4 | 86 x 56 | | 11,500 | 35 | D | 8 x 9 | 6 | 15 |
| Brent D | 3 | 70 x 70 | | | 30 | B | 8 x 3 | 10 | 11-18 |
| Cormorant A | 4 | 82 x 70 | 3600 | 9100 | 35 | A | 6 x 6 | 6 | 12-23 |
| Statfjord A | 3 | 84 x 54 | 5000 | 9350 | 55 | A | 10 x 1.6 | 6 | 12-15 |
| Statfjord B | 4 | 107 x 55 | | | 50 | D | 14 x 1.1 | 4 | 15-25 |
| TCP 2 | 3 | 72 x 51 | | | 22.5 | D | 9.6 x 1.0 | 2-10 | 10-21 |
| TP 1 | 2 | 93 x 22 | | | 17 | C | 6 x 1.2 | 5 | 22 |
| Ninian Central | | | 12,000 | 6700 | 36 | C | | | |

Load Carrying Members

Note: 3.28 ft = 1 m
10.8 ft$^2$ = 1 m$^2$
0.394 inches = 1 cm
1.1025 tons = 1 tonne

Type of member:

A. Box girder
B. Plate girder
C. Tubular trusses
D. Trusses

Table 20-2
Fractional Composition
of Deck Loading for Typical Platforms

| Platform | Structural self-weight of decks | Equipment (internal in decks) | Modules and equipment on decks | Total deck weight |
|---|---|---|---|---|
| Andoc (Dunlin A) | 0.25 | 0.09 | 0.66 | 1.00 |
| Beryl A | 0.20 | 0.09 | 0.71 | 1.00 |
| Brent B | 0.21 | 0.09 | 0.7 | 1.00 |
| Brent C | 0.33 | 0.17 | 0.5 | 1.00 |
| Cormorant A | 0.26 | 0.14 | 0.6 | 1.00 |
| Statfjord A | 0.17 | 0.13 | 0.7 | 1.00 |

**Integrated vs. Modular Design**

With integrated deck structure, most of the equipment is distributed evenly throughout the decks. In the modular design the loads represented by the modules are transferred to the deck substructure through four to six support points. These points of concentrated load require analysis to ensure sufficient local strength.

The *Beryl A* platform has an integrated steel deck.[2] In this design the crossings of the plate girders created 15 separate compartments in which practically all of the process equipment and utility systems are housed. Partially integrated systems have been used often. It is said that the partially integrated deck system has less structural self-weight and is thus less expensive than the module deck concept. Construction of integrated and partially integrated deck systems are highly dependent on the timely delivery of the equipment to be installed. Accessibility for modifying the equipment after its integration into the deck structure is very limited.

All modules need not be built by the same fabricator nor all at one construction yard. They commonly are constructed at scattered sites. There must be a planned sequence for mounting modules on the deck substructure. Sometimes it may be possible to adjust the sequence and install a module which is behind schedule at a later time—even after platform installation—although this procedure is much more expensive. Module weights range from 275 to 385 tons (250 to 350 tonnes) ordinarily. Some have weighed as much as 2200 tons (2000 tonnes).

While module deck systems are more expensive to construct, they are also recognized as more flexible. To minimize topside weight during tow-out to location, some of the uppermost modules may be towed-out separately on

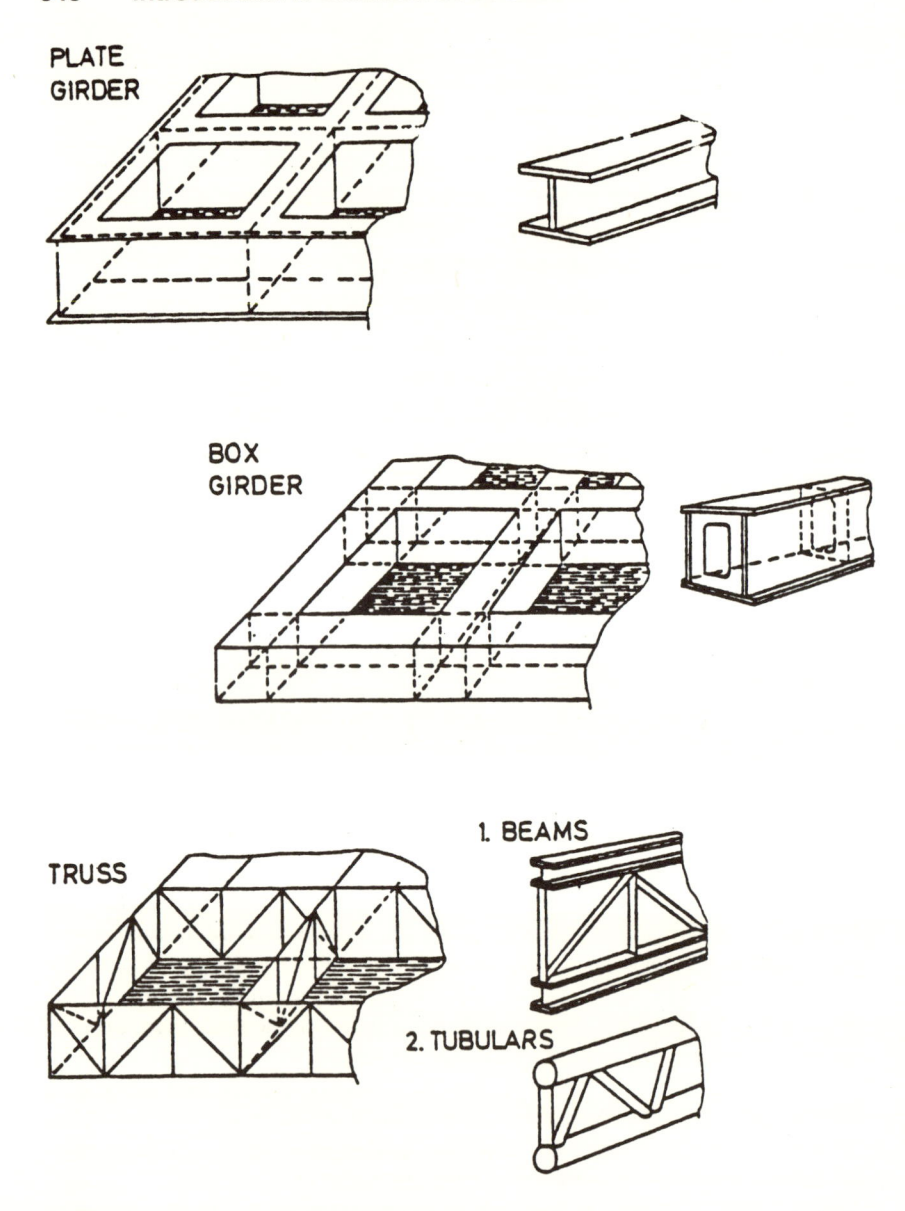

**Figure 20-1.** Main structural arrangements of steel decks on concrete platforms. (Source: O. Furnes and Ø. Løset, "Shell Structures in Offshore Platforms: Application and Design," (Det Norske Veritas). Paper presented at the International Association for Steel and Spatial Structures, Twentieth Anniversary World Congress, Madrid, September 24-28, 1979.)

**Figure 20-2.** Installation of living quarters module on the Statfjord A platform. (Copyright © 1977 Offshore Technology Conference.)

barges and lifted into place after the basic platform has been ballasted into position. Figure 20-2 shows a living quarter module being lifted into position on the *Statfjord A* platform.

## Transition Pieces

The forces resulting from waves and wind produce large deformations of the deck structure. These deformations cause rotations of the tower tops which in turn produce bending moments in the deck structural elements. These forced deformations and moments occur in addition to or in conjunction with the static loads on the decks. It has been estimated that under storm conditions the dynamic stresses in some deck elements may be as high as five times the static stresses. Steel plate thicknesses are consequently large. (See Table 20-1.) The transition pieces between the deck structure and

**350　Introduction to Offshore Structures**

the towers thus become stiff, complex three-dimensional shell structures. (See Figure 20-3a.) Fatigue behavior is an important factor in the design of all deck elements.[3]

The steel transition pieces are about 39 ft (12 m) in diameter and 20-26 ft (6-8 m) in height. These transition pieces transfer the loads from the rectangular deck substructure to the circular concrete towers. As shown in Figure 20-3b, this load transfer requires curved steel panels forming, approximately, an inverted truncated cone. The vertical stiffeners shown in Figures 20-3a and 20-3b may be channels, angles, tees, or combinations of these sections. Horizontal stiffeners often encircle the inside of the periphery also.

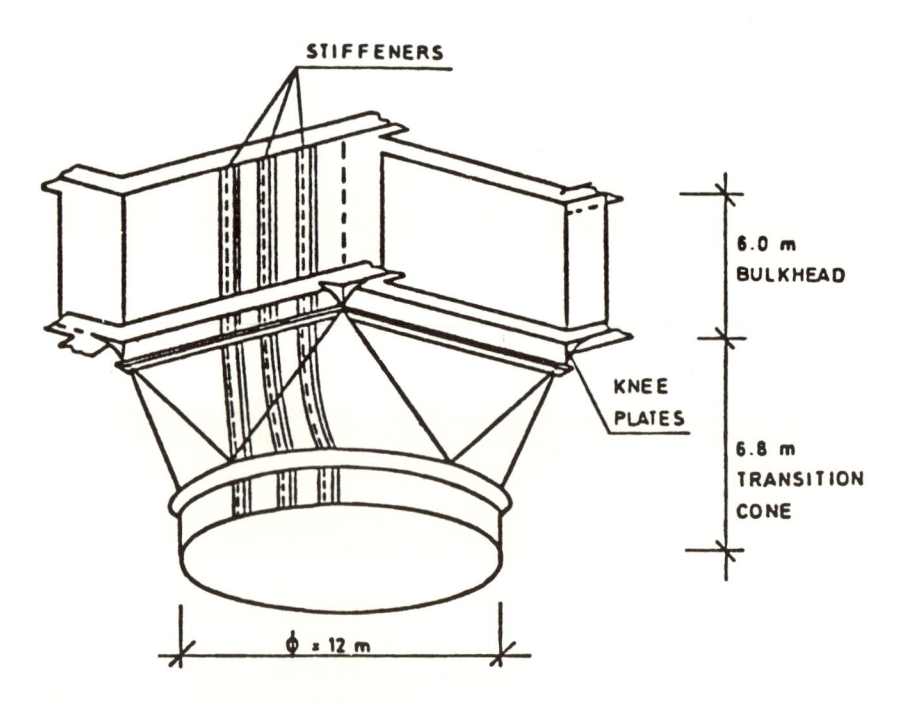

*Figure 20-3a. Steel deck transition piece to join the deck substructure to the concrete tower top. (O. Furnes and Ø. Løset, "Shell Structures in Offshore Platforms: Application and Design," (Det Norske Veritas). Paper presented at the International Association for Steel and Spatial Structures, Twentieth Anniversary World Congress, Madrid, September 24-28, 1979.)*

**Figure 20-3b.** *View underneath the deck substructure showing the steel transition pieces for connection to the concrete towers of the Brent B platform. (Courtesy Harcourt Brace Jovanovich.)*

The steel transition pieces can be seen in Figures 20-4, 20-5, and 20-6, which show the process of deck installation. In Figure 20-4 the deck structure of the *Beryl A* platform, supported on two ship-shaped barges, is approaching the platform, which is completely submerged except for the tops of the three towers. Figure 20-5 shows the barges and deck structure about to pass over the transition pieces on the tops of the towers. Figure 20-6 shows the deck structure of the *Statfjord A* platform being floated toward its three tower tops.

## Steel Grades

Two or three different grades of steel are normally employed in the deck structure. A special high-strength steel is used for the transition pieces and all the principal cruciform plate element joints. A lesser strength, so-called primary, steel is used for all other structural members, although a still lower grade secondary steel may be used for walkways, ladders, buckling stiffeners, railings, etc.

**352** **Introduction to Offshore Structures**

*Figure 20-4.* Beryl A *platform in extremely deep submerged position. (Copyright © 1976 Offshore Technology Conference.)*

*Figure 20-5. Steel deck being lined up for floating into position atop the concrete towers of the* Statfjord A *platform. (Copyright © 1977 Offshore Technology Conference.)*

**Deck Structures** 353

*Figure 20-6.* Steel deck being floated into position atop the concrete towers of the Mobil Beryl A *platform. (Source: E. Gausel and P.S. Hafskjold,* Proceedings of the First International Conference on the Behaviour of Offshore Structures, BOSS-'76, Vol. 1, *page 624.)*

### References

1. Furnes, O. and Løset, Ø.; "Shell Structures in Offshore Platforms Application and Design," Paper presented at the 20th anniversary world congress of the International Association for Steel and Spatial Structures, Madrid, September 24-28, 1979.
2. Furnes, O.; "Overview of Offshore Oil Industry With Emphasis on the North Sea," *Lectures on Offshore Engineering, 1978* (edited by Graff, W.J. and Thoft-Christensen, P.) Institute of Building Technology and Structural Engineering, Aalborg University Centre, Aalborg, Denmark.
3. Furnes, O. and Sorensen, K.T.; "The Use of Reinforced and Prestressed Concrete in Offshore Structures," Paper presented at the Institution of Engineers and Shipbuilders in Scotland annual meeting in Glasgow, December 19, 1978.

# Chapter 21

# *Construction and Installation*

### Construction

The various construction steps for a tower platform are outlined in Figure 21-1.[1] Figure 21-2 shows the steps used in constructing the Ekofisk oil storage tank typical of a manifold concrete platform.[2,3] As the two figures indicate, the construction steps are basically the same.

Construction begins in a dry dock excavated close to the sea and closed by a steel sheet-pile wall. The base slab is cast of reinforced concrete and, using slipforming, enough of the shell walls are constructed to form the base raft. The buoyancy needed for floatation in the base raft is critical to the design, since it is economically advantageous to keep the dry dock depth as small as possible. When the periphery of the base raft includes a perforated antiscour wall, the holes are closed with removable steel plugs before floating the raft out of dry dock. The dry dock is filled with water, and the sheet-pile wall is removed to permit tow-out of the raft to the deepwater construction site.

Construction of the structure exclusive of decks is then completed at the deepwater site using slipform techniques. The last step in construction is building or installing the deck structure. The deck substructures for the *Frigg CDP 1* and the *Ekofisk 1* storage tank were made of precast or cast-in-place prestressed and reinforced concrete. The *Ninian Central* platform and those of the tower type have steel deck structures which were installed by temporarily immersing the structure to a deep draft, floating the more or less completed deck structure into place, and deballasting to lift and join the components. Final uniting of the components was accomplished by welding.

After the transition pieces between the deck structure and the rest of the platform have been securely welded, the completed platform is deballasted to a minimum drag draft, towed to its final location, and ballasted onto the ocean floor.

## Construction and Installation 355

1. CONSTRUCTION OF BOTTOM STRUCTURE IN DRYDOCK.

2. THE DRYDOCK HAS BEEN FILLED WITH WATER AND THE DOCK GATE REMOVED.

 THE CONSTRUCTION IS FLOATING ON AN AIR CUSHION WITHIN THE SKIRT WELLS.

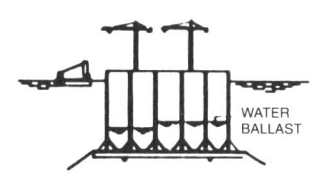

3. SLIPFORMING OF CELL WALLS.

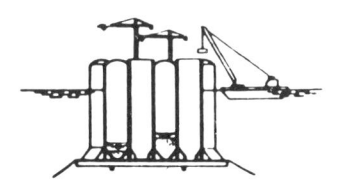

4. CONSTRUCTION OF THE UPPER DOMES, FILLING OF BALLAST SAND, AND INSTALLATION OF EQUIPMENT IN UTILITY AND RISER CELLS.

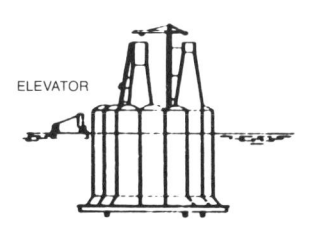

5. SLIPFORMING THE SHAFTS.

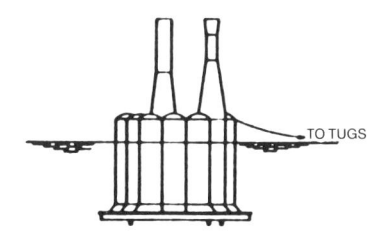

6. TOWAGE TO DEEPWATER SHELTERED SITE.

*Figure 21-1.* Construction steps for Condeep platform. (Source: Dag N. Jenssen paper "Design and Construction of the Condeep Type of Platform," Lectures on Offshore Engineering, 1978, Institute of Building Technology and Structural Engineering, Aalborg University Centre, Aalborg, Denmark.)

**356    Introduction to Offshore Structures**

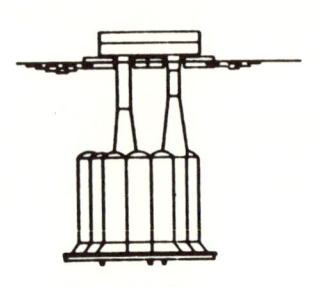

7. ERECTION OF DECK.

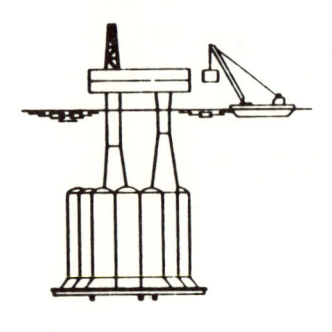

8. MOUNTING OF MODULES.

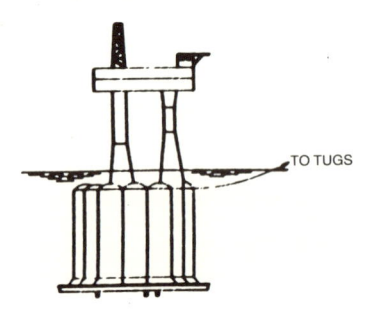

TO TUGS

9. TOWAGE TO THE FIELD.

10. PENETRATION OF THE SKIRTS INTO THE SEABED.

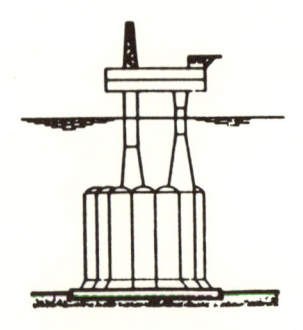

11. GROUTING—THE PLATFORM IS READY FOR OPERATION.

*Figure 21-1 continued.*

**Construction and Installation** 357

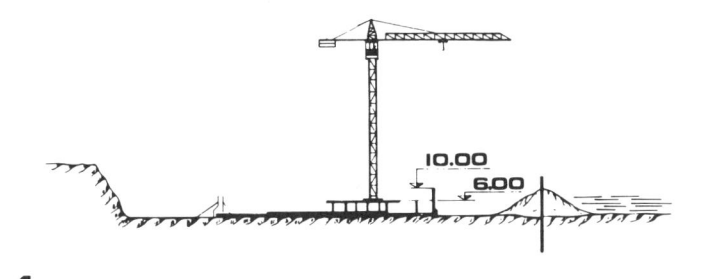

**1**
**CONSTRUCTION IN DRY DOCK**

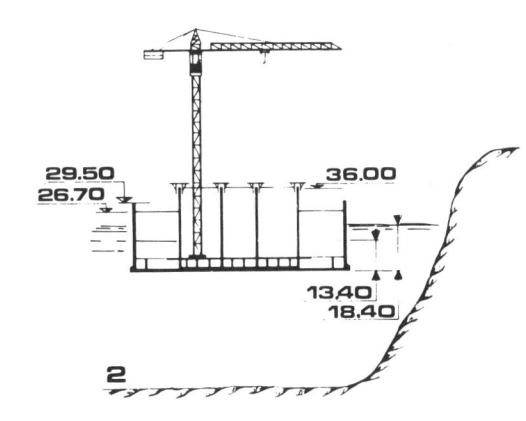

**2**
**SLIP FORM CONSTRUCTION IN FLOATING CONDITION**

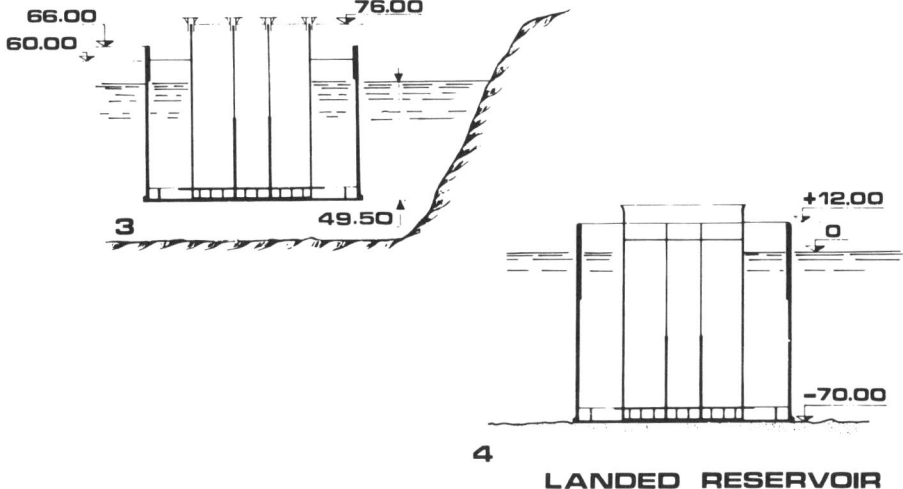

**3**

**4**
**LANDED RESERVOIR**

**Figure 21-2.** Construction steps for the Ekofisk type structure. *(Courtesy Freyssinet International.)*

## 358   Introduction to Offshore Structures

Fully automatic batching plants are used for concrete production. They are mounted on barges and are moored alongside the floating structure being built. A barge may produce 52-65 yd$^3$ (40-50 m$^3$) of concrete per hour. The water-cement ratio of the concrete mix is too low to use concrete pumps. In dry dock trucks and crane buckets convey the concrete to the forms. At the deepwater construction site, crane buckets deliver the concrete to the top of the structure where wheelbarrows are used for spreading. High-frequency vibrators operating around 12,000 rpm compact the wet concrete in the forms. In freezing weather it is necessary to heat the mixing water and insulate the forms. One basic mix is used throughout.

Because of large variations in cross sections, the heat generated by hydration and the cooling rates vary. Thus, the possibility of thermal stresses must be considered in the design stage. In some sections the maximum recorded temperatures have been as high as 150°F (65°C). Various means of reducing the temperature have been tried, including using low-heat cement, cooling the concrete mix, and cooling the sections where heat generation appears to be a problem.

It is important that geometric dimensions remain within the tolerances assumed in the design phase. To acheive a reliable reference system for measuring dimensions, a laser system is used. Deviations of the tower centerline of less than 2 inches (5 cm) have been maintained in slipforming. Variations in the 66 ft (20 m) diameter of caisson cells have generally not exceeded 0.4 inches (1.0 cm).

Detailed manuals are written to define procedures, sequences, structural consequences, and emergency actions for all possible operations during construction.

Continual inspection is conducted during the construction phase. The items monitored include: material testing, workmanship, concrete quality, placement of steel and concrete, tensioning and grouting of prestressing tendons, changes in plans and procedures, dimensions and accuracy of geometry, as-built conditions, and quality of repairs.

### Towing

The final construction steps for concrete platforms are accomplished in sheltered deepwater sites close to the onshore construction/supply yard. The towing time from this sheltered deepwater construction site to the final offshore location usually averages about three to five days.

As shown in Figures 21-3 and 21-4, several tugboats are used for towing. Four or five are positioned forward for pulling while one or two are positioned astern to provide positive lateral control. The total power required for towing ranges from 30,000 to 50,000 hp. (22,300 to 37,300 kW). The towing cables are

Construction and Installation 359

*Figure 21-3.* Towing a Condeep *platform to its offshore site. (Courtesy Freyssinet International.)*

*Figure 21-4.* Towing of the Frigg CDP 1 *platform. Notice the temporary giant ginpole derrick. This derrick was used to install the prefabricated concrete columns and the prestressed concrete girders which formed the supports for the deck structure and equipment modules. (Courtesy Freyssinet International.)*

kept as short as practical to enable quick response in the event the platform must be repositioned. The number of tugboats and the towline forces required for various towing speeds are deduced from model tests in a towing tank. The important variables are drag due to platform draft, towing speed, wave and wind resistance. Surveys of the ocean floor all along the towing route are made prior to the tow to ensure sufficient depth and no underwater peaks. The wind and wave effects are also studied from a stability viewpoint. It is important to know how the floating stability is affected by increased hydrostatic pressure and to evaluate the dynamic response of the structure to waves. Frequently, the economical towing speed is about two knots (3.7 km/hr).

Figure 21-5 shows the manifold type platform, *Frigg CDP 1*, during construction at its sheltered deepwater site. Three days were required for towing this platform from the sheltered deepwater site out to the offshore location. Thus, a three-day forecast of favorable weather was required. Good weather was needed for an additional day at the offshore site to accommodate installation. Note in Figure 21-4, which shows the *Frigg CDP 1* being towed, the temporary gin-pole crane used after the platform was placed on location to install the prefabricated, prestressed concrete columns and girders comprising the supportive structure for the steel deck system.

## Installation

Installation of a gravity platform on the ocean floor is complex and represents a critical phase in the life of the structure. Design of the structure and analysis of the geotechnical data must ensure that neither the structure foundation nor the soil beneath it become overloaded during any of the installation processes. Structural response must be monitored during the installation phase. To provide oversight, a comprehensive system of instrumentation is used. The factors that are normally monitored include:

1. Draft
2. Platform inclination
3. Water level in the cells
4. Water pressure in the skirt compartments beneath the base raft
5. Contact earth pressure at the tips of any domed cells
6. Stresses in the steel pipe dowels
7. Structural strain in any domed cells
8. Structural strain in the lower part of the foundation walls
9. Short-term settlement of the structure foundation

The installation phase is normally divided into several steps. For a tower type structure, the steps are:

**Construction and Installation** 361

*Figure 21-5.* Frigg CDP 1 *platform under construction in the Andalsnes fjord. (Courtesy Freyssinet International.)*

1. Touchdown
2. Steel pipe dowel insertion
3. Steel skirt penetration
4. Concrete skirt penetration
5. Dome seating
6. Water extraction and grouting inside the skirt compartments

For each step, the sequence of items to be monitored is outlined, contingency plans and alarm limits are developed, and written records are compared against the limits as the events occur.

By using the ballasting system in the different compartments of the structure foundation, it is possible to keep the platform level throughout the touchdown and skirt penetration steps with careful monitoring of the uneven distribution of earth pressure underneath the base raft. The gravity forces from the ballast in the various compartments of the caisson or foundation constitute a system of vertical parallel forces. By adjusting ballast water levels in different compartments and keeping the water pressures in the skirt compartments beneath the base raft within prescribed limits, the point of application of the resultant total penetration force can be controlled to virtually eliminate platform tilt. Water extraction and cement grouting injected inside the skirt compartments keeps the platform permanently level. The time required for grouting is usually three to five weeks.

On arrival at the offshore site, the platform is held in position by several tugboats usually arranged in a delta or square pattern.[4] During the early immersion portion of installation the platform is held within a 160-330 ft (50-100 m) radius circle of the desired location.

Before starting the towing operation for a platform of the manifold type, the perforations in the breakwater wall are closed and sealed with dished steel plate plugs and neoprene gaskets. At the offshore site, the platform is immersed by seawater ballasting; the descent speed is accurately controlled by submersible pumps installed on the outside of the main wall.

The tower type platform is towed to the offshore site with the tops of the caisson cells submerged under about 33 ft (10 m) of water. This submergence gives the structure much better stability than towing with the caisson cell roofs above the water surface. With the tops of the caisson cells submerged, the platform is practically insensitive to sea swells regardless of the condition of the sea.

Seawater ballasting by pumps located in the utility shaft is remotely controlled from an adjacent command ship. Water-level heights in the different compartments are continuously radioed to the command ship monitors.

## Instrumentation

Because of the need to measure so many factors during the installation phase, and the desire to monitor soil-structure performance for months after the operational life begins, the instrumentation systems built into concrete gravity platforms are comprehensive. Many of these instruments are installed during the construction of the foundation and remain dormant and submerged for months while the rest of the structure is completed.

Besides the measurements already mentioned as needed for platform installation, the kinds of data required for evaluating soil-structure performance are:

1. Horizontal, vertical, and rotational movements of the foundation of the structure.
2. Horizontal and vertical movements within the foundation soils beneath the structure.
3. Vertical contact pressures on the underside of the structure base slab.
4. Horizontal pressures or forces on the skirts.
5. Pore-water pressures within critical strata of the soil.

For complete analysis of soil-structure performance records of wave heights and periods, changes in platform structural weight and other external forces should be maintained.

## References

1. Jenssen, D.N., "Design and Construction of the Condeep Type of Platform," *Lectures on Offshore Engineering, 1978,* (edited by Graff, W.J. and Thoft-Christensen, P.) Institute of Building Technology and Structural Engineering, Aalborg University Centre, Aalborg, Denmark.
2. Clausen, C.J.; Dibiagio, E.; Duncan, J.M.; and Anderson, K.H.; "Observed Behavior of the Ekofisk Oil Storage Tank Foundation," *1975 Offshore Technology Conference Proceedings,* Paper 2373.
3. "Design, Construction Principles, and Setting of One Type of Concrete Gravity Platform Installed on Oil Fields in the North Sea," C.G. Doris do Brasil, (A paper in *Offshore Structures Engineering* edited by Carneiro, F.L. et al) Gulf Publishing Company, 1979, pp. 355-364.
4. Lindgren, J.; "Fabrication of Concrete Structures," Second WEGEMT graduate school on advanced aspects of offshore engineering, The Technical University of Aachen, Netherlands Ship Model Basin Wageningen, March 5-16, 1979.

# Index

Accident reports, 17
Accommodations. *See also* Platform, Living quarters.
Acoustic beacons, 60
Air cushion, 331
Air gap, 77, 120
Airy wave, 70, 133
American Institute of
  Steel Construction (AISC), 134, 194
American Petroleum
  Institute (API), 115, 160
American Society of
  Civil Engineers (ASCE), 160
American Welding Society (AWS), 157, 192
Anchorages (prestress), 295, 298, 330
Andoc platform, 266, 286, 320, 331
Anglo-Dutch Offshore Concrete, 331
Antiscour wall (gravity platforms), 288, 308, 354
Atlantic Richfield Company, 12
Atmospheric zone, 250
AWS fatigue curves, 192
Average wave, 98-99

Barge
  derrick, 12, 15, 28, 30
  launch, 12, 113
  launch (rocker arms), 113
  launch (skid beams), 212, 226
  load-out, movement, 226
  trim angle, 231
Barge ballasting, 226
Barge bumpers, 98, 107, 136, 203, 232
Barge movement (pitch and roll), 231
Barges (ship-shaped), 351
Base raft
  (gravity platform), 266, 277-278, 354
Base shear
  gravity platform, 270, 280
  template platform, 134, 137, 146

Base slab
  Andoc platform, 331
  manifold gravity platforms, 312-313
  Petrobras platforms, 316
Bathymetric survey, 276
Batter, 27, 76, 88, 95, 107-108, 137
Beam-columns, 110
Beam on elastic foundation, 89
Bent. *See* Jacket.
Bents, 113
*Beryl A*, 298, 320, 323, 328, 330, 347, 351
Blowout, 19
Boat landing, 15, 46, 98, 107, 136, 203
Bottles, 28, 96
Bottom slide, 66, 136
Braces, 92, 108, 149
  countoured end, 201
  diagonal, 17, 27, 29, 37, 54
  effective length coefficient, 112
  fabrication, 203
  flared end, 176
  horizontal, 27, 29, 54, 110
  ice formation, 252
  member sizing, 110
  ovalization, 161
  slenderness ratio, 110
  temporary, 237
  tie-down, 68, 222, 231, 237
Bracing
  K-type, 37
  splash zone, 252
  X-type, 17, 37
Bracing system
  functions, 110
Brazil, 261
Breakwater wall. *See also*
  Jarlan breakwater wall.
  sealing holes, 362
*Brent B*, 298, 320, 323, 328, 330

## Index 365

*Brent D,* 323
Bridge. *See also* Catwalk, Walkways.
British Petroleum Company, 17
Brittle fracture, 125, 257
Brown & Root, Inc., 5
Buckling stress, 186
Bumpers, 3
Buoyancy, 28, 108, 133, 145-146
  225, 233, 320, 354
Buoyancy tanks, 203, 205, 231-2
Buoyant members, 76

Cables, 68,100. *See also* Tendon.
California
  Santa Barbara, 4, 12, 15, 233
  University of (Berkeley), 139
Carbon content, 125, 128
Carbon equivalent, 128
Caisson (gravity platform),
  261, 266, 277, 279
  ballast forces during installation, 361
  construction (Condeep platform), 323
  overpressure, 293
  storage, 277, 293, 320
Caisson cells
  automatically filled, 277
  permanently ballasted, 277
  pressure in, 285
Can. *See* Joint can.
Cathodic protection. 205, 245-250,
  257. *See also* Corrosion.
Cathodic protection anodes, 30
  *See also* Corrosion anodes.
Catwalk, 36, 38
  cross-sectional arrangement, 39
  steel grating, 40
Cellar deck fabrication, 215
Chord
  collapse, 164
  heavy-wall, 152, 161
  torsion, 152
Clapotis, 314
*Cognac,* 2, 12, 47, 48-65, 234
  cost, 61
Column
  buckling (braces), 175
  end conditions, 112
Columns
  intermediate, 112
  long, 112
Communications
  room, 35
  tower, 345
Compressors (natural gas), 32, 34
Computer
  deterministic analysis, 134

dynamic analysis, 138-139, 145
input,
  soil conditions, 133
  wave force, 133
joint displacements, 139
member-end forces, 135, 139, 145
nonlinear analysis, 141
output options, 135
post processors, 132
preprocessors, 132
static/elastic analysis, 133, 137, 146
Computer methods, 132 ff
Computer program (oceanography), 132
Computer program (soil mechanics), 132
Computer service companies, 132
Concrete
  aggregates, 337
  durability, 336, 339
  fatigue behavior, 340
  high-strength, 298, 336
  instability failure, 337
  mix, 326, 330
  setting (retarding), 323, 329, 337
  modulus of elasticity, 337
  placement,
    tremie method, 330
    vibration, 313, 323, 330,
      338, 358
  Poisson's ratio, 337
  prestressed, 285, 336
  pumps, 338, 358
  reinforced, 285, 336
  reinforcement
    corrosion of, 340
    cover, 339
    resistance to corrosion
      and fatigue, 385
  S-N curves, 341
  seawater penetration, 340
  splash zone, 337, 339
  water/cement ratio, 336
Concrete batching plants 337, 358
Concrete construction material, 336 ff
Concrete corrosion prevention
  (low permeability), 339
Concrete structures (differences
  onshore/offshore), 285
Condeep (platform), 266, 279,
  286, 298, 320, 331
  conductors, 323
  prestressing arrangement, 298, 302, 303
Conductor, 28, 50, 97, 106, 110,
  113, 234, 274, 277, 330

guides, 203, 226
size, 235
skin friction, 277
Connection
  hybrid, 149
Connection (tubular). *See* Tubular joint.
Construction
  criteria for, 46
  inspection, 358
  schedule, 139
  steps
    manifold-type gravity
      platform, 357
    tower-type gravity
      platform, 355
  time (template platform), 47
Contact pressure (gravity platform),
  277, 278, 313, 327, 360, 362
Continental slope, 48
Control cable *(Cognac),* 60
Cook Inlet (Alaska), 28, 30, 100
Corrosion, 239 ff
  allowance, 253
  atmospheric, 127, 244
  barnacles, 256
  biological, 254
  coatings, inspection of, 253
  concentration cell, 241
  concrete reinforcement, 338
  crevice, 256
  current density
    definition of, 250
    Gulf of Mexico, 246
    North Sea, 246
  differential aeration, 241
  dissimilar metals, 242
  dry, 239
  electrochemical, 239, 242
  electrons, flow of, 249
  external circuit, 241
  fatigue, 123, 257
    limit, 257
  galvanic series, 242-243
  hydrogen evolution, 240
  internal circuit, 241
  mechanism, 240
  oxide films, 244
  oxygen absorption, 240
  pitting, 242, 255-257
  tensile stress, 190
  stress raiser, 256
  zones, 250
Corrosion anodes, 110, 226, 240
  auxiliary, 250
  GALVALUM, 245
  magnesium, 252

  sacrificial, 245, 323
  zinc, 252
Corrosion cathode, 240
Corrosion protection
  cathodic, 245-250
  coatings, selection of, 254-255
  impressed current, 247
  rectifier, 247
Corrosion protection system, 107
Corrosion recommendations
  (gravity platform), 340
Corrosive environment, 194
Cost (Cognac), 61
Cost savings (construction), 138
Crane, 3
  pedestal-mounted, 98
Crew
  drilling, 34
  travel time, 40
Current. *See also* Water current.
  bottom, 116, 308
  force, 75, 102, 110, 133, 274
  *Frigg CDP 1*, 312
  wind-driven, 274
Cumulative damage rule, 194, 342

Dead loads
  gravity platform, 270, 280
  template platform, 76, 133, 145
Deck
  beams, 119-120, 202
  cellar, 92
  definition of, 344
  design loads
    (gravity platforms), 345
  dimensions, 133
  drilling, 15, 92
  equipment, 215
  floor loads, 78
  flooring, 97, 119, 125, 215
  framing (cantilevered), 120, 215
  grating, 97
  girders, 119-120
  integrated, 344
  layout, 133
  levels, 43, 46, 97
  loading (gravity platforms), 345
  manifold, 308
  modularized, 344, 347
  piping, 215
  production, 15, 92
  steel template (types), 2
  structure *(Cognac),* 51
  topside facilities, 92
Deck components (anchorages), 345
Deck-mating (gravity platform),
  261, 285, 314

## Index

Deck substructure, 92, 97, 201, 215, 222, 226, 235
  box girders, 212, 344
  center of gravity, 237
  columns, 123, 235
  construction, 212
  design, 119
  dynamic stresses, 349
  erection, 122
  installation, 235
  integrated, 347
  legs terminating in
    jacket framing, 123
  lifting arrangement, 215
  plate girders, 344
  rectangular tubes, 152
  trusses, 123, 344
  web stiffeners, 123
Deck modules, 92, 97, 201, 215, 237
Design loads (inertial component), 345
Derrick barge 10, 47, 77, 232
  lifting capacity 215, 225, 237
Derrickman, 34
Derrick skids, 9, 97
Design criteria, 43, 152
Design loads
  (template platform), 66-78, 139
Design storm, 15, 17
Design wave, 98-99, 134, 137
Det Norske Veritas, 272, 293
Diffraction forces, 75, 274
Diffraction theory, 271-272
Directional drilling, 2
Doris, C. G., Company, 260, 288, 308
Drag force, 69, 71, 140
Driller, 34
Drilling equipment weight, 23
Drilling load, 76
Drilling mud, 23
Drilling rig (vessel)
  floating, 2
  jack-up, 2, 260
  mobile, 2, 10, 23
  semi-submersible, 2
  ship-shape, 2
  submersible, 2
Drilling rig (derrick)
  anchorage, 345
  package, 222
  skidding of, 120
Dry corrosion, 239
Dry dock (base slab
  construction), 322-333
Ductile fracture, 125
Ductility, 125, 177
Ducts (prestress), 295
Duhamel integral, 142

*Dunlin A,* 331
Dynamic amplification, 196, 272, 278
  *Cognac,* 54
Dynamic analysis (gravity platforms),
  139, 272, 286
Dynamic loading (template platforms),
  69, 190

Earthquake, 15, 43, 66, 77, 136
  139, 142, 145-146, 233
Eccentricity, 152, 157, 174
Economics, 21
Ekofisk prestressing arrangement, 298-300
Ekofisk storage tank, 260, 288, 315, 354
Elastic soil modulus, 89
Endurance limit. *See* Fatigue.
Environmental criteria, 43, 136
Environmental loads, 201, 231, 275,
  285, 288
  *Frigg CDP 1,* 312
  gravity platforms, 270 ff
  North Sea, 336
Equations of motion, 141
Explosion and fire, 19
Exxon Company, 15, 233


Fabrication *(Cognac),* 57
Fabrication ways. *See* Skid runners.
Fabrication yard
  skid runners (beams), 215, 235
Failure
  chord collapse, 164
  modes, 150
  punching shear, 164
  tubular joint (criteria), 149
Fatigue
  cumulative damage rule, 197, 342
  failure, 188
  loading rate, 190-191
  low cycle, high stress, 194
  low stress, high cycle, 194
  maximum tensile stress, 191
  mean stress, 188
  poor fit-up 191
  random stress fluctuations 188
  reinforcement, effect of, 191
  stress
    concentration 191
    distribution (long-term), 195, 341
    frequency, 190
    ratio, 188
    range, 190, 196, 342
  random, 192
  wave height distribution (long-term), 196
Fatigue analysis literature, 192
Fatigue analysis (simplified), 195

Fatigue behavior, 51
  overlapping braces, 174
Fatigue fracture mechanics, 192
Fatigue life, 152, 157, 167,
    175, 192, 194
Fatigue limit, 190
F. E. M. analysis (gravity platform), 286
Fire protection, 32
Flame-front generator, 38
Flare boom, 16
Flare tower, 32, 36, 345
Flexibility ratio, 155
Flooded members, 133, 145, 152
Fluid-structure interaction, 140-141
Forces. *See also* Wave forces,
  Wind forces.
  installation, 226
Foundation (gravity platforms)
  concrete caisson, 261
  nonlifting, 277
  rocking, 272, 275, 278
  sliding, 272, 275, 278
Foundation design, 46, 278
  problems for gravity platform, 275
Foundation failure modes
    (gravity platforms), 280-283
  bearing capacity, 282
  liquefaction, 282
  rocking, 282
  sliding, 280
Foundation stability
    (gravity platform), 277-278
Freyssinet tendons, 298, 316
*Frigg CDP 1*, 288, 208, 354, 360
*Frigg MP2*, 288, 309
Froude-Krilov hypothesis, 135

GALVALUM anodes, 245
Galvanizing, 245
Geologist, 1, 40, 46
Geology, 1
Geophysicist, 1, 5, 276
Geophysics, 1
Geotechnical design
    (gravity platforms), 275 ff
Geotechnical (soils) data, 99-100
Gradient wind, 67, 273
Grating, 202, 215, 245
Graving dock, 233
Gravity platforms
  advantages, 268
  construction, 345 ff
  design procedure, 269
  functions, 286
  installation, 285
  manifold-type, 288, 360
  material quality, 285
  oil storage, 293

  serviceability, 285
  tower-type, 286, 320 ff
Grout (prestressing ducts), 298, 338
Grouting
  template platforms, 50, 60,
    145, 234, 235
  gravity platforms, 267, 275, 279, 314,
    320-321, 330, 358, 362
Gulf of Mexico, 4, 98, 139, 196, 215, 260
Gussets, 152, 161, 171, 175
Gust wind speed, 100, 273

Habitats (dewatered), 17
Hammers, 80
Hammer-pile combination, 117
*Handbook of Ocean and
    Underwater Engineering*, 67, 109
Handrails, 203, 245
Heat-affected zone, 125, 129
Helicopters, 3, 35
  advantages, 40
  travel time, 40
Heliport, 15, 36, 42, 46, 344
Hollow structural member, 149
*Hondo*, 15
Horizontal force (at mudline).
    *See* Base shear.
Horizontal loads (gravity platforms), 280
Hot spot(s), 133, 161, 175, 191
Hotel service company, 34
Howe truss, 110
Hurricane, 9, 10, 19, 71, 76,
    92, 139, 157
  Betsy, 17, 66
  Hilda, 17, 66
  wave, 98
Hydraulic jack (prestressing), 295
Hydrogen embrittlement, 256
Hydrostatic pressure
  gravity platforms, 285, 293, 313,
    337, 340, 360
  template platforms, 108, 112, 231

Ice, 46, 66, 100, 139
  thickness, 30
Ice floes, 29, 253
Igniter, 38
Immersed zone, 250
Impact, 77
Inertial force, 69, 71
Ingot
  hot-top, 126
  piping, 126
Installation (gravity platforms), 360 ff
  soil overloading, 360
  steps, 360-361
Instrumentation (gravity platforms),
    360, 362
International Institute of Welding, 128

# Index 369

Jacket, 9, 12, 27, 79, 150
  braces, sizing of, 50, 106
  column planes, designation of, 92
  common framing plans, 110
  definition of, 106
  design, essential ingredients of, 106
  eight-leg, design of, 92 ff
  fabrication, 201
  flare, 36
  forces (launching), 201
  frame (narrow dimension), 47, 57, 202
  framing plans, 110
  gravity, center of, 231
  installation, 225
  launching, 46, 139
  lifting into water, 226
  load-out, 225-226
  loads during fabrication, 201
  lofting, 202
  members, sizing of, 50, 106
  navigational aids for ships, 107
  overall dimensions, 133
  plan view, dimensions of, 27
  sectionalized,
    horizontally connected, 233
    vertically connected, 234
  self-floating, 233
  shallow water, four leg, 212
  slip-over, 23
  stresses during launch, 231
  torsional resistance, 110
  upending, 232
  walkway, 95
  weight, distribution of, 107
  weight on ocean bottom, 232
  well-protector, 3

Jacket legs
  bottles, 28, 203
  can, 202
  grout, 28
  internal shims, 202
  lamination, effect of, 126
  launch truss collapse, 231
  numbering, 212
  oversize
    hydrostatic collapse, 233
    wave forces, 233
  piles protruding, 215
  shims, 36
  sizing, 107, 175
  spacing, 95
  thickness equation, 108
  wraps, 253

Jacket-pile system (wrenching motion), 110

Jarlan breakwater wall, 288, 308-309, 314
Joint can, 110, 161, 175-176, 192, 202

Joint nodes, 202
Joints (radiography), 129

K joint, 151
K truss, 110

Ladders (steel grade), 351
Lamellar tearing, 126, 128, 150
Lamination, 125
Laser leveling system, 323, 358
Laterally loaded pile, 88
Launch barge, 60, 95, 202, 205, 225, 232
  ballasting, 231
  rocker arms, 203, 226, 231
Launch beams. *See* Skid beams.
Launch rails (beams), 46, 47
Launch trusses, 95, 112, 136, 202, 231
Lake Maracaibo, 4
Leak detection (gas), 32
Lifting eyes, 46, 203, 222, 237
Limit state
  definition of, 288
Limit state design
  gravity platforms, 288
  safety, 292
Linear wave, 70
Liquefaction (gravity platforms), 278, 282
Live load, 76, 133, 270
Living quarters, 9, 344
Lloyd's of London Press Ltd., 19
*Lloyd's Register,* 128
Load conditions (number), 132, 134
Load factors (gravity platforms), 288
Loading conditions (gravity platforms), 285
Local buckling, 29, 112, 150
Local capacity (torsional), 155
Local yielding, 142, 161
Longitudinal planes (designation), 92
Louisiana, 4, 7, 8

Macao (Brazil), 315
Magnolia Oil Company, 8
Manifold gravity platforms, 272, 308 ff
  shallow water, 315
Material safety factors
  (gravity platforms), 288
Matrix bandwidth, 137
Maximum wave height, 9
McFaddin Beach, 6
Metallurgist, 127
Meteorologist, 46
Metering equipment, 32
Microorganisms, 254
Mobil Oil Company, 12
Modal analysis, 142, 145-146
Model tests (gravity platforms), 271-272
  tow-out, 360
  structural F. E. M., 286

# 370 Introduction to Offshore Structures

Modules
  construction, 347
  deck, 47, 273, 344
  dimensions, 222
  gravity, center of, 238
  lifting, 237
  pipe joints, 97
  placement, 349
  weight, 222, 347
Moment. *See* Overturning moment.
Moment at the mudline, 137
Monel leg wrap, 253
Mooring bitts, 3, 107
Morison equation, 66, 69,
  102, 135, 140, 271
  $C_m$ and $C_d$, 70
Motion. *See also*
  Foundation, Foundation failure modes.
Motorman, 34
Mud man, 34
Mudmats, 119, 212, 232

Natural gas (reinjection), 34
*Ninian Central*, 288, 315, 354
Node, 125, 157
N Joint, 151
Nonlinear wave, 70
North Sea, 98, 140, 157, 161
  196, 261, 278, 298, 336, 341
  British sector, 260
  Danish sector, 99
  wave period, 271
Norway, 337
Norwegian Contractors, 320
Notch
  brittleness, 125
  toughness, 127
Notches, 194

Ocean current. *See* water current.
Ocean floor topography, 279
Oceanographer, 46
Oceanographic criteria *(Cognac)*, 51
Offshore Technology Conference, 160
Oil operations
  description of, 1
  development drilling, 2
  exploration, 1
  exploratory drilling, 2
  production, 3
  worker transportation, 3
Operating waves (number), 341
Operational criteria, 43
Operator
  pump, 34
  crane, 34
Overburden pressure, 82, 84

Overturning moment
  gravity platform, 275, 277, 280
  template platform, 95, 134, 146

Paint
  deterioration of, 239
  splash zone, 252
  zinc, 253
Painting, 51
Palmgren-Miner rule, 194, 197, 342
Panama Canal, 14
Panel buckling, 185
Persian Gulf, 17
Personnel
  boat transfer of, 40
  safety, 79
Petrobras gravity platforms, 315
Petroleo Brasileiro S. A., 315
pH
  concrete, 338
  neutral, 240
  seawater, 254
Phillips Petroleum Company, 260
Pierson-Moskowitz spectrum, 140
Pigs, 32
Pile(s)
  add on, 80, 115-116
  battering, 90
  bending moment distribution, 89, 109, 135
  cap, 84
  capacity
    axial, 80
    pull-out, 96, 115
    ultimate, 85, 113
  circumferential rings for, 115
  deflection, 138
  design (remedial alternatives), 116
  diameter, 10
  diameter selection, 108
  drilling undersize hole, 117
  end bearing, 81
  factor of safety, 86
  field splice, 110
  grouted, 84, 87, 113
  guides, 28, 96, 203, 226
  high-strength steel, 225
  insert, 86
  installation, 234
  jetting, 84, 88, 117
  linear springs, 134, 137
  number of, 95, 138
  penetration, 79, 115, 117
  pilot hole, 88
  *P-y* data, 116, 146
  roundness and straightness, 225
  segments, 80, 110
  simulated springs, 109

## Index

sizing, 48, 95
skin friction, 81, 115
skirt (sleeves), 11, 17, 28, 48, 96, 113, 234
soil, upward yielding of, 88
steel, 8, 92, 175
storage on deck, 212
thickness at mudline, 116
timber, 6
undrivable, 86
wall thickness, 10, 79, 85
welding, 225
Pile analysis
difference equation, 89
lateral load, 88, 116
limit load method, 89
point of fixity method, 89
$P$-$y$ method, 89
Pile driving
dynamic formula, 84
impact stresses, 117
underwater hammer, 234
wave equation, 117
Pile follower, 234
Pile foundation, 201
Pile group, 29, 79, 90
Pile hammers
size, 235
types, 80, 84-85, 117
Pile head
rotational restraint, 116
shear displacement, 135
Pile loads
maximum, 109
$P$-$y$ analysis, 109
ratio lateral to vertical, 79
Pile tops
internal cones, 235
Piling
laterally braced, 260
weight, 225
wooden, 260
Pipeline, 4, 15, 25, 32, 38, 43, 255, 260
breaks, 19
riser, 203
Plastic strains, 141
Platform
auxiliary, 38
concrete gravity, 1, 260-261
design (aesthetics), 201
drilling/well-protector, 21
*Holly*, 12
*Hondo*, 15
locations, 260
manifold, 261
modifications in design, 276
number in existence
concrete gravity, 261
total, 1
production, 32
quarters, 17, 34, 349
reinjection, 17, 32
self-contained, 9, 12, 15, 26, 92
storage, 3
tender-type, 3, 25
tension-leg, 1
timber, 92
tower-type
concrete gravity, 271
steel template, 26, 28, 32
well-protector, 17, 21, 39
wooden, 4
Platform tilt (gravity structure), 282
Plugs, 314
Pollution, 19
Pontoon(s) (floatation), 225, 233
Pore water pressure, 278, 282, 340, 362
Post-tensioning
concrete, 293, 295, 345
Condeep tower, 327
Postwar boom, 7
Potable water, 36
Power law (wind loading), 67, 273
Pratt truss, 38, 110
Precast concrete, 298
Preheating (electrical), 129
Prestress tendon passageways, 298
Prestressing (concrete), 293-307
ducts, 313, 326, 330, 338
fatigue, 340
length, 299
tendons, 338
Pretensioning (concrete), 293
Pumps, 25, 32, 34, 106
Punching shear, 150, 164, 176, 181, 192, 195
Pure Oil Company, 5
$P$-$y$ curves, 116

Quarters platform. *See* Platform.
Quay, 202, 215, 237

Radar sidescan, 276
Railings, 46, 125, 136, 351
Random process, 144
Random waves, 140
Rayleigh distribution, 71
Rayleigh-Ritz analysis method, 141
Reinjection, 17, 32
Residual stresses, 128, 190
Ring stiffeners, 50, 151, 171, 175-176, 182, 203, 231, 233
external, 161

## 372    Introduction to Offshore Structures

internal, 161
spacing of, 176, 181
Risers, 21, 30, 97-98, 274, 277, 340
Rocker arms, 113
Rotation
  tower tops (gravity platforms), 349
Roughneck, 34
Roustabout, 34
Rust, 240

Safety, 285
  personnel, 32, 34
  risks, 138
Salvador (Brazil), 317
Sandblasting, 51
Santa Barbara (California), 4, 12, 15
Scour
  definition of, 100, 116
  gravity platform, 278
Scouring (gravity platform), 275, 320
Scour protection (gravity platform), 267, 314
Sea floor
  access, 268
*Sea Tank* (platform), 266, 286, 320
Seadeck, 27, 95
Seawater
  corrosive medium, 190
  dissolved oxygen, 251-252, 339-340
  electrolyte, 241
  pH, 254
  salinity, 252
  temperature, 251
Secondary braces, 157
Semi-probabilistic design
  (gravity platforms), 288
Septic tank, 36
Serviceability limit state, 293
Settlement, 277-278, 360.
  *See also* Soil.
Shell Oil Company, 11-12, 27, 48, 51, 331
Ship
  crew boat, 3
  LST, 10
  tanker, 3, 43, 261
  tender, 3, 10, 25
  tugboat, 232, 358, 359
  work boat, 3
Shipworm, 9
Sidesway, 76
Signal Oil and Gas Company, 4
Site investigation (gravity platform), 276
Skid beams, 58, 113, 222
Skidding (at touchdown), 279
Skid runners, 202, 212, 226
Skirt(s) (gravity platforms), 278-279, 321
  compartments, 361
  concrete, 267
  height and spacing, 280

penetration, 275
purposes, 278
steel, 267
on tower gravity platform, 320
water pressure, 361
Skirt piles. *See* Pile(s).
Sleeve
  skirt pile, packer 234
Sliding 267, 312
Slipforming, 354, 358
  rate, 299, 313, 323, 330
  scaffolds, 299
  tower construction (gravity platforms), 320
Slipforms
  size, 328
S-N curves, 190, 192, 341
Soil
  bearing pressure, 275
  cohesive shear strength, 81
  cyclic loading, 275
  importance to gravity platform, 275
  modeling (linear springs), 134
  modulus (elastic), 89
  settlement, 275
  set-up, 84
  survey, 276
Soil-structure interaction, 138
Specialist
  directional drilling, 35
  gas turbine, 35
  meteorology, 98
  mud, 35
  oceanography, 98
  soil mechanics, 98
*Specification for Design,
  Fabrication and Erection
  of Structural Steel,* 185
Sphere launchers, 32
Splash zone, 29, 250, 252
Stabbing cones, 233
Stainless steel wrap, 253
Stairs, 3, 46, 125-226, 245
Stairways, 212
*Statfjord A,* 323, 349, 351
*Statfjord B,* 266
Static design, 69
Stationary condition
  statistically, 271
Statistical regularity, 144
Steel
  carbon content, 125
  deck structure, 322
  grades, 124, 351
  grating (catwalk), 40
  high-strength, 112, 124, 351
  lamellar tearing, 126
  nonlinear material properties, 142

preheat, 128
primary, 351
reinforcement, 338, 342
reinforcing, 295
  bars, 340, 342
secondary, 351
selection, 123
short transverse
  properties, 123, 125-126
skirts (Andoc), 331
strength, 170
  ultimate, 123
  yield, 123
through-the-thickness
  properties, 123, 125-126
surface roughness, 252
Steel decks (gravity platforms)
  dimensions, 345
  substructure, 345
Steel deck structure (gravity platforms), 320, 331, 341, 344
Steel dowels (gravity platforms), 275 279, 321, 360
Steel transition joints
  fatigue behavior, 350-351
  gravity platforms, 344, 349
Stiffened steel columns (Andoc), 331
Stiffeners, 125, 345, 351
  longitudinal, 176, 182, 233
Stochastic analysis, 144
Stokes wave, 100-101, 133, 140
Storm wave, 69-70, 77, 99, 312, 341.
  *See also* Design wave.
Strain hardening, 177, 182
Strain
  through-the-thickness, 127
Strand. *See* Tendon.
Stream function wave theory, 70
Stress
  maximum principal, 168
  punching shear, 170, 174
  transverse shear (instability), 188
Stresses
  chord (membrane), 161
  elastic, 166-167
  hot spot, 166, 194
  redistribution, 150, 161, 190
  residual, 157
  tubular joint, 149
    interaction method, 177
Stress concentration, 133, 161, 166, 175, 191, 194
Stress corrosion, 123, 256
Stress range, 190, 342
Stress ratio
  applied to allowable 137-138, 146
Stress relieving, 128, 157

Structural analysis
  deterministic, 142
  equations of motion, 140
  finite element method, 140
  initial hand calculations, 136
  manual, 132
  matrix bandwidth, 146
  natural frequencies, 141
  solution techniques (comments), 142
  stiffness method, 135
  stochastic, 142
Structural damping coefficient, 141
Structural design
  *Cognac,* 54
  gravity platforms, 285 ff
  physical dimensions, 133
  template platforms, 46
Structural stiffness matrix, 141
Structural Welding Code, 160
STRUDL program, 133
Submerged Lands Act, 11
Summerland Oil field, 4
Superior Oil Company, 5, 9
Sustained wind speed, 68, 273

Tangent modulus, 112
Template. *See* Jacket.
Temporary loading conditions, 285
Tendon (prestress), 293, 295
Tenneco Corporation, 12
Teredo, 9
Tidal current, 274
Time-history analysis, 142
T joint, 151
Tool pusher, 34
Topside facilities, 100, 102.
  *See also* Decks.
Tower
  flare, 32
Tower platform
  leg grout, 29
  gravity structure, 320 ff
Tow-out
  *Cognac,* 58
  dynamic response, 360
  economical speed, 360
  favorable weather, 360
  gravity platform, 347, 358
    manifold platform, 313
    tower platform (draft), 362
Transition joint (gravity platform), 266-267, 354
Transportation barge, 47, 95, 113, 215, 222, 234, 237
Transverse planes
  designation, 92
Treatment facilities, 98
Treatment. *See* Platform.

# 374  Introduction to Offshore Structures

Tremie method, 330
Tubes
  advantages, 149
  heavy-walled, 201
  out-of-roundness, 108
  rectangular, 152
    effective width, 155
  stiffened, 182, 188
    axial compression, 182
    buckling, 182
    hydrostatic pressure, 185
  thin and thick, designation, 112
Tubular joint(s), 108, 125, 149, 191
  allowable stress, 171
  can, 152, 157
  collapse, 167, 176, 180
  complex, 155, 157
  configurations, 150
  diameter ratio, 161
  diaphragms, 176
  eccentric, 152
  fatigue behavior, 188 ff
  gamma ratio, 164
  gussets, 151
  initial cracking, 166-167
    yielding, 166
  in-plane, 150, 155, 161
  load redistribution, 177
  multiplane, 157
  nonoverlapping, 175
  offset, 152
  overlapping, 152, 157, 174
  penetrating braces, 176
  plastic behavior, 176-177
  plastic zone, 169
  plug, 152, 167
  rectangular section, 152, 155
  reinforcement (methods), 176
  reserve capacity, 166, 171
  saddle, 152
  safety factor, 180
Tugboats, 232, 362

Ubarana Oilfield, 315
Umbilical *(Cognac)*, 60
U.S. Geological Survey, 19, 47

Valves
  Christmas tree, 25
Venezuela, 4
Vortex shedding, 68, 100, 274

Walkways, 95, 125, 203, 212, 245, 351
Warren truss, 38, 92, 97, 110
Water-cement ratio, 358
Water current, 66, 99, 270
  *Cognac*, 54
  North Sea, 100
Water particle velocity, 102, 274

Wave celerity, 102, 274
Wave forces, 10, 43, 69, 110,
  139-140, 142, 272, 275, 285
  computer generation, 137
  distribution, 102
    drag pressure, 102, 104
    inertial pressure, 102, 105
  maximum, 274
  phase relation, 102
  shielding, 71
Wave height
  design, 71
  distribution, 270, 277
  operating, 196
  significant, 51, 71, 98
  storm, 29, 43, 66
Wave loads, 139, 201, 270, 280
  design wave method, 270
  spectral analysis method, 271
  number, 340
Wave pressure (appurtenances), 75
Wave profile (steepening), 75, 274
Wave spectrum, 140, 271
Wave theory, 102
Weld, 47, 87
  defects, 129, 194
  inspection methods, 129
  metal shrinkage, 126
  passes (sequencing), 128-129
  penetration, 191
  porosity, 191
  post heat, 129
  throat thickness, 150
  toe, 170
Weldability, 127
Welders
  crew members, 34
  prequalified, 129
  qualification, 160, 191
  wind protection for, 205
Welding, 17, 46, 119, 123,
  149, 152, 155, 157, 190
  accessibility, 128-129
  beveling, 202
  dry habitats, 234
  engineer, 127
  field, 97
  heat-affected zone (HAZ), 125
  inspectors, 129
  procedures, 129, 191
  specifications, 129, 160
  transition joints (gravity platforms),
    354
Welds
  filler metal, 150
  fillet, 150
  full-penetration groove, 17, 150, 160,
    174, 182, 203, 225, 234-235

# Index

length, 120
pile, 80
Wells
 number, 43, 97
Well-kill system, 25
Well-protector jacket, 3
Wet corrosion, 239
Wind forces, 43, 66, 102, 120, 133, 142, 201, 280, 285
Wind loads
 calculation, 66, 272
 coefficients $C_D$ and $C_L$, 67, 273
 gravity platforms, 270
Wind loading, 98, 139, 345
Wind pressure design, 68

Wind speed
 fastest-mile-velocity, 68, 273
 Gulf of Mexico, 99
 gust, 68, 273
 North Sea, 100, 274
 sustained, 68, 273
Wind shield factors, 69
Wire. *See* Tendon.
Wire rope, 51
Wooden derricks, 4
Work day, 34
Work period, 35
Worker transportation, 3

X joint, 151

Y joint, 151